Differential-Algebraic Systems

Analytical Aspects and
Circuit Applications

Differential-Algebraic Systems

Analytical Aspects and Circuit Applications

Ricardo Riaza
Universidad Politécnica de Madrid, Spain

NEW JERSEY · LONDON · SINGAPORE · BEIJING · SHANGHAI · HONG KONG · TAIPEI · CHENNAI

Published by
World Scientific Publishing Co. Pte. Ltd.
5 Toh Tuck Link, Singapore 596224
USA office: 27 Warren Street, Suite 401-402, Hackensack, NJ 07601
UK office: 57 Shelton Street, Covent Garden, London WC2H 9HE

British Library Cataloguing-in-Publication Data
A catalogue record for this book is available from the British Library.

DIFFERENTIAL-ALGEBRAIC SYSTEMS
Analytical Aspects and Circuit Applications

Copyright © 2008 by World Scientific Publishing Co. Pte. Ltd.

All rights reserved. This book, or parts thereof, may not be reproduced in any form or by any means, electronic or mechanical, including photocopying, recording or any information storage and retrieval system now known or to be invented, without written permission from the Publisher.

For photocopying of material in this volume, please pay a copying fee through the Copyright Clearance Center, Inc., 222 Rosewood Drive, Danvers, MA 01923, USA. In this case permission to photocopy is not required from the publisher.

ISBN-13 978-981-279-180-1
ISBN-10 981-279-180-9

Printed in Singapore.

A la memoria de mi madre

Preface

Differential-algebraic equations (DAEs) have been the object of increasing attention in the last three decades. Nowadays, they provide a valuable tool for system modeling and analysis within different fields, including nonlinear electric and electronic circuit theory, constrained mechanics, and control theory, among others.

The first part of this book addresses analytical properties of such differential-algebraic systems. The few existing monographs on DAEs are mainly focused on numerical aspects and, in most cases, they are restricted to a specific approach, structured around the differentiation, geometric, tractability, perturbation or strangeness indices, respectively. By contrast, the present book attempts to discuss a variety of analytical frameworks for the study of DAEs. The emphasis will be on projector methods based upon the tractability index, and also on reduction techniques supported on the geometric index. The differentiation index will also be briefly examined; note that it has received comparatively more attention in the DAE literature than the projector-based and reduction frameworks, in spite of the many benefits displayed by these approaches.

Projector-based methods, introduced for linear DAEs in Chapter 2, allow for precise input-output functional descriptions and explicit solution characterizations in terms of the original problem variables: this holds for linear time-varying DAEs with arbitrary index, under mild smoothness requirements. These methods have been mainly developed by Roswitha März and, accordingly, the material in Chapter 2 is crucially based on her work. Nevertheless, some recent or new contributions can also be found in this Chapter, concerning e.g. the so-called Π-projectors, a simplification of the decoupling of DAEs with properly stated leading term, or a detailed characterization of standard form linear problems.

Reduction methods, based on the research of Rabier, Rheinboldt and other authors, define a powerful framework for the analysis of nonlinear DAEs. Chapter 3 is mainly focused on these techniques, introducing in particular a local approach for quasilinear DAEs in settings where the global assumptions of Rabier and Rheinboldt do not hold, and paving the way for subsequent analyses of singular problems. The differentiation index is also discussed in this Chapter; cf. Sections 3.1, 3.2 and 3.7.

From a dynamical point of view, the essential differences between DAEs and explicit ordinary differential equations (ODEs) arise in so-called singular problems, which lead to new dynamic phenomena such as those displayed at impasse points or singularity-induced bifurcations. Recent results on the classification and analysis of singularities are extensively discussed in Chapter 4. The topics covered on singularities of linear time-varying problems are the result of recent research, whereas the material on singular points of quasilinear DAEs is completely new.

The second part of the present monograph is focused on the analysis of DAEs arising in electrical circuit theory, emphasizing modeling aspects and considering both linear and nonlinear problems. Chapter 5 discusses nodal analysis methods, widely used in circuit simulation programs, with special attention to index characterizations. Chapter 6 addresses so-called branch-oriented analysis methods. The models arising from these techniques, which are not as well-suited as nodal ones for numerical simulation purposes, provide however several advantages from an analytical point of view. In particular, they make it possible to frame in the DAE context the state formulation problem for electrical circuits, including normal tree methods used to tackle it, and also to analyze different qualitative aspects of circuit dynamics.

I have tried to write this book in a self-contained manner, making it accessible to as many interested readers as possible. Some background material has been added with this aim; this includes regular matrix pencil theory, Schur complements and Cauchy-Binet expansions from matrix analysis, several elementary aspects of differential geometry, and rather detailed introductions to digraphs and to circuit theory fundamentals. See, specifically, Sections 2.1.1, 3.3 and 5.1. With this material, the prerequisites for reading this book virtually amount to undergraduate courses on differential calculus, linear algebra and ordinary differential equations. However, readers with more background on mathematical analysis, differential equations, and circuit and system theory will probably get a deeper insight into the contents of this monograph.

In the same spirit, technicalities have been intended to be kept at a minimum. In particular, the local properties discussed in Chapter 3 are not stated in terms of germs in order to make these results available to readers unfamiliar with this language. Examples have been chosen to be as simple as possible, being just focused on the ideas that they are intended to illustrate. In Part II, controlled sources are excluded from most analyses for the sake of simplicity, although subsection 6.2.6 indicates how to extend the results to circuits including a broad family of these sources. Notational and terminological issues are explained or recalled at the points where they show up; this should make the reading easier, specially regarding the different variable splittings appearing throughout.

This book can be used in a graduate course on DAEs. Additionally, different reader profiles might benefit from its contents. Some hints are given in the section *How to read this monograph* on page 22. In particular, the book may be of interest for applied mathematicians and analysts. It is worth indicating in this regard that, very often, the main goal in DAE analyses is to unveil their behavior in terms of some kind of related ODE; from this perspective, DAEs pose many problems which can be better framed within linear algebra or mathematical analysis contexts, rather than in the theory of differential equations. The book can also be helpful within numerical mathematics, covering the analysis which usually precedes the numerical simulation of differential and differential-algebraic equations.

Regarding applications, the present book may be useful for scientists and engineers interested in singular system theory and modeling issues. Part II may be of help in electrical and electronic circuit analysis and design. On the other hand, the general theory discussed in Part I can be used in other applications of DAEs arising for instance in mechanics, chemical processes or control theory. Other scientists aiming to get a glimpse on DAEs would profit from certain parts of the book.

Unfortunately, other aspects of DAE theory which might be of interest for some readers are beyond the scope of this monograph. These topics include numerical issues, normal form theory, over- and underdetermined problems, partial differential-algebraic equations (PDAEs), stochastic DAEs, singularity-induced bifurcations, or impasse phenomena and singularities in electrical circuits. Hopefully, the references cited on these topics will be of some help for the interested readers.

I wish to thank Ms Anna Tong, Commissioning Editor of World Scientific, for her kind invitation to undertake this project, and Ms Lai Fun Kwong for her support in the book production process. I am greatly indebted to Roswitha März not only for her comments and suggestions on this material, but also for the continued and pleasant research collaboration on these topics. Actually, I have a very nice memory of my research stays with the DAEs group at the Humboldt University of Berlin, where I also benefited from many fruitful discussions with René Lamour and Caren Tischendorf. The help of Diana Estévez-Schwarz is gratefully acknowledged, as well as that of Uri Ascher and Steve Campbell, who guided my first steps on DAEs. I am also thankful to Clara Ministral for many useful English tips. The efficient bibliographic support offered by the personnel of the Library of the ETSI Telecomunicación (UPM) has been very much appreciated. Finally, I want to express my gratitude to my wife, Raquel, for her warm understanding, patience and support during the writing of this book.

Ricardo Riaza
Madrid, December 2007

Contents

Preface vii

1. Introduction 1
    1.1 Historical remarks: Different origins, different names ... 2
    1.2 DAE analysis 4
        1.2.1 Indices 5
        1.2.2 Dynamics and singularities 8
        1.2.3 Numerical aspects 10
    1.3 State vs. semistate modeling 11
    1.4 Formulations 12
        1.4.1 Input-output descriptions 13
        1.4.2 Leading terms 15
        1.4.3 Semiexplicit, semilinear and quasilinear DAEs .. 16
    1.5 Contents and structure of the book 20

Analytical aspects of DAEs 23

2. Linear DAEs and projector-based methods 25
    2.1 Linear time-invariant DAEs 26
        2.1.1 Matrix pencils and the Kronecker canonical form 27
        2.1.2 Solving linear time-invariant DAEs via the KCF . 28
        2.1.3 A glance at projector-based techniques 29
    2.2 Properly stated linear time-varying DAEs 35
        2.2.1 On standard form index one problems 36
        2.2.2 Properly stated leading terms 38
        2.2.3 P-projectors: Matrix chain and the tractability index 39
        2.2.4 The Π-framework 43
        2.2.5 Decoupling 51
        2.2.6 A tutorial example 65
        2.2.7 Regular points 74

2.3	Standard form linear DAEs		75
	2.3.1	The tractability index of standard form DAEs	75
	2.3.2	Decoupling	78
	2.3.3	Time-invariant problems revisited	79
2.4	Other approaches for linear DAEs: Reduction techniques		81

3. Nonlinear DAEs and reduction methods   83

3.1	Semiexplicit index one DAEs		85
3.2	Hessenberg systems		88
3.3	Some notions from differential geometry		90
3.4	Quasilinear DAEs: The geometric index		93
	3.4.1	The framework of Rabier and Rheinboldt	94
	3.4.2	Index zero and index one points	97
	3.4.3	Higher index points	103
	3.4.4	Manifold sequences and locally regular DAEs	108
	3.4.5	Local equivalence	111
	3.4.6	Examples	118
	3.4.7	Nonautonomous problems	123
3.5	Dynamical aspects		130
3.6	Reduction methods for fully nonlinear DAEs		133
3.7	The differentiation index and derivative arrays		134

4. Singularities   137

4.1	What is a singular DAE?		137
4.2	Singularities of properly stated linear time-varying DAEs		139
	4.2.1	Classification of singular points	140
	4.2.2	Decoupling	145
4.3	Singularities of standard form linear time-varying DAEs		154
	4.3.1	Classification	154
	4.3.2	Decoupling	156
	4.3.3	Analytic problems	157
4.4	Singularities of autonomous quasilinear DAEs		160
	4.4.1	Quasilinear ODEs and impasse phenomena	162
	4.4.2	Singular points of quasilinear DAEs	168
	4.4.3	A reduction framework for singular problems	171
	4.4.4	Dynamical aspects	179
	4.4.5	Singular semiexplicit index one DAEs	181
	4.4.6	Examples	184

Semistate models of electrical circuits　　　　　　191

5. Nodal analysis　　　　　　　　　　　　　　　　　　193

　5.1　Background on graphs and electrical circuits . . . .　195
　　　5.1.1　Graphs and digraphs . . . .　196
　　　5.1.2　Elementary aspects of circuit theory . . . .　203
　5.2　Formulation of nodal models . . . .　212
　　　5.2.1　Node Tableau Analysis . . . .　213
　　　5.2.2　Augmented Nodal Analysis . . . .　214
　　　5.2.3　Modified Nodal Analysis . . . .　215
　5.3　Index analysis: Fundamentals . . . .　216
　　　5.3.1　Structural form of nodal models . . . .　216
　　　5.3.2　On the tractability index of quasilinear DAEs . .　218
　5.4　Index analysis: Passive circuits . . . .　219
　　　5.4.1　Tableau equations and Augmented Nodal Analysis　220
　　　5.4.2　Modified Nodal Analysis . . . .　232
　5.5　Index analysis: Tree methods for non-passive circuits . . .　236
　　　5.5.1　Augmented Nodal Analysis . . . .　237
　　　5.5.2　Modified Nodal Analysis . . . .　243

6. Branch-oriented methods　　　　　　　　　　　　　　255

　6.1　Branch-oriented semistate models . . . .　257
　　　6.1.1　The basic model . . . .　258
　　　6.1.2　Tree-based formulations . . . .　259
　　　6.1.3　The state formulation problem . . . .　262
　6.2　Geometric index analysis and reduction of branch models　264
　　　6.2.1　Operating points . . . .　265
　　　6.2.2　Implicitly described resistors and strict passivity .　265
　　　6.2.3　Multiport reduction . . . .　271
　　　6.2.4　Index characterization . . . .　279
　　　6.2.5　State space reduction . . . .　283
　　　6.2.6　Controlled sources . . . .　289
　6.3　Qualitative properties . . . .　293
　　　6.3.1　Equilibria of DC circuits . . . .　295
　　　6.3.2　Nonsingularity . . . .　297
　　　6.3.3　Hyperbolicity and exponential stability . . . . . .　300

Bibliography 309

Index 325

Chapter 1

Introduction

Many systems in science and engineering can be modeled by an explicit ordinary differential equation (ODE) of the form

$$u' = f(u, t), \quad (1.1)$$

where $f \in C^1(\Omega, \mathbb{R}^r)$, the set Ω being open in $\mathbb{R}^{r+1}$. This equation is often said to define a *state space model* of the system, and u is referred to as a *state variable*. Explicit ODEs such as (1.1) have a long-term mathematical history, and a large number of analytical and numerical tools have been developed for their study.

However, in many cases such an explicit state space model for the dynamics of a given system is not available. The system may instead be described by an *implicit* ODE

$$F(t, x, x') = 0, \quad (1.2)$$

where $F : \Omega \to \mathbb{R}^m$ is defined on an open set $\Omega \subseteq \mathbb{R}^{2m+1}$. For reasons detailed later, the equation (1.2) is called a *semistate model* of the system and, accordingly, x is said to be a *semistate variable*.

If the algebraic (in the sense of non-differential) problem $F(t, x, p) = 0$ arising from (1.2) can be globally and uniquely solved for p, a global state model for the system can be derived, and the implicit formulation would not make any difference with the explicit ODE setting. From a local point of view, if F is a C^1 mapping and the matrix of partial derivatives $F_p(t, x, p)$ is nonsingular at a given zero of F, then by the implicit function theorem the set defined by $F(t, x, p) = 0$ can be locally described as $p = f(x, t)$ for a C^1 map f, so that $x' = f(x, t)$ defines a *local* state space model for (1.2).

Broadly speaking, the attention in this book will be mainly focused on the implicit ODE (1.2) under the assumption that the derivative $F_p(t, x, p)$ is everywhere singular. In many cases, some of the relations within (1.2) do not involve at all the time derivative x', hence being purely algebraic equations. This motivates calling (1.2) a *differential-algebraic equation* (DAE) or a *differential-algebraic system*. From the modeling perspective, algebraic constraints within the DAE (1.2) make some components of the m-dimensional semistate variable x redundant; the elimination of this redundancy may lead to a state space model of the form (1.1), in which the dimension of the state variable u is $r < m$.

For the sake of completeness, explicit ODEs such as (1.1), as well as implicit systems (1.2) in which F_p is everywhere nonsingular, are framed in the differential-algebraic context as *index zero* DAEs. Certain problems in which F_p is singular only on a subset (typically, a hypersurface) of the semistate space will also be addressed. In this setting, the points where F_p is singular are particular instances of what will be called a *singularity*; note, however, that the notion of a singularity will also appear in problems with an everywhere singular matrix F_p.

In this Chapter we present an overview of DAE theory. Some notes on the origin of DAEs, from a mathematical perspective but also regarding application fields where they show up, can be found in Section 1.1. Section 1.2 introduces several frameworks developed for the analysis of DAEs. Dynamical features and, in less detail, numerical aspects are also discussed. The role of DAEs in system modeling is examined in Section 1.3. Several issues concerning DAE formulations, as well as the structural forms that arise more often in applications, are compiled in Section 1.4. All this background will make it possible to define, in Section 1.5, the goals and contents of this monograph in a more detailed manner.

1.1 Historical remarks: Different origins, different names

The origins of DAE theory can be traced back to the work of K. Weierstrass and L. Kronecker on parametrized families of bilinear forms [144, 301]. In terms of matrices, these *pencils* were applied to the analysis of linear systems of ordinary differential equations with a possibly singular leading coefficient matrix by F. R. Gantmacher [90, 91]; see specifically Section 7 in Chapter XII of [91].

Another milestone is the work of P. Dirac on generalized Hamiltonian systems [79–81]. The key ideas supporting what nowadays is known as the differentiation index of a semiexplicit DAE can be found in these references. A geometric approach to the study of so-called *constrained sys-*

1.1. Historical remarks: Different origins, different names 3

tems stemmed from the work of Dirac, mainly motivated by applications in mechanics [17, 101, 102, 170, 171, 205, 208, 261–263]. From different perspectives, mechanical systems have driven much investigation on DAEs; cf. additionally [83, 122, 163, 227] and the bibliography therein.

A large amount of research on differential-algebraic equations has also been motivated by applications in circuit theory. The differential-algebraic form of circuit equations is naturally due to the combination of differential equations coming from reactive elements with algebraic (non-differential) relations modeling Kirchhoff laws and device characteristics. The term 'algebraic-differential system' was already used in the circuit context by Brown in 1963 [32]. In the electrical circuit literature, these models are often referred to as *semistate systems*. The word 'semistate', which is due to Dziurla (see p. 31 in [136]), appeared for the first time in the joint work of Dziurla and Newcomb [82]; see also [210]. In spite of their different origins, the terms 'semistate' and 'differential-algebraic' can be understood as synonyms hereafter. DAEs are nowadays pervasive in nonlinear circuit analysis and design, specially because of their appearance in nodal analysis methods used to set up network equations in circuit simulation programs [85, 87, 112–116, 194, 253, 292, 293]. The reader is referred to Chapters 5 and 6 for extensive discussions of the role of DAEs in circuit modeling.

Control theory has also been the focus of considerable attention in the DAE literature. Differential-algebraic equations are often called in this context *descriptor systems*, after the seminal work of Luenberger [168, 169]. Linear and nonlinear, possibly time-varying control systems have been the focus of an increasing interest from the differential-algebraic perspective in the last three decades, optimal control problems playing an important role; see [12, 14, 30, 50, 53, 73, 97, 109, 146, 150–154, 282, 306, 307] and references therein.

In the 1970s, a mathematical approach to differential-algebraic systems, somehow independent of specific application fields, began to flourish. It is worth mentioning at this point the work of Gear [93], Takens [288] and, in the early 1980s, the books of Campbell [39, 40] as well as the papers by Petzold [216], Gear and Petzold [95], and Rheinboldt [238]. Numerical aspects were emphasized in many of these references. A variety of approaches to DAE analysis began to be developed in that period, through the work of the above-mentioned authors as well as Griepentrog, Hairer, März, Rabier, and Reich, among others; see subsection 1.2.1 below. After 1990, the attention has also been directed to so-called *singular DAEs*, extending in different ways the seminal work of Rabier [219] and Chua and Deng [61, 62] on impasse points. Singular problems will be presented in subsection 1.2.2 and extensively discussed in Chapter 4. Much recent research is focused on *partial differential-algebraic equations* (PDAEs) and stochastic DAEs, beyond the scope of this book; as a sample of the related literature, see [2, 25, 111, 172, 176, 232, 264, 293] and [259, 305], respectively.

Besides the ongoing mathematical interest on analytical and numerical aspects of differential-algebraic systems, quite a lot of attention on DAEs has remained associated with applications in the above-mentioned fields of mechanics, electrical circuits and control theory; applications of DAEs are also found in chemistry [146], power systems [11, 134, 175, 296, 298], magnetohydrodynamics [48, 49, 174], neural networks [55, 211, 256], fault diagnosis [138, 295], model identification and observer design [57, 212], and robotics [46, 211, 278, 279], among other fields. DAEs also arise from problems in other branches of mathematics, such as the discretization of PDEs [30, 167], root-finding [245, 247, 254, 255] or optimization [260].

Other names for DAEs, not linked with specific applications, are 'implicit systems', 'singular systems' and, restricted to quasilinear problems (cf. subsection 1.4.3 below) 'generalized vector fields'. Finally, DAEs should not be confused with 'algebraic differential equations', that is, problems of the form $x' = p(x)$ in which p is a polynomial mapping.

1.2 DAE analysis

A C^1-*solution* of the differential-algebraic system (1.2) is a C^1 mapping $x : \mathcal{I} \to \mathbb{R}^m$, with $\mathcal{I} \subseteq \mathbb{R}$ an open interval, such that $(t, x(t), x'(t)) \in \Omega$ and $F(t, x(t), x'(t)) = 0$ for all $t \in \mathcal{I}$. Other types of solutions for DAEs will arise in different contexts, for instance when using properly stated formulations, or in the presence of impasse points. For the moment the term 'solution' will mean 'C^1-solution'.

A solution $x(t)$ which is *a priori* required to take a prescribed value $x(t_0) = x_0$ at a fixed time t_0 is said to have *initial value* x_0. Contrary to explicit ODEs (cf. (1.1)) defined by a C^1 map f, for DAEs it is usually the case that not every point admits a solution through it. A point x_0 for which there is indeed a solution of (1.2) with $x(t_0) = x_0$ is said to be a *consistent initial value* at t_0, and the set of pairs (x_0, t_0) admitting a solution is called the *solution set*. Additionally, we say that the DAE has a unique solution through x_0 at t_0 if any two solutions $x(t)$ and $\tilde{x}(t)$ (defined on the open intervals $\mathcal{I}$ and $\tilde{\mathcal{I}}$, respectively, both of them comprising t_0)

1.2. DAE analysis

such that $x(t_0) = \tilde{x}(t_0) = x_0$ verify $x(t) = \tilde{x}(t)$ for all $t \in \mathcal{I} \cap \tilde{\mathcal{I}}$. The solution is said to be locally unique if the condition $x(t_0) = \tilde{x}(t_0) = x_0$ implies $x(t) = \tilde{x}(t)$ on some neighborhood of t_0.

For autonomous DAEs (that is, problems of the form (1.2) with F independent of t), the consistency of an initial value does not depend on the choice of t_0, and the solution set can be thought of as lying on x-space. In many cases, the solution set has a manifold structure (cf. Section 3.3), solutions are uniquely defined through every point on this manifold, and they yield a C^1-flow (see e.g. [3]) on it. This has led to the *solution manifold* concept and to several definitions in the literature around the notions of a *solvable DAE* and a *regular DAE*.

However, in many other problems the above-mentioned properties are not met. For instance, solutions though a given point may not be unique; examples can be found in (4.35) on p. 167, and (4.61), p. 185. Furthermore, the solution set may not be a manifold, as the same examples illustrate. To accommodate solutions at the backward impasse points discussed in subsection 4.4.1 we would need to deal with C^0-semiflows. Note that these phenomena are not due to weak smoothness assumptions, since all of them can be displayed even in analytic problems. There are too many situations to handle, so that the discussion of notions such as that of the solution manifold or the one of a regular DAE is postponed until Chapter 3. At this introductory level, it is enough for the moment to have in mind the above-presented definition of a C^1-solution and the notion of the solution set.

1.2.1 Indices

The Kronecker index of regular linear time-invariant DAEs

Linear time-invariant DAEs are systems of the form

$$Ax' + Ex = q(t), \qquad (1.3)$$

where A and E belong to $\mathbb{R}^{m \times m}$ ($\mathbb{R}^{n \times p}$ denoting throughout the set of real matrices with n rows and p columns), and $q : \mathcal{J} \to \mathbb{R}^m$ is a sufficiently smooth mapping defined on an open interval $\mathcal{J} \subseteq \mathbb{R}$. As detailed in Section 2.1, the solutions of (1.3) can be completely characterized in terms of the Weierstrass-Kronecker canonical form of the matrix pencil $\{A, E\}$, provided that this pencil is a regular one; the *Kronecker index* (also called the *nilpotency index*) of the pencil will play a key role in this analysis. Needless to say, our interest will be directed to cases in which the leading matrix A is singular, since otherwise the problem trivially amounts to a linear explicit ODE with constant coefficients.

Linear time-varying and nonlinear DAEs

More difficult is the treatment of the linear time-varying counterpart of (1.3), namely

$$A(t)x'(t) + E(t)x(t) = q(t), \qquad (1.4)$$

as well as that of nonlinear problems of the form $F(t, x, x') = 0$ depicted in (1.2). From an analytical point of view, usually the solutions are not explicitly sought in terms of (1.2) or (1.4). Instead, most strategies aim at constructing a related mathematical object from which the solutions of the DAE can be described; this object may be an explicit ODE, a vector field on a manifold, or a canonical form for the equation. These strategies have led to different analytical frameworks, organized around different index notions which generalize the nilpotency index of a regular matrix pencil. These approaches are commented on below.

The differentiation index

The idea behind the *differentiation index* framework is, roughly speaking, to define the index of (1.2) as the number of differentiations needed to write x' in terms of (x, t). As illustrated in Sections 3.1 and 3.2, the constraints are differentiated in order to realize an explicit *underlying ODE* for which the solution set of the DAE is an invariant manifold. A more general discussion can be found in Section 3.7. When applied in particular to a linear time-invariant equation of the form (1.3) having a regular pencil, this notion yields the nilpotency index of the pencil.

The differentiation index approach has been developed mainly by S. Campbell; important contributions are also due to C. W. Gear and L. Petzold, among other authors. The reader is referred to the book [30] for a detailed introduction to this framework; see also [10, 42, 43, 45, 46, 54, 94].

The tractability index and projector-based methods

A different generalization of the nilpotency index of a regular matrix pencil comes from the *tractability index* notion, which is the key concept in the projector-based framework developed by R. März. Within this approach, the behavior of linear DAEs such as (1.3) or (1.4) is unveiled by means

1.2. DAE analysis

of an *inherent ODE* which, speaking again in general terms, is obtained from the projection of the equation and the problem variables into certain characteristic subspaces. Allowing for mild smoothness assumptions, this framework provides explicit solution characterizations for DAEs with arbitrarily high index, as well as precise functional input-output descriptions of the system behavior in the original problem setting.

The main references concerning projector techniques directed to linear time-varying problems of the form (1.4) are [107, 108, 179, 182, 185]; recently, a different formulation for DAEs, discussed in subsection 1.4.2 below, has allowed for a substantial improvement of this approach, regarding both linear and nonlinear problems; see [16, 188, 189, 191, 192, 190, 193] and references therein, as well as the forthcoming title [157]. Projector techniques for linear DAEs are extensively discussed in Chapter 2; Chapter 4 addresses singularities of these linear problems, whereas nodal models of electrical circuits are analyzed via projector-based methods in Chapter 5.

The geometric index and reduction methods

Reduction methods, based on the so-called *geometric index*, describe the behavior of an autonomous DAE in terms of a vector field defined on the solution set, provided that it has a manifold structure. This vector field defines a differential equation on this manifold and induces a flow on it. Using local parametrizations, the ODE on the solution manifold locally leads to an explicit *reduced ODE* on an open subset of $\mathbb{R}^r$; note that, in contrast to the underlying ODE arising in the differentiation index framework, the state dimension of this reduced ODE is strictly lower than that of the original DAE. In the original problem coordinates, the reduction process can be roughly described as the elimination of certain variables by solving the constraints. The solution manifold and the reduced ODE are computed in an iterative manner, and the name of iteration steps needed for the algorithm to stabilize defines the index.

The geometric reduction approach has a key reference in the paper [238] by W. Rheinboldt, and has been later developed by S. Reich [229–231] and in the joint research of P. Rabier and Rheinboldt [220, 221, 228]. Reduction methods for nonlinear problems are the focus of Chapter 3; special attention will be paid to quasilinear DAEs (cf. subsection 1.4.3 below), including nonautonomous cases. Their application to linear time-varying DAEs (1.4) is briefly addressed in Section 2.4. Singularities of quasilinear problems are tackled via reduction methods in Chapter 4 (Section 4.4).

Reduction techniques will be applied to the analysis of certain electrical circuit models in Chapter 6.

Perturbation, strangeness and structural indices

A salient feature of linear time-invariant DAEs (1.3) with an index $\nu \geq 2$ regular pencil is that solutions will depend explicitly on the derivatives of the excitation $q(t)$ up to order $\nu - 1$. This dependence, which is displayed (in a more subtle way) also by linear time-varying and nonlinear problems, supports the *perturbation index* concept [121], which reflects the sensitivity of the DAE solutions to perturbations. See also [45, 122].

The reader is referred to the book [151] for an extensive analysis of the *strangeness index* notion and related canonical forms for DAEs introduced by P. Kunkel and V. Mehrmann. Other references on this topic are [147–150, 152]; some relations with the tractability and geometric indices are discussed in [157] and [228], respectively.

Finally, different aspects of the so-called *structural index* are addressed in [215, 237].

The paragraphs above give an overview of the main frameworks developed for the analysis of DAEs. This book will be mainly focused on projector-based and reduction methods, supported on the tractability and geometric indices, respectively. In these contexts, when the assumptions supporting the index notion are met the DAE is called *regular*. The failing of these assumptions will lead to *singular* problems, presented below.

1.2.2 Dynamics and singularities

As sketched above, when the conditions supporting the tractability or geometric index notions are met, the behavior of the DAE (1.2) can be described in terms of a related (inherent or reduced, respectively) explicit ODE. For the moment, we will informally refer to these systems as *regular*. In spite of the mathematical challenges raised by the unveiling of such an explicit ODE, from a dynamical point of view a regular autonomous DAE does not lead to new qualitative phenomena. For instance, in problems with a well-defined geometric index the dynamical behavior of the DAE is given by the flow induced by a vector field on a lower-dimensional manifold. Locally, the solutions of a reduced ODE are embedded onto those of the DAE, and therefore the flows of both are locally diffeomorphic.

1.2. DAE analysis

Things change substantially in the presence of *singularities*. Singular points of a DAE will be defined as those where the assumptions supporting the definition of an index fail, and indeed yield several types of dynamical behavior not displayed by explicit ODEs. The best-known phenomenon associated with singular DAEs is the one occurring at *impasse points* [61, 62, 219, 222, 223, 233]. Other singular phenomena in autonomous DAEs have been addressed only for particular structures, mainly quasilinear problems with a dense subset of regular points [165, 202, 203, 236, 254, 276, 277, 309] and semiexplicit DAEs [22, 23, 173, 236, 249]. Last-step singularities (cf. Chapter 4) were considered by Rabier and Rheinboldt in [228]. From a different perspective, several results concerning singularities of scalar problems have been discussed in [7, 34, 72, 289]. Singular bifurcations are tackled in [19–21, 166, 241–243, 246, 275, 296–298]. In the linear time-varying context singularities have been studied in [196, 197]; some results can also be found in [139, 225, 228].

A systematic approach to the local analysis of singular DAEs is presented in Chapter 4; no previous framework for the general study of singularities in differential-algebraic systems is known to the author. An outline of the main ideas follows. After providing invariant definitions of singular points in linear time-varying and quasilinear autonomous DAEs, the "regular" analysis frameworks will be modified in order to accommodate singular problems. In both contexts, this will result in an implicit ODE which can be rewritten in explicit form on an open dense subset of its domain, the leading term of which captures the singularities of the original DAE. In particular, this approach extends the scope of the results of Rabier and Rheinboldt for quasilinear systems. Broadly speaking, the working scenarios allowing for this replace constant rank assumptions by the requirement that certain characteristic spaces admit a continuation through the singularity, thereby relaxing the hypotheses which support the analysis in regular settings.

Qualitative properties of DAEs

Qualitative results for DAEs are of major importance in different applications. For regular autonomous DAEs, even though the local dynamics amount to that of explicit ODEs, it is necessary to have tools allowing one to assess qualitative properties directly in the differential-algebraic setting, since an explicit state space description of the system dynamics may not be feasible in many practical cases. The qualitative theory of DAEs makes such an explicit ODE description unnecessary. In this regard it is important to examine, in DAE terms, stability properties of invariants such as equilibrium points or periodic trajectories, as well as the associated bifurcations characterizing the dependence of these properties on system parameters.

Although a general discussion of qualitative aspects of differential-algebraic systems would exceed the scope of this book, we will address some issues in this direction. Linear stability properties of regular equilibria in quasilinear DAEs will be characterized in terms of matrix pencils in Section 3.5; from different points of view, this type of results has been discussed in [92, 183, 184, 186, 228, 231, 240, 281, 291]. The reduction framework seems to be well-suited for characterizing the situations in which the matrix pencil which results from the DAE linearization describes the stability properties of equilibria. The electrical counterpart of these results is tackled in Section 6.3, where a key point will be to distinguish the topological conditions characterizing the hyperbolicity or exponential stability of equilibrium points from those allowing for the formulation of a state space model for the network, a distinction which is not always clear in the circuit literature.

Other qualitative aspects of differential-algebraic systems have been analyzed in the last two decades. Local normal forms are discussed in [21–23, 67, 68, 202, 236, 277, 309]. Regarding periodic DAEs and periodic solutions, several results concerning Floquet theory for DAEs can be found in [158, 159]; related aspects involving the analysis of nonlinear oscillations in semistate models of electrical circuits are addressed in [74]. Finally, the reader is referred to Chapter 4 and the bibliography cited there for the discussion of qualitative phenomena at singularities, in particular at singular equilibria and pseudoequilibria.

1.2.3 Numerical aspects

The numerical treatment of differential-algebraic systems has driven a great deal of research on this field. Most monographs on DAEs are partially (or, in some cases, totally) devoted to numerical methods: see [10, 30, 107, 121, 122, 151, 157, 228]. Essentially, all the frameworks discussed in subsection 1.2.1 have a numerical counterpart. This book will be focused on analytical properties of DAEs, and hence the reader is referred to the above-mentioned titles for detailed discussions of computational issues. Without any attempt to be exhaustive, other important references are [8, 38, 44, 47, 51, 52, 96, 129–133, 140, 141, 182, 216, 217, 267, 270].

1.3. State vs. semistate modeling

Regarding the problem of computing consistent initial values, see [30, 33, 86, 97, 124, 126, 156, 162, 215] and the bibliography therein. Numerical techniques for the analysis of singular DAEs are discussed in [9, 199, 207, 218, 223, 228, 275, 304], among other references.

1.3 State vs. semistate modeling

The dichotomy between DAEs and explicit ODEs is also relevant from the point of view of system modeling. In fact, much attention has been directed to DAEs because of its chance to model easily systems in which certain constraints are imposed among the variables appearing within a given set of differential equations. This is very clear for instance in circuit theory, where time-domain models of electrical networks combine differential equations arising from capacitors and inductors with algebraic ones coming from Kirchhoff laws and the characteristics of devices (cf. Sections 5.2 and 6.1).

These constraints mean that there exist some *redundancy* among the model variables. For this reason differential-algebraic models are sometimes called *semistate models* [82, 210]; the idea is that, in contrast to *state space models* based on explicit ODEs of the form (1.1), where an initial value can be freely assigned to every component of the state variable u, in a DAE such as (1.2) an initial value can be specified only for some components of the vector x for a solution to be well-defined. The dynamic degree of freedom of a given DAE, which is a local quantity, will be called the (local) state dimension of the problem. More details in this regard can be found in the reduction framework of Chapter 3.

From the modeling perspective, the elimination of redundant variables can be seen as a *model reduction* process. This includes, in particular, the derivation of a state space description of a given system by means of the reduction of an initial DAE model; in this direction, the state formulation problem for electrical circuits will be tackled in Chapter 6 as a reduction of certain semistate models. On the other hand, in many practical cases only some of the variables are eliminated from a given semistate equation, yielding another DAE which may be easier to analyze while still capturing the essential features of the system. Again, the electrical circuits discussed in Chapters 5 and 6 will provide several examples of this, since for instance augmented nodal analysis or multiport models can be seen as intermediate formulations between a tableau or a branch-oriented model, in which no variable has been eliminated and therefore redundancy is maximal, and a state space model, displaying no redundancy among variables.

It is important to notice that a reduction process leading to a state space model of a given system is not always feasible in practice. When this is the case, the thoroughly discussed mathematical tools for the analysis of dynamical systems formulated in terms of explicit ODEs become unavailable. These tools include, in particular, qualitative theory and numerical integration methods. In these cases the DAE-oriented techniques referred to in subsections 1.2.2 and 1.2.3 above, which apply without recourse to a state space reduction, become very relevant in system analysis and simulation.

Finally, a particularly important role in system modeling is played by so-called *semiexplicit* DAEs (cf. subsection 1.4.3 below), having the form

$$y' = h(y, z, t) \tag{1.5a}$$
$$0 = g(y, z, t). \tag{1.5b}$$

These systems arise in applications when algebraic (that is, non-differential) constraints are explicitly added to a set of differential relations, but also as the reduced equation of the *singular perturbation* problem

$$y' = h(y, z, t) \tag{1.6a}$$
$$\varepsilon z' = g(y, z, t) \tag{1.6b}$$

where, typically, $0 < \varepsilon \ll 1$. Conversely, the way from (1.5) to (1.6) is usually called a *regularization* of the DAE [10, 30, 142, 143, 214]. Semiexplicit DAEs also arise as the *enlarged* system

$$x' = p \tag{1.7a}$$
$$0 = F(t, x, p) \tag{1.7b}$$

of the fully implicit problem (1.2).

1.4 Formulations

When using DAEs to describe physical systems, several options arise in the formulation of the model. In particular, there are different ways to capture the variables which are required to be differentiated. If the derivatives of all the components of the semistate vector x are involved in the model, as it happens in explicit ODEs, we are faced with a *standard form DAE*. Nevertheless, this presents certain disadvantages which are discussed, in terms of input-output system descriptions, in subsection 1.4.1 below. This

1.4. Formulations

will motivate the introduction of so-called *properly stated DAEs*, presented in subsection 1.4.2. Finally, in subsection 1.4.3 we discuss some structural forms for DAEs which are often displayed in applications.

1.4.1 Input-output descriptions

Explicit ODEs

Let us consider from a functional point of view the linear constant coefficient ODE

$$x' + Ex = q(t), \qquad (1.8)$$

where E is an $m \times m$ real matrix, together with an initial condition

$$x(t_0) = x_0 \in \mathbb{R}^m. \qquad (1.9)$$

The initial value problem (IVP) defined by (1.8) and (1.9) is known to yield a *smoothing* operator

$$\begin{aligned} C^k(\mathcal{J}, \mathbb{R}^m) &\longrightarrow C^{k+1}(\mathcal{J}, \mathbb{R}^m) \\ q(t) &\longrightarrow x(t), \end{aligned}$$

where $x(t)$ is the unique solution of the ODE (1.8) which satisfies (1.9), in the understanding that $t_0 \in \mathcal{J}$. This means that the initial value problem induces a mapping between the functional spaces $C^k(\mathcal{J}, \mathbb{R}^m)$ and $C^{k+1}(\mathcal{J}, \mathbb{R}^m)$, for any $k \geq 0$; this point of view is important for instance in system theory or in electrical engineering, where $q(t)$ and $x(t)$ are viewed as input and output signals of a system defined by the matrix E and with initial state x_0.

Moreover, the ODE (1.8) alone defines a *bijection*

$$\begin{aligned} C^k(\mathcal{J}, \mathbb{R}^m) \times \mathbb{R}^m &\longrightarrow C^{k+1}(\mathcal{J}, \mathbb{R}^m) \\ (q(t), x_0) &\longrightarrow x(t), \end{aligned}$$

with an inverse given by $x(t) \to (x'(t) + Ex(t), x(t_0))$. The bijective nature of this mapping somehow indicates that C^k-spaces are the right ones to accommodate excitations and solutions of the problem. Note that the same remarks apply if we replace the constant matrix E in (1.8) by a time-varying one $E(t) \in C^k(\mathcal{J}, \mathbb{R}^{m \times m})$.

Linear time-invariant DAEs

In the differential-algebraic context one is often interested in obtaining similar functional descriptions of the solution behavior and, specially, in providing functional spaces where excitations and solutions are mapped into one another bijectively. As discussed below, and unlike the ODE case, neither the DAE formulations handled so far nor the use of C^k spaces are appropriate in this regard.

Indeed, consider the linear time-invariant DAE (1.3), and assume that the matrix pencil $\{A, E\}$ is regular with Kronecker index one (cf. Section 2.1). It will shown that in this situation there exist nonsingular matrices G, H such that the solutions of (1.3) can be obtained from those of

$$u' + Wu = \tilde{q}_1(t) \tag{1.10a}$$
$$v = \tilde{q}_2(t), \tag{1.10b}$$

for a certain matrix W, by means of the transformation $x = Hw$. Here the variable w stands for (u, v), whereas the excitation $\tilde{q}(t) = (\tilde{q}_1(t), \tilde{q}_2(t))$ in (1.10) is defined by $\tilde{q}(t) = Gq(t)$.

Let us assume that $q(t)$ (and thereby $\tilde{q}(t)$) is continuous. System (1.10) certainly has a solution $w(t) = (u(t), v(t))$, defined by a C^1-solution $u(t)$ of (1.10a) and a C^0-mapping $v(t)$ given by (1.10b). But if the excitation $q(t)$ (or, more precisely, $\tilde{q}_2(t)$) is not differentiable, then neither are $v(t)$, $w(t)$, nor $x(t) = Hw(t)$; in this situation it does not make sense to differentiate x in (1.3). This raises the problem of the sense in which $x(t) = Hw(t)$ can be considered a solution of (1.3) in cases in which $q(t)$ is just a continuous map. We are additionally faced with the characterization of the functional spaces of (i) solutions onto which continuous excitations are mapped, and (ii) excitations yielding C^1-solutions.

We might agree to restrict the attention to C^1-excitations $q(t)$. This is however unsatisfactory, not only because continuous excitations seem to be acceptable in the light of (1.10), but also when looking for the space of solutions $x(t)$ onto which C^1-excitations are mapped.

This introductory digression shows that, for index one DAEs, there is no hope for a bijective transformation $C^k \to C^{k+1}$ or $C^k \to C^k$ of excitations onto solutions; the cases considered above correspond to $k = 0, 1$, but the same is true for $k \geq 2$. Note that for the sake of simplicity we are deliberately vague concerning initial values.

A solution can be devised by rewriting $Ax' = (Ax)'$ and then considering, instead of (1.3), the reformulation

$$(Ax)' + Ex = q(t), \tag{1.11}$$

which comprises all C^1-solutions of (1.3) but allows for the search of additional ones in the larger space

$$C_A^1(\mathcal{J}, \mathbb{R}^m) = \{x \in C^0(\mathcal{J}, \mathbb{R}^m) \ / \ Ax \in C^1(\mathcal{J}, \mathbb{R}^m)\}.$$

Continuous excitations $q(t)$ can now be shown to yield a solution in the space C_A^1, whereas C^k excitations lead to solutions within

$$C_A^{k+1}(\mathcal{J}, \mathbb{R}^m) = \{x \in C^k(\mathcal{J}, \mathbb{R}^m) \ / \ Ax \in C^{k+1}(\mathcal{J}, \mathbb{R}^m)\}.$$

This type of reformulations will also lead to a characterization of the excitations which are bijectively mapped onto C^k-solutions.

The discussion above is restricted to linear time-invariant DAEs with nilpotency index one. Using the Kronecker canonical form (see Section 2.1), these remarks can be extended to linear time-invariant problems with higher index, under stronger smoothness requirements on $q(t)$. However, in the linear time-varying framework defined by (1.4), and also in nonlinear contexts, this kind of results are more involved. These issues can be tackled in time-varying and/or nonlinear settings by means of a recently proposed formulation for the leading term of DAEs, detailed below. The reader is referred to subsections 2.1.3.2 and 2.2.1 for additional details concerning input-output functional descriptions.

1.4.2 Leading terms

As indicated above, rewriting the leading term of (1.3) as $(Ax)'$ yields several advantages from the analytical point of view. The same can be achieved by recasting the leading term as $A(Px)'$, where P is a projector along $\ker A$, based on the identity $A = AP$ (note that $A(I - P) = 0$).

The key remark at this point is that these ideas can be extended to the linear time-varying context. Indeed, several benefits will follow from the consideration of DAEs of the form

$$A(t)(D(t)x)' + B(t)x = q(t), \tag{1.12}$$

instead of (1.4); see the seminal work [16] and [157, 188, 189, 191, 192] together with the references therein. The leading term of (1.12) will be said to be *properly stated* when the matrix mappings $A(t)$ and $D(t)$ are well-matched in a certain sense (cf. Definition 2.2 on p. 38). The above-mentioned input-output functional descriptions are better tackled in the framework defined by (1.12), which allows for a precise formulation of smoothness requirements on the excitation $q(t)$ and the matrix mappings $A(t)$, $D(t)$ and $B(t)$. Additionally, the leading term of (1.12) arises in this form in different circuit and control applications and, in particular, displays nice symmetry properties when considering adjoint formulations [15, 16, 187]. Several advantages are also met from the numerical point of view [130–132, 157, 293].

Certainly, a lot of DAE literature is directed to the form (1.4). We will refer to these problems as *standard form* linear DAEs. Under mild requirements on the leading matrix $A(t)$, a standard form DAE can be rewritten in the properly stated form (1.12), and therefore the results coming from the properly stated framework can be applied to standard form problems; see, in this regard, Sections 2.3 and 4.3.

Properly stated formulations can be also used in nonlinear settings; some instances can be actually found in subsection 5.3.1. This involves, however, some additional complexities, and most readers will be unfamiliar with this type of formulation in the nonlinear context. Due to the intrinsic difficulties of nonlinear problems, in their analysis we will try to keep at a minimum certain technicalities. Therefore, the attention in Chapter 3 will be restricted to standard form DAEs, and precise smoothness requirements will be disregarded; that is, all objects will simply be assumed to be smooth enough, and the C^∞ setting will be often invoked for that. In any case, we believe that the results of Chapter 2 should be enough to give a reader a solid understanding of the benefits of properly stated formulations, and we refer him/her to the forthcoming title [157] for an extensive discussion of nonlinear, properly stated DAEs.

1.4.3 Semiexplicit, semilinear and quasilinear DAEs

Many DAEs in applications display some kind of structure. In most practical cases the equation is linear in the derivative x'; depending on the specific form of the matrix map in front of this derivative, these *quasilinear* or *linearly implicit* systems have received several names in the literature. The semilinear and quasilinear problems here presented can be thought of as being located between linear DAEs and the "nonlinear" system (1.2). Note that these families are not mutually exclusive; quite on the contrary, linear DAEs are a particular case of semilinear problems (mind, in this regard, that semiexplicit problems may well be linear), which in turn are instances of quasilinear DAEs with a leading matrix independent of x. Similarly, the form (1.2) comprises quasilinear problems. Actually, referring to (1.2) as a 'nonlinear DAE' is a terminological abuse since it may stand in particular for linear cases. System (1.2) is often called a *fully implicit* DAE.

1.4. Formulations

1.4.3.1 Semiexplicit and semilinear DAEs

In an autonomous setting, *semiexplicit* DAEs are defined by a system of the form

$$y' = h(y, z) \tag{1.13a}$$
$$0 = g(y, z), \tag{1.13b}$$

where for the moment we disregard smoothness requirements on the maps $h : W_0 \to \mathbb{R}^r$ and $g : W_0 \to \mathbb{R}^p$. Here W_0 is an open set in $\mathbb{R}^{r+p}$. The form depicted in (1.13) makes it possible to distinguish the r-dimensional vector y of *dynamic variables* from the p-dimensional one z of *algebraic* ones. The nonautonomous counterpart of (1.13) is displayed in (1.5).

It is worth remarking at this point that the term 'differential-algebraic equation' is sometimes used in the literature to mean the semiexplicit equation (1.13), and even to refer to (1.13) under the assumption that the matrix of partial derivatives g_z is everywhere nonsingular. The nonsingularity of g_z will be later defined as an *index one* requirement.

Autonomous *semilinear* DAEs are problems of the form

$$Ax' = f(x), \tag{1.14}$$

where A is an $m \times m$ real matrix and $f : W_0 \to \mathbb{R}^m$ is a sufficiently smooth mapping, W_0 being open in $\mathbb{R}^m$. Semilinear equations comprise in particular the semiexplicit DAE (1.13), since the latter can be written as

$$\begin{pmatrix} I_r & 0 \\ 0 & 0 \end{pmatrix} \begin{pmatrix} y' \\ z' \end{pmatrix} = \begin{pmatrix} h(y, z) \\ g(y, z) \end{pmatrix}$$

which is a system of the form (1.14) with $x = (y, z)$, $f = (h, g)$. Conversely, semilinear DAEs may written in semiexplicit form: denoting $\operatorname{rk} A = r$, there exist nonsingular matrices G, H such that [90]

$$GAH = \begin{pmatrix} I_r & 0 \\ 0 & 0 \end{pmatrix}.$$

Now, if we premultiply (1.14) by G and perform the coordinate change $x = H\tilde{x}$, the semilinear system reads

$$\begin{pmatrix} I_r & 0 \\ 0 & 0 \end{pmatrix} \tilde{x}' = Gf(H\tilde{x})$$

which, via the splitting $\tilde{x} = (y, z)$, is easily checked to be a semiexplicit equation.

The semiexplicit DAE (1.13) can be written in properly stated form as

$$\begin{pmatrix} I_r \\ 0 \end{pmatrix} \left[\begin{pmatrix} I_r & 0 \end{pmatrix} \begin{pmatrix} y \\ z \end{pmatrix} \right]' = \begin{pmatrix} h(y,z) \\ g(y,z) \end{pmatrix}, \qquad (1.15)$$

where

$$A = \begin{pmatrix} I_r \\ 0 \end{pmatrix}, \quad D = \begin{pmatrix} I_r & 0 \end{pmatrix},$$

define a well-matched pair of matrices yielding a properly stated leading term; cf. subsection 2.2.2 for additional details in this regard.

Finally, by allowing the above-introduced operators A and f depend explicitly on the time t, we obtain the nonautonomous analog of (1.14), namely,

$$A(t)x' = f(x,t).$$

1.4.3.2 Hessenberg DAEs

In several applications, the mappings h and g within the semiexplicit DAE (1.5) depend only on certain variables. This is often the case in mechanics or, as will be detailed in Chapter 6, in electrical circuit theory. In this direction, we present below so-called *Hessenberg DAEs* directly in the nonautonomous context.

When the constraints g do not depend on the algebraic variable z, we are led to a *Hessenberg DAE of size two*. These systems can be written as

$$y' = h(y, z, t) \qquad (1.16\text{a})$$
$$0 = g(y, t). \qquad (1.16\text{b})$$

For simplicity, we assume that $h : \mathbb{R}^{r+p+1} \to \mathbb{R}^r$ and $g : \mathbb{R}^{r+1} \to \mathbb{R}^p$, although the domains of these mapping may be assumed to be open sets $W_0 \subseteq \mathbb{R}^{r+p+1}$ and $\hat{W}_0 \subseteq \mathbb{R}^{r+1}$. Labeling (1.16) as a 'size two' Hessenberg DAE without additional requirements comprises a terminological abuse, since this term is usually reserved to systems of the form (1.16) in which the product $g_y h_z$ is nonsingular [30, 70]. Later on we will elaborate further on the nonsingularity of $g_y h_z$, which will define an *index two* condition for (1.16); note that this requires, in particular, that $r \geq p$.

If, additionally, in the Hessenberg DAE (1.16) the mapping g does only depend on t and some of the dynamic variables y (to be denoted by $y_2 \in \mathbb{R}^{r_2}$, with $r_2 < r$), and the corresponding components of h do not depend on z,

1.4. Formulations

we are led to a *Hessenberg DAE of size three*, which may be written as

$$y_1' = h_1(y_1, y_2, z, t) \tag{1.17a}$$
$$y_2' = h_2(y_1, y_2, t) \tag{1.17b}$$
$$0 = g(y_2, t). \tag{1.17c}$$

Analogously, a *Hessenberg DAE of size k* has the structure

$$y_1' = h_1(y_1, y_2, \ldots, y_{k-1}, z, t) \tag{1.18a}$$
$$y_i' = h_i(y_{i-1}, \ldots, y_{k-1}, t), \quad 2 \leq i \leq k-1 \tag{1.18b}$$
$$0 = g(y_{k-1}, t), \tag{1.18c}$$

where $y_i \in \mathbb{R}^{r_i}$ for $i = 1, \ldots, k-1$, $z \in \mathbb{R}^p$. This DAE will be index k if the product

$$\frac{\partial g}{\partial y_{k-1}} \frac{\partial h_{k-1}}{\partial y_{k-2}} \cdots \frac{\partial h_2}{\partial y_1} \frac{\partial h_1}{\partial z} \tag{1.19}$$

is nonsingular. This implies that $r_i \geq p$ for $i = 1, \ldots, k-1$. The above-mentioned terminological abuse applies also here, since the nonsingularity of (1.19) is usually required in order to call (1.18) a 'size k' Hessenberg DAE; this holds in particular for the case $k = 3$ displayed in (1.17).

The term *Hessenberg DAE* stands for a DAE which has a Hessenberg structure of size k for some $k \geq 2$. Sometimes the semiexplicit DAE (1.13) itself is also considered as a Hessenberg DAE [10].

1.4.3.3 Quasilinear DAEs

Finally, *quasilinear* autonomous DAEs are systems of the form

$$A(x)x' = f(x), \tag{1.20}$$

where $A(x)$ and $f(x)$ are sufficiently smooth mappings $W_0 \to \mathbb{R}^{m \times m}$ and $W_0 \to \mathbb{R}^m$, respectively; the set W_0 is open in $\mathbb{R}^m$. The pair (A, f) is sometimes called a *generalized vector field* on W_0 [67, 68, 202, 203].

Most cases of interest in applications will be defined by an everywhere singular matrix mapping $A(x)$. However, in the reduction framework of Chapter 3 an important role will be played by problems of the form (1.20) in which $A(x)$ is singular only on a codimension-one submanifold of W_0. These cases have received considerable attention in the literature [165, 166, 202, 203, 236, 240, 254, 276, 277, 309]; we will reserve the term *quasilinear ODE* to refer to problems of the form (1.20) in which $A(x)$ is nonsingular on a dense subset of W_0.

The nonautonomous counterpart of a quasilinear DAE is defined by the assumption that A and/or f depend on t, yielding an equation of the form
$$A(x,t)x' = f(x,t).$$
These systems arise for instance when setting up electrical circuit models using Modified Nodal Analysis (MNA); cf. Chapter 5.

1.5 Contents and structure of the book

With the background presented in previous Sections, we are now in a position to define the goals of the present book, and also to provide a more detailed description of its contents and structure. Broadly speaking, the main goal is to present several frameworks for the analysis of differential-algebraic systems and, subsidiarily, of semistate circuit models. Thus, the two parts of the book are respectively devoted to the discussion of analytical aspects of DAEs in general, and to different issues arising in the use of DAEs in electrical circuit modeling.

In Part I, we discuss in detail several analysis methods for DAEs. The main focus is on projector-based techniques and reduction methods. Both approaches certainly apply to linear time-varying and nonlinear DAEs; nevertheless, a detailed discussion of both frameworks in these two settings would be excessively long. For this reason, projector methods will be discussed in the context of linear time-varying systems, whereas reduction techniques will be addressed for nonlinear DAEs; special emphasis will be put on quasilinear problems. However, reduction techniques for linear problems are briefly presented in Section 2.4, whereas some remarks on the use of projector methods for the analysis of quasilinear DAEs arising in circuit modeling can be found in Section 5.3. More details can be found in [228] and [157], respectively.

Therefore, Chapters 2 and 3 address linear and nonlinear DAEs via projector-based and reduction techniques, respectively. Undoubtedly, the reader will find a certain gap between both approaches, and also with respect to the differentiation index framework (cf. Section 3.7) but, from the author's point of view, the different perspectives should provide a richer knowledge of the fundamentals of DAE analysis. In spite of the salient differences between the projector and reduction frameworks, the main ideas somehow go in parallel. In both cases we examine in detail the assumptions supporting the (tractability and geometric) index notions in which the methods are supported, arriving at two results (Theorems 2.3 and 3.2 on

1.5. Contents and structure of the book

pages 51 and 107, respectively) which characterize the solutions of DAEs in regular contexts. The failing of these assumptions will lead to the analysis of singularities carried out in Chapter 4.

Part II tackles different analytical issues in electrical circuit theory using semistate (differential-algebraic) models. The use of time-domain formulations make the results applicable both to linear and nonlinear problems. Two model families are considered; those based on nodal analysis are considered in Chapter 5, whereas branch-oriented models define the scope of Chapter 6. In both cases, the focus is placed on index analyses.

As detailed in Chapter 5, nodal models have attracted quite a lot of attention in the DAE literature, specially regarding Modified Nodal Analysis (MNA) systems, widely used in circuit simulation programs. The tractability index and projector methods have succeeded in providing an accurate characterization of the index of these models in passive contexts; we will discuss these results and present as well several index characterizations for non-passive problems by means of tree-based techniques.

The branch-oriented models in Chapter 6 have been comparatively overlooked from the DAE point of view. We attempt to illustrate that the differential-algebraic formalism is certainly of interest also in the analysis of these circuit models. In particular, the geometric index framework makes it possible to recast the state formulation problem and the normal tree method as a reduction of a branch-oriented circuit model. Several advantages, regarding for instance qualitative properties of lumped circuits, will be derived from this approach.

We aim to discuss all these results in a self-contained manner, by means of the inclusion of background material on different topics. From a mathematical point of view, we compile some basic results coming from linear algebra (involving matrix pencils, projectors or Schur complements, as well as specific results such as the Cauchy-Binet formula), differential geometry and graph theory, which virtually reduce the technical prerequisites for reading this book to basic courses on differential calculus, linear algebra and ordinary differential equations.

Regarding circuit theory, Section 5.1 presents a detailed introduction to the fundamentals of electrical circuit analysis. Subsection 5.1.1 may be useful for readers interested in getting an introduction to elementary aspects of graph theory, regardless of specific applications. The material in Section 5.1, together with that of Sections 5.2 and 6.1, should allow readers without a background in circuit analysis to profit from the results discussed in Chapters 5 and 6.

How to read this monograph

Some readers will be mainly interested on certain parts of this book. There are several ways to go through the material here presented; some of them are proposed below.

Readers who aim to get an introductory vision on DAEs, or maybe trying to get a glimpse at the topic before going into details, should probably read this Chapter and then Sections 2.1.1, 2.1.2, 3.1, 3.2 and 4.1, before proceeding further.

Mathematically-oriented readers without a substantial background on DAEs may find of interest, from an analytical point of view, Sections 2.2 and 2.3 for linear time-varying DAEs, Sections 3.4 and 3.5 for quasilinear problems, and Chapter 4 for the study of singularities in both contexts.

Expert readers interested on specific approaches will find projector methods and the tractability index in Chapter 2, focused on linear systems. Sections 4.2 and 4.3 adapt the projector-based framework to singular linear time-varying problems, whereas some ideas concerning the tractability index of quasilinear DAEs arising in circuit theory can be found in Section 5.3. Reduction methods and the geometric index are discussed for linear problems in Section 2.4 and for nonlinear systems in Chapter 3; singularities of quasilinear DAEs are then addressed in Section 4.4. The differentiation index notion is presented in Section 3.7, after some introductory remarks in Sections 3.1 and 3.2.

Finally, readers interested on applications of DAEs will find a thorough discussion of differential-algebraic circuit models in Chapters 5 and 6, many parts of which can be read independently of the rest of the book.

PART I
Analytical aspects of DAEs

Chapter 2

Linear DAEs and projector-based methods

In this Chapter we undertake the study of linear differential-algebraic equations, in both the time-invariant and time-varying contexts. In a time-invariant setting, linear DAEs read

$$Ax' + Ex = q(t), \qquad (2.1)$$

where A, $E \in \mathbb{R}^{m \times m}$ and $q(t) \in C(\mathcal{J}, \mathbb{R}^m)$, $\mathcal{J} \subseteq \mathbb{R}$ being an open interval. Smoothness requirements on q will be discussed later. The problems of interest will be those in which A is a singular matrix, since otherwise the system trivially amounts to an explicit linear constant-coefficient ODE; these particular cases are accommodated in the differential-algebraic framework as *index zero* problems. The linear time-invariant DAE (2.1) will be tackled in terms of the *matrix pencil* $\{A, E\}$ [90, 91, 144, 301], as detailed in Section 2.1. Under an assumption of regularity on the pencil, a key role will be played by its *nilpotency index*.

The time-varying analog of (2.1) can be written as

$$A(t)x'(t) + E(t)x(t) = q(t), \quad t \in \mathcal{J}, \qquad (2.2)$$

with $A(t)$, $E(t) \in C(\mathcal{J}, \mathbb{R}^{m \times m})$, $q(t) \in C(\mathcal{J}, \mathbb{R}^m)$. This will be called the *standard form* for a linear time-varying DAE.

We will also consider linear DAEs of the more general form

$$A(t)(D(t)x(t))' + B(t)x(t) = q(t), \quad t \in \mathcal{J}, \qquad (2.3)$$

where $A(t) \in C(\mathcal{J}, \mathbb{R}^{m \times n})$, $D(t) \in C(\mathcal{J}, \mathbb{R}^{n \times m})$, $B(t) \in C(\mathcal{J}, \mathbb{R}^{m \times m})$ and $q(t) \in C(\mathcal{J}, \mathbb{R}^m)$. Under certain conditions specified later, this DAE will be said to be *properly stated*. Note that, under mild assumptions, the standard form DAE (2.2) can be rewritten in the form (2.3). Both forms will be extensively discussed in Sections 2.2 and 2.3, where in particular we detail smoothness requirements on A, E, D, B and q.

Different approaches have been proposed in the last decades for the analysis of linear time-varying DAEs. They can be roughly classified into four groups which are structured around different index concepts, namely the differentiation, tractability, geometric, and strangeness indices. A glimpse at these notions can be found in Chapter 1 (pp. 6-8). The frameworks based on the differentiation and the strangeness indices are addressed in detail in [30] and [151], respectively. The reader is referred to these books but also to [41, 42, 54] and [147, 148] for extensive discussions of these approaches; in particular, so-called *standard canonical forms* can be found in [30, 41, 42]. The differentiation index will be briefly analyzed for fully implicit problems in Section 3.7.

In this Chapter we will focus our attention mainly on projector-based methods developed around the tractability index and, in less detail, on reduction methods supported upon the geometric index. Since the main interest will be directed to projector techniques, we detail here the structure of Sections 2.1.3, 2.2 and 2.3, in which we present this framework. In order to provide the reader with a friendly introduction to the tractability index approach, we discuss projector methods first in the time-invariant context of the DAE (2.1) in subsection 2.1.3, where index one problems are given special attention; pages 30-34 may be useful as a first reading on projector methods. The transition to time-varying problems proceeds via index one DAEs of the form (2.2) in 2.2.1. Properly stated linear time-varying DAEs (2.3) with arbitrary index will be then analyzed in subsections 2.2.2 to 2.2.5. The results presented there will be illustrated by means of an example discussed in detail in 2.2.6. In subsection 2.2.7 we introduce a local restatement of this approach aimed at the analysis of singular problems performed in Chapter 4. This framework is particularized to standard form DAEs (2.2) in Section 2.3.

Finally, Section 2.4 briefly presents reduction methods for linear problems; further details can be found in [224–226, 228]. These techniques will be extensively discussed for nonlinear DAEs in Chapter 3.

2.1 Linear time-invariant DAEs

Consider the linear time-invariant homogeneous DAE

$$Ax' + Ex = 0, \qquad (2.4)$$

with $A, E \in \mathbb{R}^{m \times m}$, A possibly singular. As in the classical theory of ODEs, the search for solutions of (2.4) having the form $e^{\lambda t}x_0$ naturally

2.1. Linear time-invariant DAEs

leads to the generalized eigenvalue problem defined by

$$\det(\lambda A + E) = 0,$$

and therefore drives the analysis of homogeneous linear time-invariant DAEs to a *matrix pencil* setting.

2.1.1 Matrix pencils and the Kronecker canonical form

Given a pair of matrices A, E in $\mathbb{R}^{m \times m}$, the *matrix pencil* $\{A, E\}$ is defined as the one-parameter family $\{\lambda A + E : \lambda \in \mathbb{C}\}$. We will sometimes denote the pencil simply by the expression $\lambda A + E$. If there exists some $\lambda \in \mathbb{C}$ such that $\lambda A + E$ is a nonsingular matrix, i.e. if $\det(\lambda A + E)$ does not vanish identically, the matrix pencil is called *regular*; otherwise it is said to be a *singular* pencil [91].

Two matrix pencils $\{A, E\}$ and $\{\tilde{A}, \tilde{E}\}$ will be said to be *equivalent* if there exist nonsingular matrices G, $H \in \mathbb{R}^{m \times m}$ such that $\tilde{A} = GAH$ and $\tilde{E} = GEH$. Note that this use of the term 'equivalent' stands for 'strictly equivalent' in [91]. A regular pencil $\{A, E\}$ is equivalent to a pencil $\{\tilde{A}, \tilde{E}\}$ in which $\tilde{A}$ and $\tilde{E}$ have the form (cf. Theorem XII.3 in [91])

$$\tilde{A} = GAH = \begin{pmatrix} I_s & 0 \\ 0 & N \end{pmatrix}, \quad \tilde{E} = GEH = \begin{pmatrix} W & 0 \\ 0 & I_{m-s} \end{pmatrix}, \quad (2.5)$$

where $W \in \mathbb{R}^{s \times s}$ for some nonnegative $s \leq m$, and $N \in \mathbb{R}^{(m-s) \times (m-s)}$ is a nilpotent matrix with index $\nu \leq m - s$, that is, $N^\nu = 0$, $N^{\nu-1} \neq 0$. Without loss of generality N and W can be assumed to be in Jordan form. Note that it may be $s = 0$, meaning that $\tilde{A} = N$, $\tilde{E} = I_m$, and also $s = m$, which yields $\tilde{A} = I_s = I_m$, $\tilde{E} = W$.

This defines the *Weierstrass-Kronecker canonical form* of the pencil [91]; this result is due to Weierstrass [301], and was extended later by Kronecker to singular matrix pencils [144]. In the DAE literature this normal form is often referred to simply as the *Kronecker canonical form* and will be henceforth abbreviated as 'KCF'.

The nilpotency index ν is called the *Kronecker index* of the matrix pencil, although again the term 'Weierstrass-Kronecker index' would possibly be more accurate. It is also called the *nilpotency index* of the pencil itself. The matrix pencil is said to have index zero if $s = m$, which amounts to require that A is a nonsingular matrix. The index one case is defined by a null matrix N of dimension $m - s > 0$. In particular, if A vanishes the DAE (2.4) amounts to $Ex = 0$; if E is nonsingular, then N is a null matrix of dimension m and the pencil has index one.

The *spectrum* of a regular matrix pencil is
$$\sigma(\{A,E\}) = \{\lambda \in \mathbb{C} : \det(\lambda A + E) = 0\}. \qquad (2.6)$$
Regardless of the index, a regular pencil $\{A, E\}$ satisfies
$$\sigma(\{A,E\}) = \sigma(-W), \qquad (2.7)$$
since
$$\begin{aligned}\det(\lambda A + E) &= (\det G \ \det H)^{-1}\det(\lambda GAH + GEH)\\ &= (\det G \ \det H)^{-1}\det(\lambda I_s + W) \qquad (2.8)\end{aligned}$$
because of the fact that $\det(\lambda N + I_{m-s}) = 1$. The identity depicted in (2.8) implies that $\det(\lambda A + E) = 0 \Leftrightarrow \det(\lambda I_s + W) = 0$, as stated in (2.7).

This means that the spectrum of the matrix pencil has exactly s eigenvalues (counted with multiplicity) or, equivalently, that $\det(\lambda A + E)$ is a polynomial in λ with degree s. The value s is sometimes called the *core-rank* of the pencil [30, 39], and, for the reasons stated in subsection 2.1.2, can be also called the *dynamical degree of freedom* (cf. [179]) or the *state dimension* of the DAE. A regular matrix pencil with index $\nu \geq 1$ is said to have an infinite eigenvalue of multiplicity $m - s$, with m and s defined above. The identity (2.7) will also play a role in the stability analysis of equilibria in DAEs; cf. Section 3.5.

2.1.2 Solving linear time-invariant DAEs via the KCF

Supposed that the matrix pencil $\{A, E\}$ is regular, not only the homogeneous DAE (2.4) but also the inhomogeneous linear problem (2.1) can be tackled via the Weierstrass-Kronecker canonical form. The regularity of the pencil is actually equivalent to the solvability of the linear time-invariant DAE, in the sense specified in [30]; see also [91, 107, 151]. If the index of the regular pencil is ν, then both DAEs (2.1) and (2.4) are said to be index ν as well.

Indeed, premultiplying (2.1) by G and using the linear coordinate change $x = Hw$ we transform the DAE into
$$\begin{pmatrix} I_s & 0 \\ 0 & N \end{pmatrix} w' + \begin{pmatrix} W & 0 \\ 0 & I_{m-s} \end{pmatrix} w = \begin{pmatrix} \tilde{q}_1(t) \\ \tilde{q}_2(t) \end{pmatrix} \qquad (2.9)$$
with $\tilde{q}(t) = Gq(t)$. Splitting $w = (u, v)$ with $u \in \mathbb{R}^s$, $v \in \mathbb{R}^{m-s}$, equation (2.9) reads
$$u' + Wu = \tilde{q}_1(t) \qquad (2.10a)$$
$$Nv' + v = \tilde{q}_2(t). \qquad (2.10b)$$

Equation (2.10a) is an explicit linear constant coefficient ODE for $u \in \mathbb{R}^s$, not involving the v component. An initial value problem is well-defined by any $u_0 \in \mathbb{R}^s$ and therefore this equation has s dynamical degrees of freedom.

In turn, (2.10b) is decoupled from (2.10a) since it does not involve the u component. Following [30] (see also [91]), equation (2.10b) can be rewritten as

$$(ND + I)v = \tilde{q}_2(t)$$

with $D \equiv d/dt$, and then

$$v = (ND + I)^{-1}\tilde{q}_2(t) = \sum_{j=0}^{\nu-1}(-1)^j(ND)^j\tilde{q}_2(t),$$

since $N^j = 0$ for $j \geq \nu$. This means that $v \in \mathbb{R}^{m-s}$ has no degree of freedom since it is completely determined from $\tilde{q}_2(t)$ via the relation

$$v = \tilde{q}_2(t) - N\tilde{q}_2'(t) + \ldots + (-1)^{\nu-1}N^{\nu-1}\tilde{q}_2^{(\nu-1)}(t). \qquad (2.11)$$

Note that in this framework $\tilde{q}(t)$ (or equivalently $q(t)$) must be in $C^{\nu-1}(\mathcal{J}, \mathbb{R}^m)$, and that in higher ($\nu \geq 2$) index DAEs solutions will depend explicitly on the derivatives of the excitation $q(t)$.

In the homogeneous case (2.4), obtained by setting $q(t) = 0$ in (2.1), the explicit ODE (2.10a) reads

$$u' + Wu = 0,$$

whereas making $\tilde{q}(t) = Gq(t) = 0$ in (2.11) we get

$$v = 0.$$

In this case, solutions are only defined on the s-dimensional linear space defined by the algebraic restriction $v = 0$, that is, $w_{s+1} = \ldots = w_m = 0$. The variables u (i.e. $w_1, \ldots, w_s$) can be understood to yield a parametrization of this linear subspace, the dynamical behavior on it being described by the explicit equation $u' = -Wu$.

2.1.3 A glance at projector-based techniques

The Kronecker canonical form and the decoupling (2.10a)-(2.11) based on it provides much insight into the linear time-invariant DAEs (2.1) and (2.4). Nevertheless it also displays some limitations, the need to compute the canonical form being an obvious one. Additionally, the solution description is given in terms of the "transformed" variables $w = H^{-1}x$. In particular, the smoothness of the different solution components would be stated in terms of w, somehow obstructing a precise functional description of the system behavior. Similarly, the smoothness demands on $q(t)$ must be stated in terms of the transformed excitation $\tilde{q}(t) = Gq(t)$; if we rewrite (2.11) as

$$v = \tilde{q}_2(t) - (N\tilde{q}_2)'(t) + \ldots + (-1)^{\nu-1}(N^{\nu-1}\tilde{q}_2)^{(\nu-1)}(t),$$

it is clear that the $C^{\nu-1}$-smoothness demand on $\tilde{q}(t)$ can be relaxed by directing the requirements to the appropriate components of $\tilde{q}(t)$.

Proper formulations and projector methods attempt to overcome these drawbacks, allowing additionally for an extension of the results to the time-varying context. These techniques provide an index characterization in terms of the original problem description. They yield a precise functional description of the solution behavior under mild smoothness requirements directed to the appropriate components of q. This approach will be discussed in detail in Section 2.2; we present below an introduction for linear time-invariant problems. This way the reader can get a glimpse at projector-based methods without the technicalities arising in the linear time-varying context. Special emphasis is put on index one systems.

2.1.3.1 Index one characterization via projectors

Definition 2.1. A square matrix $Q \in \mathbb{R}^{m \times m}$ is called a *projector* if $Q^2 = Q$.

From the relation $Q^2 = Q$ it is easy to check that $\ker Q \cap \operatorname{im} Q = \{0\}$ and therefore $\mathbb{R}^m = \ker Q \oplus \operatorname{im} Q$. Note that all vectors within $\operatorname{im} Q$ are invariant, that is, $Qv = v$ if (and, actually, only if) $v \in \operatorname{im} Q$. The projector Q is said to project *along* $\ker Q$ *onto* $\operatorname{im} Q$. If I denotes the identity in $\mathbb{R}^{m \times m}$, then $P = I - Q$ is a projector along $\operatorname{im} Q$ onto $\ker Q$, and therefore $PQ = QP = 0$. When the spaces $\ker Q$ and $\operatorname{im} Q$ are orthogonal to each other, Q is said to be an *orthogonal* projector.

Assume now that A is a singular matrix, and let Q be *any* projector onto $\ker A$. Maybe the key seminal result in the tractability index framework is the characterization of an index one pencil depicted below.

Proposition 2.1. *Let $A \in \mathbb{R}^{m \times m}$ be a singular matrix, and Q a projector onto $\ker A$. Then the matrix pencil $\{A, E\}$ is regular with Kronecker index one if and only if the matrix $A_1 = A + EQ$ is nonsingular.*

The proof of this result can be found in Theorem A.13 of [107] and, in a more general context, in Theorem 3 of [108].

2.1.3.2 Decoupling of linear time-invariant index one DAEs

Besides the index characterization depicted above, the matrix A_1 allows for a decoupling of the DAE (2.1), if it is index one, in terms of the projections Px and Qx of the semistate vector x. Recall that Q is a projector onto $\ker A$ and $P = I - Q$. Indeed, premultiplying (2.1) by A_1^{-1} we get

$$A_1^{-1}Ax' + A_1^{-1}Ex = A_1^{-1}q(t)$$

which, decomposing $x = Px + Qx$, yields

$$A_1^{-1}Ax' + A_1^{-1}EPx + A_1^{-1}EQx = A_1^{-1}q(t). \tag{2.12}$$

This equation can be simplified via the relations

(1) $A_1^{-1}A = P$;
(2) $A_1^{-1}EQ = Q$.

The identity depicted in (1) follows from $A_1 P = (A + EQ)P = AP = A$, where we have used $QP = 0$ and $AP = A$ since $AQ = 0$. In turn, (2) is due to $A_1 Q = (A + EQ)Q = EQ$.

Using these relations, (2.12) reads

$$Px' + A_1^{-1}EPx + Qx = A_1^{-1}q(t). \tag{2.13}$$

Premultiplying this equation by P we get

$$Px' + PA_1^{-1}EPx = PA_1^{-1}q(t), \tag{2.14}$$

which, denoting $u = Px$, can be written as

$$u' + PA_1^{-1}Eu = PA_1^{-1}q(t). \tag{2.15}$$

This is called an *inherent ODE* for the linear time-invariant index one DAE (2.1). It is very important at this stage to understand the nature of this ODE. The variable u takes values on $\mathbb{R}^m$ and the equation itself is defined on the whole of $\mathbb{R}^m$. Now, the linear space $\operatorname{im} P$ is invariant for this equation; indeed, an initial condition $u_0 \in \operatorname{im} P$ verifies $Qu_0 = 0$, and multiplying (2.15) by Q we get that the projection onto $\operatorname{im} Q$ of the solution $u(t)$ verifies $Qu'(t) = 0$, meaning that $Qu(t) = 0$ (i.e. $u(t) \in \operatorname{im} P$) for all t. Solutions of the DAE will be described (cf. (2.18) below) in terms of a solution u of (2.15) *lying on the invariant space* $\operatorname{im} P$.

On the other hand, premultiplying (2.13) by Q we obtain

$$QA_1^{-1}EPx + Qx = QA_1^{-1}q(t).$$

Denoting $v = Qx$, we can rewrite this as

$$v = -QA_1^{-1}Eu + QA_1^{-1}q(t). \tag{2.16}$$

Equations (2.15) and (2.16) yield a decoupling of the DAE (2.1) in terms of the projections $u = Px$, $v = Qx$. More precisely, using $A = AP$ let us reformulate (2.1) as

$$A(Px)' + Ex = q(t). \qquad (2.17)$$

This makes it possible to seek for solutions within the space

$$C_P^1(\mathcal{J}, \mathbb{R}^m) = \{x \in C(\mathcal{J}, \mathbb{R}^m) \;/\; Px \in C^1(\mathcal{J}, \mathbb{R}^m)\} \supset C^1(\mathcal{J}, \mathbb{R}^m).$$

Provided that $q(t)$ is continuous, and replacing Px' by $(Px)'$ in (2.13) and (2.14), the reasoning above shows that a given mapping $x(t) \in C_P^1(\mathcal{J}, \mathbb{R}^m)$ is a solution of (2.17) if and only if it can be written as

$$x(t) = u(t) + v(t), \qquad (2.18)$$

with $u(t) \in C^1(\mathcal{J}, \mathbb{R}^m)$ a solution of (2.15) in the invariant space $\operatorname{im} P$ and $v(t) \in C(\mathcal{J}, \mathbb{R}^m)$ given by (2.16).

The advantages of this approach over the KCF-based method discussed in 2.1.2 are worth a digression. First, via the reformulation (2.17), the projector framework provides a bijective input-output functional description of the solutions. Indeed, every continuous excitation $q(t)$ together with an initial condition x_0 satisfying $Ex_0 - q(t_0) \in \operatorname{im} A$, yields a unique solution $x(t) \in C_P^1$ of the DAE, and an inverse mapping is naturally defined by $x(t) \to (A(Px)' + Ex, x(t_0))$; additionally, the excitations $q(t)$ which lead to C^1 solutions are $\{q(t) \in C(\mathcal{J}, \mathbb{R}^m) \;/\; QA_1^{-1}q \in C^1(\mathcal{J}, \mathbb{R}^m)\}$. The key aspect here is that these ideas can be extended to the linear time-varying setting, as indicated in subsection 2.2.1.

Additionally, within the projector approach we avoid the need to compute the Kronecker canonical form; in an index one context, the KCF computation is replaced by the introduction of a projector Q onto $\ker A$ and the index is characterized by the elementary matrix operations arising in Proposition 2.1 (cf. (2.21) and Theorem 2.1 below for higher index cases). Note also that there is no change of coordinates involved in the projector-based solution description, in contrast to the framework based on the Kronecker canonical form. Finally, the projector approach will admit an extension to linear time-varying DAEs with arbitrary index, as detailed in Section 2.2.

2.1.3.3 *Geometrical remarks*

From a geometrical point of view, the index one notion for the DAE (2.4) can be expressed in terms of the transversality of certain characteristic spaces of the problem, as stated below (cf. Theorem A.13 in [107]).

2.1. Linear time-invariant DAEs

Proposition 2.2. *Let $A \in \mathbb{R}^{m \times m}$ be a singular matrix. The matrix pencil $\{A, E\}$ is regular with index one if and only if the spaces $N = \ker A$ and*

$$S = \{x \in \mathbb{R}^m \ / \ Ex \in \operatorname{im} A\}$$

verify $N \cap S = \{0\}$ or, equivalently, $\mathbb{R}^m = N \oplus S$.

Any solution of the homogeneous DAE (2.4) must lie on S. For regular pencils with index one, the solutions actually fill this space, in contrast to higher index cases for which the solution set would be a proper subspace of S. In index one problems, the projector onto $N = \ker A$ can be chosen *canonically* as the one which projects along the space S, which by Proposition 2.2 is transversal to N; this canonical projector Q_c can be computed as $QA_1^{-1}E$ from any projector Q onto N, cf. Lemma A.14 in [107] or Lemma 2.1 in [182].

With this choice, the decoupled system (2.15)-(2.16) for the inhomogeneous problem (2.1) with $A_1 = A + EQ_c$ amounts to

$$u' + P_c A_1^{-1} E u = P_c A_1^{-1} q(t) \tag{2.19a}$$
$$v = Q_c A_1^{-1} q(t), \tag{2.19b}$$

the space $S = \operatorname{im} P_c$ being now invariant for (2.19a). The fact that u is not present in (2.19b) makes the decoupling (2.19a)-(2.19b) a *complete* one [192]; cf. Remark 2.8 on page 52.

In any case, it is worth emphasizing that this canonical choice gives a geometrical insight into the solution behavior, but it is *not* necessary to compute solutions via the decoupling (2.19). This idea can be illustrated via the homogeneous problem (2.4), for which one has

$$u' + P A_1^{-1} E u = 0 \tag{2.20a}$$
$$v = -Q A_1^{-1} E u \tag{2.20b}$$

for an arbitrary choice of the projector Q onto N, and

$$u' + P_c A_1^{-1} E u = 0$$
$$v = 0$$

for the canonical choice Q_c, with $A_1 = A + EQ_c$ in this case. Both $\operatorname{im} P$ and S are transversal spaces to N, and the reader can think of (2.20b) (or, more precisely, of the mapping $u \to u - QA_1^{-1}Eu$) as a correction which drives the solutions from $\operatorname{im} P$ onto the "true" solution space S.

For the inhomogeneous DAE (2.1), similar ideas apply when S is replaced pointwise by $\{x \in \mathbb{R}^m \ / \ Ex - q(t) \in \operatorname{im} A\}$

2.1.3.4 Higher index problems

Given a matrix pencil $\{A, E\}$, if the above-constructed matrix $A_1 = A + EQ$ is singular then the pencil may well be a regular one with higher index. This can be assessed via the following matrix chain construction, originally introduced in [179]. Set $A_0 = A$, $E_0 = E$ and, for $i \geq 0$,

$$A_{i+1} = A_i + E_i Q_i, \quad E_{i+1} = E_i P_i, \tag{2.21}$$

where Q_i is a projector onto $N_i = \ker A_i$, and $P_i = I - Q_i$. For the proof of the following result we refer the reader to Theorem 3 in [108].

Theorem 2.1. *A matrix pencil $\{A, E\}$ is regular with Kronecker index ν if and only if the matrices A_i constructed in (2.21) are singular for $i < \nu$ and A_ν is nonsingular.*

Note that the chain becomes stationary if a nonsingular A_ν is reached since in this case $Q_i = 0$, $P_i = I$ for $i \geq \nu$.

If the pencil $\{A, E\}$ is regular, then as detailed in [108, 179] it is possible to choose the projectors Q_i in a way such that

$$Q_i Q_j = 0 \text{ if } j < i. \tag{2.22}$$

This choice, which will be said to define an *admissible* projector sequence, makes it possible to decouple a linear time-invariant DAE with arbitrary index, generalizing the ideas that led to (2.15)-(2.16). We omit the details since this can be derived as a particular case of the more general framework discussed in Sections 2.2 and 2.3 below: see specifically equations (2.153) and (2.154) on page 80.

2.1.3.5 Some auxiliary properties of the projectors P_i and Q_i

We compile here some properties of the projectors P_i and Q_i which hold under the admissibility requirement (2.22) and will be useful later. The reader may skip this technical subsection in a first reading. These relations apply to time-invariant problems but will also hold (pointwise) in the time-varying context of Section 2.2.

First, from the condition $Q_i Q_j = 0$ for $i > j$ stated in (2.22) we have

$$P_i Q_j = Q_j, \quad \text{if } i > j$$

since $P_i Q_j = (I - Q_i) Q_j = Q_j$. In turn, writing $P_i P_j = P_i (I - Q_j)$ we get

$$P_i P_j = P_i - Q_j, \quad \text{if } i > j.$$

Note also that, if $j > k$, $Q_j P_k = Q_j(I - Q_k) = Q_j$ if $j > k$, and then

$$P_i P_j P_k = P_i - Q_j - Q_k, \quad \text{if } i > j > k. \tag{2.23}$$

In the decoupling of index ν problems discussed in subsection 2.2.5 we will often use these products, as well as those of the form $Q_k P_{k+1} P_{k+2} \cdots P_{\nu-1}$. From the relations depicted above we can derive

$$P_{k+1} P_{k+2} \cdots P_{\nu-1} P_i = \begin{cases} P_{k+1} P_{k+2} \cdots P_{\nu-1} - Q_i & \text{if } i \leq k \\ P_{k+1} P_{k+2} \cdots P_{\nu-1} & \text{if } k < i \leq \nu - 1. \end{cases}$$

Therefore, for nonnegative $i, k \leq \nu - 1$,

$$Q_k P_{k+1} \cdots P_{\nu-1} P_i = Q_k P_{k+1} \cdots P_{\nu-1} \text{ if } i \neq k, \tag{2.24}$$

whereas

$$Q_k P_{k+1} \cdots P_{\nu-1} P_k = Q_k(P_{k+1} \cdots P_{\nu-1} - I), \tag{2.25}$$

and then

$$Q_k P_{k+1} \cdots P_{\nu-1} P_0 \cdots P_i = \begin{cases} Q_k P_{k+1} \cdots P_{\nu-1} & \text{if } i < k \\ Q_k(P_{k+1} \cdots P_{\nu-1} - I) & \text{if } i = k \\ Q_k(P_{k+1} \cdots P_{\nu-1} - P_{k+1} \cdots P_i) & \text{if } i > k. \end{cases} \tag{2.26}$$

2.2 Properly stated linear time-varying DAEs

The extension of the framework discussed in subsection 2.1.3 for the linear time-invariant DAE (2.1) to time-varying problems of the form (2.2) offered substantial difficulties. With a big technical effort, März succeeded in extending the results to linear time-varying DAEs up to index three [180–182]. However, the complete characterization of solutions of linear time-varying systems with arbitrary index remained open.

A framework accommodating the general case was devised in [16] by driving the attention to the formulation (2.3). When the matrix maps $A(t)$, $D(t)$ in this equation are well-matched in the sense specified in Definition 2.2 (p. 38), the leading term of the DAE (2.3) is said to be *properly stated*. See the above-mentioned reference [16] but also [130, 188] and specially [189, 191, 192], where linear DAEs with arbitrary index are fully characterized. Some preliminary ideas concerning this formulation had been already examined by März in the 1980s [177, 178].

The DAE (2.2) can be rewritten in the form (2.3) if there exists a C^1 projector $P(t)$ along $\ker A(t)$ for $t \in \mathcal{J}$; cf. Remark 2.1 below. In this case, since $A(t) = A(t)P(t)$, the product $A(t)x'(t)$ can be written as

$$A(t)P(t)x'(t) = A(t)(P(t)x(t))' - A(t)P'(t)x(t),$$

and therefore (2.2) takes the form

$$A(t)(P(t)x(t))' + (E(t) - A(t)P'(t))x(t) = q(t). \qquad (2.27)$$

Denoting $D(t) = P(t)$, $B(t) = E(t) - A(t)P'(t)$, equation (2.27) is seen to be a particular instance of (2.3) satisfying the requirements stated in Definition 2.2 below. This way the properly stated framework applies in particular to the standard form (2.2).

Remark 2.1. The existence of a C^1 projector defined on an interval $\mathcal{J} \subseteq \mathbb{R}$ along (or onto) a given time-dependent linear subspace of $\mathbb{R}^m$ with constant dimension is equivalent to the existence of a basis for this space defined by maps from $C^1(\mathcal{J}, \mathbb{R}^m)$ [108]. Such a space will be called a C^1-*space*, and has naturally associated a (C^1) *vector bundle* structure (cf. for instance [1]). In particular, $\ker A(t)$ and $\text{im}\, A(t)$ are C^1-spaces if $A(t)$ is in $C^1(\mathcal{J}, \mathbb{R}^{m \times m})$ and has constant rank [107]. For C^∞-analogs of these claims see e.g. [228].

The present Section, and specifically subsections 2.2.2, 2.2.3 and 2.2.5, essentially compile the analysis of linear DAEs with the properly stated form (2.3) as developed in [189, 191]. Nevertheless, some recent contributions involving so-called Π-projectors [251], discussed in 2.2.4, provide some improvements and simplifications. The working conditions within all the subsections mentioned above are assumed to hold on the whole interval $\mathcal{J}$; the local results of 2.2.7, on the contrary, open the way to the analysis of singular points carried out in Chapter 4.

2.2.1 On standard form index one problems

In order to undertake the analysis of general linear DAEs, the reader may profit from the introductory digression on standard form index one problems here presented. It may be also useful to compare the results with those of 2.1.3.2, focused on index one linear time-invariant DAEs. We proceed via the reformulation (2.27) of (2.2) assuming that $\ker A(t)$ is a C^1 space on $\mathcal{J}$ and denoting $B(t) = E(t) - A(t)P'(t)$, which yields

$$A(t)(P(t)x(t))' + B(t)x(t) = q(t). \qquad (2.28)$$

2.2. Properly stated linear time-varying DAEs

In this time-varying setting, the (tractability) index one condition will be stated as the nonsingularity on $\mathcal{J}$ of

$$A_1(t) = A(t) + B(t)Q(t), \tag{2.29}$$

with $Q(t) = I - P(t)$. We remove in the sequel dependences on t for the sake of notational simplicity. The reader can check that the identities

$$A_1^{-1} A = P \tag{2.30a}$$
$$A_1^{-1} BQ = Q \tag{2.30b}$$

of page 31 still hold in the current time-varying context. Premultiplying (2.28) by A_1^{-1}, using (2.30) and decomposing $x = Px + Qx$ we transform (2.28) into

$$P(Px)' + A_1^{-1} BPx + Qx = A_1^{-1} q. \tag{2.31}$$

Now, premultiplication of (2.31) by P leads to

$$P(Px)' + PA_1^{-1} BPx = PA_1^{-1} q. \tag{2.32}$$

Using the relation $P(Px)' = (Px)' - P'Px$, which owes to the identity $(Px)' = (PPx)' = P(Px)' + P'Px$, equation (2.32) can be recast as

$$(Px)' + (-P' + PA_1^{-1}B)Px = PA_1^{-1} q.$$

Writing $u = Px$, we obtain the inherent, time-varying ODE

$$u' + (-P' + PA_1^{-1} B)u = PA_1^{-1} q, \tag{2.33}$$

where it is noteworthy the additional term $-P'u$ with respect to the corresponding equation (2.15) derived in the time-invariant setting. The (time-varying) space $\operatorname{im} P(t)$ can be checked to be invariant for (2.33).

In turn, premultiplying (2.31) by Q we obtain

$$QA_1^{-1} BPx + Qx = QA_1^{-1} q,$$

which gives an algebraic relation for $v = Qx$ in terms of $u = Px$, namely

$$v = -QA_1^{-1} Bu + QA_1^{-1} q. \tag{2.34}$$

Except for the fact that the operators are now time-dependent, (2.34) is formally identical to (2.16).

Akin to the time-invariant context, this shows that a given mapping $x(t) \in C_P^1(\mathcal{J}, \mathbb{R}^m)$ solves (2.28) if and only if it can be written as

$$x(t) = u(t) + v(t), \tag{2.35}$$

where $u(t) \in C^1(\mathcal{J}, \mathbb{R}^m)$ is a solution of the inherent (2.33) lying on the invariant space $\operatorname{im} P(t)$, and $v(t) \in C(\mathcal{J}, \mathbb{R}^m)$ satisfies the relation depicted in (2.34). This result is a particular case, with $\nu = 1$, of Theorem 2.5 in Section 2.3; see specifically Remark 2.11 on p. 79. Note that the key assumptions supporting this result in the time-varying setting are the index one condition, stated as the nonsingularity of the matrix $A_1(t)$ in (2.29), and the constant dimension and C^1-structure of $\ker A(t)$; the latter is the only smoothness requirement needed on the continuous maps $A(t)$, $E(t)$.

The decoupling (2.33)-(2.34) paves the way for an input-output functional description of solutions entirely analogous to the one discussed in 2.1.3.2 for time-invariant problems. A continuous excitation $q(t)$, together with an initial condition x_0 verifying $E(t_0)x_0 - q(t_0) \in \operatorname{im} A(t_0)$, leads to a unique solution $x(t) \in C_P^1$ of the reformulated DAE (2.28). An inverse mapping is given by $x(t) \to (A(Px)'(t) + Bx(t), x(t_0))$. Moreover, for sufficiently smooth maps $A(t)$, $E(t)$, the excitations $q(t)$ yielding C^1 solutions are $\{q(t) \in C(\mathcal{J}, \mathbb{R}^m) \ / \ QA_1^{-1}q \in C^1(\mathcal{J}, \mathbb{R}^m)\}$. Similar characterizations can be derived for higher index cases, along the lines discussed in 2.2.5.

2.2.2 Properly stated leading terms

The reformulation (2.28) of the DAE (2.2) plays a fundamental role in the ideas sketched above for index one linear time-varying problems. In this and the forthcoming subsections we extend this approach to DAEs with arbitrary index, using the more general form (2.3). Standard form DAEs (2.2) will be revisited in Section 2.3.

The matrix-valued mapping $D(t)$ in (2.3) is intended to capture the components of the semistate vector x which actually need to be differentiated; the leading term arises in the form depicted there in different circuit and control applications, including adjoint formulations. See [12, 14–16, 130–132, 154, 157, 187–190, 293]. It makes it possible to consider solutions within the space

$$C_D^1(\mathcal{J}, \mathbb{R}^m) = \{x \in C(\mathcal{J}, \mathbb{R}^m) \ / \ Dx \in C^1(\mathcal{J}, \mathbb{R}^n)\}.$$

Note that $D(t)$ in (2.3), unlike $P(t)$ in (2.28), is not assumed to be a projector, not even needs to be a square matrix. Nevertheless, it must be well-matched with the matrix mapping $A(t)$ in the sense specified below.

Definition 2.2. The leading term of the DAE (2.3) is said to be *properly stated* on the interval $\mathcal{J} \subseteq \mathbb{R}$ if $A(t)$ and $D(t)$ satisfy

$$\ker A(t) \oplus \operatorname{im} D(t) = \mathbb{R}^n \tag{2.36}$$

for all $t \in \mathcal{J}$, and both $\ker A(t)$ and $\operatorname{im} D(t)$ are C^1-spaces.

2.2. Properly stated linear time-varying DAEs

From the requirements defining a properly stated leading term it follows that not only the matrices $A(t)$ and $D(t)$ but also the product $A(t)D(t)$ (to be denoted by $G_0(t)$) have constant rank on $\mathcal{J}$. They also imply that $\ker A(t)D(t) = \ker D(t)$.

Additionally, the transversality condition (2.36) together with the C^1 assumption on $\ker A(t)$ and $\operatorname{im} D(t)$ supports the existence of a projector mapping $R(t) \in C^1(\mathcal{J}, \mathbb{R}^{n \times n})$ onto $\operatorname{im} D(t)$ and along $\ker A(t)$ for all $t \in \mathcal{J}$. This is a consequence of a more general property, stating that there exists a C^1 projector along $X(t)$ and onto $Y(t)$, both spaces having constant dimension and verifying $X(t) \oplus Y(t) = \mathbb{R}^n$, if and only if $X(t)$ and $Y(t)$ are C^1-spaces (cf. Remark 2.1 above); if $\{e_1(t), \ldots, e_r(t)\}$ and $\{e_{r+1}(t), \ldots, e_n(t)\}$ are C^1 basis maps spanning $Y(t)$ and $X(t)$, respectively, and we denote by $M(t)$ the $n \times n$ matrix with columns $e_1(t), \ldots, e_n(t)$, the projector onto $Y(t)$ along $X(t)$ can be easily checked to be given by

$$M(t) \begin{pmatrix} I_r & 0 \\ 0 & 0 \end{pmatrix} M(t)^{-1}.$$

Instances of properly stated leading terms can be found in (1.15) or in the example discussed in subsection 2.2.6: see specifically (2.124) on p. 66.

2.2.3 P-projectors: Matrix chain and the tractability index

2.2.3.1 Matrix chain

As detailed in [189], the matrix chain construction which supports the tractability index for properly stated linear time-varying DAEs (2.3) proceeds through the introduction of a reflexive generalized inverse [24] of $D(t)$ on $\mathcal{J}$, that is, a matrix-valued mapping $D^- : \mathcal{J} \to \mathbb{R}^{m \times n}$ verifying

$$D(t)D^-(t)D(t) = D(t), \quad D^-(t)D(t)D^-(t) = D^-(t). \tag{2.37}$$

A matrix-valued map $D^-(t)$ satisfying (2.37) is called a $\{1,2\}$-inverse of $D(t)$ [24]. We will additionally require that

$$D(t)D^-(t) = R(t), \tag{2.38}$$

for all $t \in \mathcal{J}$, with $R(t)$ defined above. The matrix mapping $D^-(t)$ is not uniquely defined by (2.37)-(2.38); however, if we let P_0 be any continuous projector along $\ker G_0(t) = \ker A(t)D(t) = \ker D(t)$ and require also that

$$D^-(t)D(t) = P_0(t), \tag{2.39}$$

then the joint conditions (2.37), (2.38) and (2.39) uniquely determine a mapping $D^-(t)$, which additionally is continuous on $\mathcal{J}$ (cf. [16, 188, 189]).

Equivalently, it is possible to begin the construction with the choice of a continuous $D^-(t)$ satisfying (2.37)-(2.38) and let the projector $P_0(t)$ along $\ker G_0(t)$ be defined by (2.39).

With this equipment, let us now build the matrix chain which extends (2.21) and supports the tractability index definition for (2.3). We remove explicit dependences on t for the sake of simplicity. To begin with, define

$$G_0 = AD, \qquad (2.40)$$

which has constant rank (denoted hereafter by r_0) on $\mathcal{J}$ if the leading term is properly stated. Let N_0 stand for $\ker G_0 = \ker D$, assume that P_0 is the continuous projector along N_0 defined by (2.39), and set $Q_0 = I - P_0$. Write

$$B_0 = B \qquad (2.41a)$$
$$G_1 = G_0 + B_0 Q_0. \qquad (2.41b)$$

The construction can be iterated for $i \geq 1$ if the following conditions hold:

(a) G_i has constant rank (to be denoted by r_i) on $\mathcal{J}$;
(b) $N_i = \ker G_i$ verifies $(N_0 \oplus \ldots \oplus N_{i-1}) \cap N_i = \{0\}$ for all $t \in \mathcal{J}$.

As detailed in Remark 2.2 below, conditions (a) and (b) make it possible to choose a continuous projector Q_i onto $N_i = \ker G_i$ such that

(b') $Q_i Q_j = 0$ if $0 \leq j < i$, for all $t \in \mathcal{J}$.

Set $P_i = I - Q_i$, and assume additionally that

(c) the product $DP_0 P_1 \cdots P_i D^-$ is in $C^1(\mathcal{J}, \mathbb{R}^{n \times n})$.

Provided that these assumptions hold, define

$$B_i = B_{i-1} P_{i-1} - G_i D^- (DP_0 \cdots P_i D^-)' DP_0 \cdots P_{i-1} \qquad (2.42a)$$
$$G_{i+1} = G_i + B_i Q_i. \qquad (2.42b)$$

Remark 2.2. Writing $N_0 \oplus \ldots \oplus N_{i-1}$ as a direct sum in (b) for $i \geq 2$ is supported on the trivial intersections $(N_0 \oplus \ldots \oplus N_{j-1}) \cap N_j = \{0\}$ assumed for $1 \leq j < i$. Additionally, the existence of a projector Q_i satisfying $Q_i Q_j = 0$ for $j < i$, as stated in (b'), relies on the fact that condition (b) makes it possible to choose the projector Q_i onto N_i in a way such that

$$N_0 \oplus \ldots \oplus N_{i-1} \subseteq \ker Q_i. \qquad (2.43)$$

Since Q_j projects onto N_j for $0 \leq j < i$, the identity $Q_i Q_j = 0$ follows.

2.2. Properly stated linear time-varying DAEs

The concept of an admissible projector sequence presented below was introduced in [191, 192]; for the notions of an algebraically nice and a nice DAE in Definition 2.4 we refer the reader to [195]. For brevity, in Definitions 2.3 and 2.4 the conditions are implicitly assumed to hold for all $t \in \mathcal{J}$.

Definition 2.3. If the DAE (2.3) is properly stated, and D^- and the continuous projector P_0 satisfy (2.37), (2.38) and (2.39), then both P_0 and $Q_0 = I - P_0$ are said to be *admissible*.

A projector sequence $\{Q_0, \ldots, Q_k\}$ (resp. $\{P_0, \ldots, P_k\}$), with $k \geq 1$, is said to be *preadmissible* up to level k if Q_0 is admissible, conditions (a) and (b') above hold for $1 \leq i \leq k$, and condition (c) is met for $1 \leq i < k$.

Should the smoothness condition (c) be met also for $i = k$, the sequence $\{Q_0, \ldots, Q_k\}$ (resp. $\{P_0, \ldots, P_k\}$) is called *admissible* up to level k.

The verification of conditions (a) and (b) above at a given level k does not depend on the choice of the projectors $\{Q_0, \ldots, Q_{k-1}\}$ as long as this sequence is admissible up to level $k - 1$: see Theorem 2.3 in [191]. Specifically, $r_k = \mathrm{rk}\, G_k$ and $\dim(N_0 \oplus \ldots \oplus N_{k-1}) \cap N_k$ are proved in Proposition 2 of [195] to be independent of the (admissible) choice of the projectors $Q_0, \ldots, Q_{k-1}$. This shows that the definition of an algebraically nice DAE displayed below does not depend on the actual choice of the admissible sequence $\{Q_0, \ldots, Q_{k-1}\}$, hence capturing an intrinsic property of the equation.

Definition 2.4. The linear DAE (2.3) is called *nice* at level 0 if the leading term is properly stated.

It is said to be *algebraically nice* at level $k \geq 1$ if it is nice at level $k - 1$ and conditions (a) and (b) above hold for $i = k$, for some (hence any) admissible up to level $k - 1$ sequence $\{Q_0, \ldots, Q_{k-1}\}$.

It is called *nice* at level $k \geq 1$ if it is algebraically nice at level k and there exists an admissible choice of Q_k.

Here, an 'admissible choice' of Q_k means that the projector sequence $\{Q_0, \ldots, Q_{k-1}, Q_k\}$ meets condition (b') (that is, $N_0 \oplus \ldots \oplus N_{k-1} \subseteq \ker Q_k$, which is allowed by (b) as indicated in Remark 2.2) as well as the smoothness requirement (c) for $i = k$. Thereby, the DAE will be algebraically nice (resp. nice) at level k if and only if it admits a preadmissible (resp. admissible) sequence up to level k. The ranks $r_i = \mathrm{rk}\, G_i$, for $i = 0, \ldots, k$, are said to be *characteristic values* of a nice at level k DAE.

For algebraically nice DAEs of the form (2.3) with sufficiently smooth coefficients $A(t)$, $D(t)$ and $B(t)$, the smoothness requirement depicted in

(c) can be always met. This means that, for sufficiently smooth problems, the existence of an admissible projector sequence and the notion of a nice DAE rely only on conditions (a) and (b), as stated below (cf. Proposition 3 in [195]).

Proposition 2.3. *Assume that the coefficients $A(t)$, $D(t)$, $B(t)$ in the DAE (2.3) are in C^k, $k \geq 1$. If the DAE is algebraically nice at level k, then it is nice at level k; equivalently, the existence of a preadmissible sequence up to level k guarantees that there exists an admissible sequence up to level k.*

Proof. The result follows from the fact that Q_0 can be taken from C^k, so that D^- and $G_1 = G_0 + B_0 Q_0$ are also in C^k; the latter implies that Q_1 can be chosen from C^k in an admissible manner. With this choice, B_1 and hence G_2 are in C^{k-1} (see (2.42)) and, subsequently, for $2 \leq i \leq k$ the projector Q_i can be taken from C^{k-i+1} to yield an admissible sequence $\{Q_0, \ldots, Q_k\}$. □

2.2.3.2 The tractability index of regular linear DAEs

If there exists a minimum nonnegative i for which the matrix G_i (constructed according to (2.40), (2.41) and (2.42)) is nonsingular on the whole working interval, the linear DAE (2.3) is said to be *regular*; in this case, this integer is called the *tractability index* of the DAE. These notions are stated precisely below.

Definition 2.5. Let the leading term of (2.3) be properly stated on $\mathcal{J}$. The DAE (2.3) is said to be *regular with tractability index zero* on $\mathcal{J}$ if both A and D (or, equivalently, G_0 in (2.40)) are invertible on $\mathcal{J}$.

It is said to be *regular with tractability index* $\nu \geq 1$ on $\mathcal{J}$ if there exists an admissible projector sequence $\{Q_0, \ldots, Q_{\nu-1}\}$ for which the G_i-matrices in (2.40), (2.41b), (2.42b) are singular for $0 \leq i < \nu$ and G_ν is nonsingular, for all $t \in \mathcal{J}$.

Finally, it is called *regular* on $\mathcal{J}$ if it is regular with any nonnegative tractability index.

Equivalently, the tractability index is ν if the DAE is nice up to level $\nu - 1$, the matrix mappings G_i are singular for $i < \nu$, and G_ν is nonsingular on $\mathcal{J}$. Due to the increasing dimension of the direct sum $N_0 \oplus \ldots \oplus N_{i-1}$ arising from condition (b) on page 40, it is easy to check that the index cannot exceed m.

2.2. Properly stated linear time-varying DAEs

Remark 2.3. As shown in Proposition 2.10 of [189], the tractability index notion of Definition 2.5 does not depend on the actual choice of the admissible projector sequence $Q_0, \ldots, Q_{\nu-1}$. The same is true for the *characteristic values* $r_0, r_1, \ldots, r_\nu$ defined by the ranks of the G_i matrices.

Remark 2.4. Along the lines of Proposition 2.3, the requirement that $A(t)$, $D(t)$ and $B(t)$ are in C^{m-1} is enough to satisfy all the smoothness conditions in the chain construction. For index ν problems it suffices to assume that the operators are in $C^{\nu-1}$.

The matrix chain constructed via (2.40), (2.41) and (2.42), together with the tractability index notion of Definition 2.5 supported on it, extends to linear time-varying properly stated DAEs the chain (2.21) characterizing the Kronecker index of linear time-invariant systems. Its importance relies on the fact that it allows for a complete description of the system solutions via the decoupling discussed in subsection 2.2.5 (see specifically Theorem 2.3 on page 51), which generalizes the solution decomposition depicted in (2.35) for index one DAEs. The reader may jump at this point to 2.2.5. However, the results in subsection 2.2.4 will simplify somewhat the projector framework and clarify the decoupling presented in 2.2.5.

2.2.4 The Π-framework

The reader may have noticed the presence of certain projector products of the form $P_0 \cdots P_i$ in the matrix chain constructed according to (2.42); see specifically the expression defining B_i in (2.42a). Also, products of the form $P_0 \cdots P_{i-1} Q_i$ arise when computing $B_i Q_i$ in (2.42b). An important role is played as well by these products in the coefficients of the DAE decoupling as developed in [189, 191]. The product $P_0 \cdots P_i$ displays additionally the important feature stated below (cf. Theorem 2.3 in [191] and Proposition 1 in [195]).

Proposition 2.4. *Let the DAE (2.3) be nice at level $i \geq 0$. Then*

$$\ker P_0 \cdots P_i = N_0 \oplus \ldots \oplus N_i. \tag{2.44}$$

Moreover, the space $N_0 \oplus \ldots \oplus N_i$ is independent of the choice of the admissible projectors $\{P_0, \ldots, P_i\}$.

By contrast, the individual spaces $N_i = \ker G_i$ will certainly depend on the choice of the projectors P_j for $j < i$. These remarks suggest that there might be a way to perform the chain construction just in terms of the products $P_0 \cdots P_i$ and the spaces $N_0 \oplus \ldots \oplus N_i$, somehow disregarding the individual projectors P_i and spaces N_i. More precisely, it seems feasible that certain projectors axiomatically defined might capture the properties of these products in a way which could simplify the analysis. This is indeed the case, as shown in [251] and detailed in this subsection.

Briefly, the products $P_0 \cdots P_i$ and $P_0 \cdots P_{i-1}Q_i$ will be replaced by certain projectors Π_i and M_i based upon the spaces $N_0 \oplus \ldots \oplus N_i$, leading to an equivalent chain construction which does not involve the individual projectors P_i. The projectors P_i within the framework of subsection 2.2.3 provide a particular way to compute Π_i and M_i, and in this setting Π_i and M_i can be understood as abbreviations for $P_0 \cdots P_i$ and $P_0 \cdots P_{i-1}Q_i$, respectively. However, other options will arise; notably, the construction can be naturally performed by choosing Π_i as the *orthogonal* projector along the characteristic space $N_0 \oplus \ldots \oplus N_i$. This simple choice is not possible in the P-framework of 2.2.3 since e.g. the orthogonal projector P_1 along N_1 need not satisfy, with $Q_1 = I - P_1$, the requirement $N_0 \subseteq \ker Q_1$ supporting the relation $Q_1Q_0 = 0$. The construction based on Π-projectors will allow for a nice characterization of the index of standard form DAEs (cf. Theorem 2.4, p. 77) and will also display several benefits in the analysis of singularities performed in Chapter 4; see specifically Remark 4.1 on p. 148.

2.2.4.1 Alternative chain construction

Let the leading term of the DAE (2.3) be properly stated on $\mathcal{J}$ (cf. Definition 2.2 on p. 38). Assuming, without explicit references to t, that the forthcoming conditions hold for all $t \in \mathcal{J}$ define, as in (2.40),

$$G_0 = AD. \tag{2.45}$$

It follows from the proper statement of the DAE that G_0 has constant rank r_0 on $\mathcal{J}$, and therefore we can choose a continuous projector Π_0 along the space $N_0 = \ker G_0$. This projector stands for P_0 within the P-framework of 2.2.3. Assume that $D^-(t)$ is now given by (2.37) and (2.38) together with

$$D^-(t)D(t) = \Pi_0(t). \tag{2.46}$$

Set $M_0 = I - \Pi_0$, and define

$$B_0 = B \tag{2.47a}$$

$$G_1 = G_0 + B_0M_0. \tag{2.47b}$$

Note that M_0 amounts to the previously introduced Q_0 projector, and so far (2.47b) just involves a notational change with respect to (2.41b).

For $i \geq 1$, the construction can be iterated provided that

2.2. Properly stated linear time-varying DAEs

(a) G_i has constant rank (denoted by r_i), and that
(b) $N_i = \ker G_i$ verifies $(N_0 \oplus \ldots \oplus N_{i-1}) \cap N_i = \{0\}$.

If both conditions hold, proceed by choosing a continuous projector

$$\Pi_i \text{ along } N_0 \oplus \ldots \oplus N_i, \text{ with } \operatorname{im} \Pi_i \subseteq \operatorname{im} \Pi_{i-1}, \qquad (2.48)$$

where the direct sum is iteratively supported on condition (b). Note additionally that the requirement $\operatorname{im} \Pi_i \subseteq \operatorname{im} \Pi_{i-1}$ can be met because of the relation $N_0 \oplus \ldots \oplus N_{i-1} \oplus \operatorname{im} \Pi_{i-1} = \mathbb{R}^m$, which guarantees the existence of a space transversal to $N_0 \oplus \ldots \oplus N_{i-1} \oplus N_i$ within $\operatorname{im} \Pi_{i-1}$.

Let us then complete the construction with

$$M_i = \Pi_{i-1} - \Pi_i, \qquad (2.49)$$

and assuming that

(c) $D\Pi_i D^-$ is in $C^1(\mathcal{J}, \mathbb{R}^{n \times n})$,

define

$$B_i = (B_{i-1} - G_i D^- (D\Pi_i D^-)' D)\Pi_{i-1} \qquad (2.50\text{a})$$
$$G_{i+1} = G_i + B_i M_i. \qquad (2.50\text{b})$$

Remark 2.5. The requirement $\operatorname{im} \Pi_i \subseteq \operatorname{im} \Pi_{i-1}$ in (2.48) is satisfied automatically if Π_i is chosen as the *orthogonal* projector along $N_0 \oplus \ldots \oplus N_i$, since in this case

$$\operatorname{im} \Pi_i = (N_0 \oplus \ldots \oplus N_i)^\perp \subseteq (N_0 \oplus \ldots \oplus N_{i-1})^\perp = \operatorname{im} \Pi_{i-1}.$$

If the smoothness requirement (c) is satisfied for these orthogonal projectors, this provides a simple criterion for the choice of projectors in the tractability index construction.

The identities $\operatorname{im} \Pi_i \subseteq \operatorname{im} \Pi_{i-1}$ and $\ker \Pi_{i-1} \subseteq \ker \Pi_i$ following from (2.48) imply (cf. Proposition 2.5 below) that $\Pi_{i-1}\Pi_i = \Pi_i$, $\Pi_i\Pi_{i-1} = \Pi_i$; it is easy to check that these properties make M_i in (2.49) a projector.

2.2.4.2 Equivalence of the P- and Π-chains

The matrix chain construction supported on the Π-projectors is equivalent to the P-based one of subsection 2.2.3 in the sense specified in Theorem 2.2 below (cf. also Corollary 2.1 on page 48). See [251]. These results will proceed via the following admissibility notion for a Π-sequence, which naturally parallelizes Definition 2.3.

Definition 2.6. If the DAE (2.3) is properly stated and (2.37), (2.38), (2.46) hold, then the continuous projector Π_0 is said to be *admissible*.

A projector sequence $\{\Pi_0, \ldots, \Pi_k\}$, $k \geq 1$, is said to be *preadmissible* up to level k if Π_0 is admissible, conditions (a), (b) above hold for $1 \leq i \leq k$, Π_i satisfies (2.48) for $1 \leq i \leq k$, and condition (c) is met for $1 \leq i < k$.

The sequence $\{\Pi_0, \ldots, \Pi_k\}$ is called *admissible* up to level k if, additionally, the smoothness condition (c) is met for $i = k$.

For the sake of notational brevity, the word *admissible* below stands for *admissible up to level k* for some $k \geq 0$, i ranging from 0 to k. Note that the result holds trivially for $k = 0$.

Theorem 2.2. *An admissible P-projector sequence defines an admissible Π-sequence leading to the same matrix chain by setting $\Pi_i = P_0 \cdots P_i$.*

Conversely, an admissible Π-sequence defines an admissible P-sequence when the projector P_i is defined via

$$\ker P_i = N_i, \quad \operatorname{im} P_i = N_0 \oplus \ldots \oplus N_{i-1} \oplus \operatorname{im} \Pi_i. \tag{2.51}$$

This yields the same matrix chain, and the following relations hold:

$$\Pi_i = P_0 \cdots P_i, \quad M_i = P_0 \cdots P_{i-1} Q_i. \tag{2.52}$$

Proof. For a given admissible P-sequence we obtain an admissible Π-sequence via $\Pi_i = P_0 \cdots P_i$ due to Proposition 2.4 and the identity $\operatorname{im} P_0 \cdots P_{i-1} P_i \subseteq \operatorname{im} P_0 \cdots P_{i-1}$, which imply (2.48).

Moreover, $M_i = P_0 \cdots P_{i-1} - P_0 \cdots P_i = P_0 \cdots P_{i-1}(I - P_i) = P_0 \cdots P_{i-1} Q_i$. We then need to check that, in the light of the identities $\Pi_i = P_0 \cdots P_i$ and $M_i = P_0 \cdots P_{i-1} Q_i$, the matrix chain defined by (2.50) coincides with the one constructed by means of (2.42). In this regard, note that B_i from (2.42a) can be written as

$$B_i = B_{i-1} P_{i-1} - G_i D^- (D P_0 \cdots P_i D^-)' D P_0 \cdots P_{i-1}$$
$$= (B_{i-2} P_{i-2} - G_{i-1} D^- (D P_0 \cdots P_{i-1} D^-)' D P_0 \cdots P_{i-2}) P_{i-1}$$
$$- G_i D^- (D P_0 \cdots P_i D^-)' D P_0 \cdots P_{i-1}$$

and eventually as

$$B_i = (B_0 - G_1 D^- (D P_0 P_1 D^-)' D - \ldots - G_i D^- (D P_0 \cdots P_i D^-)' D) P_0 \cdots P_{i-1}.$$

Using this expression it is not difficult to check that the identity

$$B_i = (B_{i-1} - G_i D^- (D P_0 \cdots P_i D^-)' D) P_0 \cdots P_{i-1} \tag{2.53}$$

holds, meaning that the expression for B_i given in (2.50a) equals the one in (2.42a) if $\Pi_i = P_0 \cdots P_i$.

2.2. Properly stated linear time-varying DAEs

Additionally, from (2.53) it follows that $B_i P_0 \cdots P_{i-1} Q_i = B_i Q_i$. Therefore, the identity $M_i = P_0 \cdots P_{i-1} Q_i$ implies that $B_i M_i = B_i Q_i$ and thus the computation of G_i via (2.50b) coincides with the one based on (2.42b).

Now, assume that an admissible Π-sequence is given and let $P_0 = \Pi_0$, $Q_0 = I - P_0 = M_0$. For $i \geq 1$ we construct P_i as indicated in (2.51), allowed by the decomposition $N_0 \oplus \ldots \oplus N_{i-1} \oplus N_i \oplus \operatorname{im} \Pi_i = \mathbb{R}^m$ which follows from the definition of Π_i. Let us then set $Q_i = I - P_i$, and note that the admissibility condition $N_0 \oplus \ldots \oplus N_{i-1} \subseteq \ker Q_i$ follows immediately from (2.51) and $\ker Q_i = \operatorname{im} P_i$.

With this definition of P_i, the identity

$$\Pi_{i-1} P_i = \Pi_i \tag{2.54}$$

holds. In order to prove this statement we show below that $\Pi_{i-1} P_i$ is a projector along $N_0 \oplus \ldots \oplus N_i$ onto $\operatorname{im} \Pi_i$, thereby equaling Π_i.

Indeed, $\Pi_{i-1} P_i$ is a projector: since Π_{i-1} projects along $N_0 \oplus \ldots \oplus N_{i-1}$, $I - \Pi_{i-1}$ projects onto $N_0 \oplus \ldots \oplus N_{i-1} \subseteq \operatorname{im} P_i$. Therefore $P_i(I - \Pi_{i-1}) = I - \Pi_{i-1}$ (cf. item (i) of Proposition 2.5 on page 49) or, equivalently, $P_i \Pi_{i-1} = P_i + \Pi_{i-1} - I$. Hence

$$\Pi_{i-1} P_i \Pi_{i-1} P_i = \Pi_{i-1}(P_i + \Pi_{i-1} - I) P_i$$
$$= \Pi_{i-1} P_i + \Pi_{i-1} P_i - \Pi_{i-1} P_i = \Pi_{i-1} P_i.$$

We now show that the identity $\ker \Pi_{i-1} P_i = N_0 \oplus \ldots \oplus N_i$ holds. First, let $v \in \ker \Pi_{i-1} P_i$ and decompose $v = P_i v + Q_i v$. Since $\Pi_{i-1} P_i v = 0$, we have $P_i v \in \ker \Pi_{i-1} = N_0 \oplus \ldots \oplus N_{i-1}$ and this together with $Q_i v \in N_i$ shows that

$$\ker \Pi_{i-1} P_i \subseteq N_0 \oplus \ldots \oplus N_{i-1} \oplus N_i = N_0 \oplus \ldots \oplus N_i.$$

On the other hand, if $v \in N_0 \oplus \ldots \oplus N_{i-1} \subseteq \operatorname{im} P_i$, we have $P_i v = v$ and therefore $\Pi_{i-1} P_i v = \Pi_{i-1} v = 0$ since $v \in N_0 \oplus \ldots \oplus N_{i-1} = \ker \Pi_{i-1}$. Additionally $N_i = \ker P_i \subseteq \ker \Pi_{i-1} P_i$ and, altogether,

$$N_0 \oplus \ldots \oplus N_i = N_0 \oplus \ldots \oplus N_{i-1} \oplus N_i \subseteq \ker \Pi_{i-1} P_i.$$

We still need to check that $\Pi_{i-1} P_i$ projects onto $\operatorname{im} \Pi_i$. Because of the relation $\ker \Pi_{i-1} P_i = N_0 \oplus \ldots \oplus N_i = \ker \Pi_i$ we know that $\operatorname{rk} \Pi_{i-1} P_i = \operatorname{rk} \Pi_i$ and thereby it suffices to check that $\operatorname{im} \Pi_i \subseteq \operatorname{im} \Pi_{i-1} P_i$. Let $w \in \operatorname{im} \Pi_i$. Since $\operatorname{im} \Pi_i \subseteq \operatorname{im} \Pi_{i-1}$ and also $\operatorname{im} \Pi_i \subseteq \operatorname{im} P_i$, we can easily see that $\Pi_{i-1} P_i w = \Pi_{i-1} w = w$, so that $w \in \operatorname{im} \Pi_{i-1} P_i$. This means that $\operatorname{im} \Pi_{i-1} P_i = \operatorname{im} \Pi_i$, and therefore completes the proof of (2.54).

From (2.54) we get $\Pi_i = P_0 \cdots P_i$ and this implies in turn that $M_i = P_0 \cdots P_{i-1} Q_i$ since $M_i = \Pi_{i-1} - \Pi_i = P_0 \cdots P_{i-1} - P_0 \cdots P_i = P_0 \cdots P_{i-1} Q_i$.

The identity depicted in (2.54) then yields $B_i \Pi_i = B_i \Pi_{i-1} P_i = B_i P_i$ since $B_i \Pi_{i-1} = B_i$ because of (2.50a). This implies that the construction of the B_i matrices via (2.42a) amounts to the one defined by (2.50a). Additionally, $G_i + B_i Q_i = G_i + B_i \Pi_{i-1} Q_i = G_i + B_i M_i$ and therefore the expression for G_i in (2.42b) equals the definition given in (2.50b). □

Corollary 2.1. *Let the leading term of the linear DAE (2.3) be properly stated on $\mathcal{J}$. Then (2.3) is regular with tractability index $\nu \geq 1$ on $\mathcal{J}$ if and only if there exists an admissible sequence $\{\Pi_0, \ldots, \Pi_{\nu-1}\}$ for which the G_i-matrices in (2.45), (2.47b), (2.50b) are singular for $0 \leq i < \nu$ and G_ν is nonsingular, for all $t \in \mathcal{J}$.*

Along the same lines, the DAE (2.3) will be algebraically nice or nice at level k (cf. Definition 2.4) on $\mathcal{J}$ if and only if there exists a preadmissible or an admissible up to level k Π-sequence on $\mathcal{J}$. Additionally, from Remark 2.3 and Theorem 2.2 it follows that the computation of the tractability index is independent of the actual choice of admissible Π-projectors.

Remark 2.6. If the DAE (2.3) is regular with index ν, the Q_i and P_i projectors can be explicitly computed from an admissible Π-sequence via $Q_0 = M_0$, $P_0 = \Pi_0$ and, for $1 \leq i < \nu$,

$$Q_i = G_\nu^{-1} B_i M_i, \quad P_i = I - Q_i. \tag{2.55}$$

Indeed, from (2.50b) we can write

$$G_\nu = G_i + \sum_{k=i}^{\nu-1} B_k M_k.$$

Multiplying this relation by Q_i, we derive (2.55) from the resulting identity $G_\nu Q_i = B_i M_i$; the latter follows from $G_i Q_i = 0$ and the relations $M_k Q_i = 0$ if $k > i$, $M_i Q_i = M_i$, which in turn are due to (2.52) and the admissibility condition $Q_k Q_i = 0$ for $k > i$.

Remark 2.7. As shown in [251], the B_i matrices can be replaced by the somehow simpler ones H_i, defined as

$$H_0 = B \tag{2.56}$$

$$H_i = H_{i-1} - G_i D^-(D \Pi_i D^-)' D, \quad \text{for } i \geq 1. \tag{2.57}$$

Using

$$G_{i+1} = G_i + H_i M_i \tag{2.58}$$

instead of (2.50b), the reader can check that the relation $B_i = H_i \Pi_{i-1}$ holds and that the G_i-chains coincide in both constructions.

2.2. Properly stated linear time-varying DAEs

2.2.4.3 Some properties of the projectors Π_i and M_i

The following result will be helpful in order to characterize several properties of the projectors Π_i and M_i. We will make systematic use of these properties within the decoupling discussed in subsection 2.2.5 below.

Proposition 2.5. *Let T, $S \in \mathbb{R}^{m \times m}$ be projectors.*

a) If $\operatorname{im} S \subseteq \operatorname{im} T$, then

  (i) $TS = S$;
  (ii) ST is a projector onto $\operatorname{im} S$ along $\ker T \oplus (\ker S \cap \operatorname{im} T)$.

b) If $\ker S \subseteq \ker T$, then

  (iii) $TS = T$;
  (iv) ST is a projector onto $\operatorname{im} S \cap (\ker S \oplus \operatorname{im} T)$ along $\ker T$.

Proof. For item (i), note that for any $v \in \mathbb{R}^m$ it is $Sv \in \operatorname{im} S \subseteq \operatorname{im} T$ and therefore $TSv = Sv$.

Regarding the statement in (ii), ST is a projector since, by item (i), $TS = S$ and thereby $STST = SST = ST$.

In order to show that $\operatorname{im} ST = \operatorname{im} S$ we just need to check the inclusion $\operatorname{im} S \subseteq \operatorname{im} ST$. Assume that $v \in \operatorname{im} S$, i.e. $v = Sv$. Since $\operatorname{im} S \subseteq \operatorname{im} T$ we have $v = Tv$ and therefore $v = Sv = STv$, that is, $v \in \operatorname{im} ST$.

Now let us prove that $\ker ST = \ker T \oplus (\ker S \cap \operatorname{im} T)$. On the one hand, $\ker T \oplus (\ker S \cap \operatorname{im} T) \subseteq \ker ST$: indeed, $\ker T \subseteq \ker ST$, and if $v \in \ker S \cap \operatorname{im} T$ then $Tv = v$ and $0 = Sv = STv$, i.e. $v \in \ker ST$, meaning that $\ker S \cap \operatorname{im} T \subseteq \ker ST$. Hence $\ker T \oplus (\ker S \cap \operatorname{im} T) \subseteq \ker ST$. In turn, the inclusion $\ker ST \subseteq \ker T \oplus (\ker S \cap \operatorname{im} T)$ can be proved as follows. Let $v \in \ker ST$, i.e. $STv = 0$; writing $v = (I-T)v + Tv$, it is very easy to check that $(I-T)v \in \ker T$ and $Tv \in \ker S \cap \operatorname{im} T$.

For item (iii), recast the condition $\ker S \subseteq \ker T$ as $\operatorname{im}(I - S) \subseteq \operatorname{im}(I - T)$ and use (i) to write $(I-T)(I-S) = I - S$, that is, $I - S - T + TS = I - S$ and then $T = TS$.

Item (iv) is largely analogous to (ii). Indeed, ST is a projector because $STST = STT = ST$. Additionally, the identity $\ker ST = \ker T$ holds since $STv = 0$ means $Tv \in \ker S \subseteq \ker T$ and then $Tv = 0$. The inclusion $\operatorname{im} ST \subseteq \operatorname{im} S \cap (\ker S \oplus \operatorname{im} T)$ follows easily from the decomposition of a given vector $v \in \operatorname{im} ST$ as $(I - T)v + Tv$. Finally, in order to show that $\operatorname{im} S \cap (\ker S \oplus \operatorname{im} T) \subseteq \operatorname{im} ST$ we write $v \in \operatorname{im} S$ as $v_1 + v_2$ with $v_1 \in \ker S$, $v_2 \in \operatorname{im} T$, and use the identities $v = Sv = Sv_2 = STv_2$ to check that $v \in \operatorname{im} ST$. □

Many properties of the projectors Π_i and M_i can be derived from Proposition 2.5. Due to Theorem 2.2, these properties will hold for the products $P_0 \cdots P_i$ and $P_0 \cdots P_{i-1}Q_i$ in the P-framework of subsection 2.2.3. In particular, using the identity $M_i = \Pi_{i-1}Q_i$ we can characterize the image and the kernel of the projector M_i as follows; since

$$\ker \Pi_{i-1} = N_0 \oplus \ldots \oplus N_{i-1} \subseteq \operatorname{im} P_i = \ker Q_i,$$

cf. (2.51), in the light of item (iv) above we get

$$\operatorname{im} M_i = \operatorname{im} \Pi_{i-1} \cap (N_0 \oplus \ldots \oplus N_i), \tag{2.59a}$$
$$\ker M_i = \ker Q_i = N_0 \oplus \ldots \oplus N_{i-1} \oplus \operatorname{im} \Pi_i. \tag{2.59b}$$

From these relations and the fact that $\operatorname{im} \Pi_{j-1} \subseteq \operatorname{im} \Pi_i$ if $j > i$, it is easy to prove that

$$M_i M_j = \begin{cases} 0 & \text{if } i \neq j \\ M_i & \text{if } i = j. \end{cases} \tag{2.60}$$

Using items (i) and (iii) from Proposition 2.5 together with (2.48), (2.51) and (2.59), the reader can also check that the following identities hold.

$$\Pi_i \Pi_j = \begin{cases} \Pi_i & \text{if } i \geq j \\ \Pi_j & \text{if } i < j \end{cases} \tag{2.61}$$

$$\Pi_i M_j = \begin{cases} 0 & \text{if } i \geq j \\ M_j & \text{if } i < j \end{cases} \tag{2.62}$$

$$M_i \Pi_j = \begin{cases} M_i & \text{if } i > j \\ 0 & \text{if } i \leq j. \end{cases} \tag{2.63}$$

Some other properties such as

$$M_i Q_j = 0 \text{ for } i > j, \quad M_i Q_i = M_i, \tag{2.64}$$

already used in 2.2.4.2, can also be seen as a consequence of (2.59); note that

$$\operatorname{im} Q_i = N_i, \quad \ker Q_i = N_0 \oplus \ldots \oplus N_{i-1} \oplus \operatorname{im} \Pi_i, \tag{2.65}$$

in the light of (2.51) and $Q_i = I - P_i$. We will also make future use of the relation

$$Q_i M_i = Q_i \tag{2.66}$$

which follows as well from $\ker Q_i = \ker M_i$ and item (iii) above.

2.2.5 Decoupling

The matrix chain defining the tractability index, in either one of the equivalent forms discussed in subsections 2.2.3 and 2.2.4, is aimed at the forthcoming characterization of the solutions of regular linear DAEs with arbitrary index. Such a characterization is given in Theorem 2.3 below for DAEs with properly stated leading term (2.3), and in Theorem 2.5 within Section 2.3 for the standard form DAE (2.2).

A preliminary digression is however necessary in order to understand the meaning of the symbols arising in Theorem 2.3, specially for readers who might have skipped subsection 2.2.4 in a first approach. The matrix chain can be certainly constructed via P_i and Q_i projectors, as detailed in subsection 2.2.3, and in this case the symbols Π_i and M_i (for $i = 0, \ldots, \nu-1$) below should be understood as abbreviations for the products $P_0 \cdots P_i$ and $P_0 \cdots P_{i-1} Q_i$, respectively. Nevertheless, the reader should be aware of the properties derived in 2.2.4.3 for these products, which will be systematically used below.

Conversely, when the matrix chain is supported on the (equivalent) framework of subsection 2.2.4, based on Π_i and M_i projectors, the individual projectors P_i and Q_i are well-defined by Theorem 2.2 and can be computed via the relations depicted in (2.55). In this setting, the reader may profit from the chance to choose the Π_i projectors naturally as the *orthogonal* ones along $N_0 \oplus \ldots \oplus N_i$, as stated in Remark 2.5; in this case, later appearances of the space $\operatorname{im} \Pi_i$ can be understood as $(N_0 \oplus \ldots \oplus N_i)^\perp$, which by Proposition 2.4 (p. 43) is a characteristic space of the DAE, not depending on any choice of projectors.

Within the following statement, it is worth emphasizing that a straightforward simplification follows in cases in which $N_0 \oplus \ldots \oplus N_{\nu-1} = \mathbb{R}^m$, with G_ν nonsingular. In this situation it is $\Pi_{\nu-1} = 0$. The decoupling does not involve an inherent ODE since there is no dynamic degree of freedom, and the solution decomposition (2.67) amounts to $x = v_{\nu-1} + \ldots + v_0$.

Theorem 2.3. *Assume that the DAE (2.3) is regular with tractability index ν on $\mathcal{J}$, and that $q(t)$ meets the smoothness requirements stated in Remark 2.9 below. Then,*

$$x(t) \in C_D^1(\mathcal{J}, \mathbb{R}^m) = \{x \in C(\mathcal{J}, \mathbb{R}^m) \;/\; Dx \in C^1(\mathcal{J}, \mathbb{R}^n)\},$$

solves (2.3) if and only if it can be written as

$$x = D^- u + v_{\nu-1} + \ldots + v_1 + v_0, \tag{2.67}$$

where $u \in C^1(\mathcal{J}, \mathbb{R}^n)$ is a solution of the inherent ODE

$$u' - (D\Pi_{\nu-1}D^-)'u + D\Pi_{\nu-1}G_\nu^{-1}BD^-u = D\Pi_{\nu-1}G_\nu^{-1}q \qquad (2.68)$$

lying on the invariant space $\operatorname{im} D\Pi_{\nu-1}D^- = \operatorname{im} D\Pi_{\nu-1}$, whereas

$$v_k = -\mathcal{K}_k D^- u + \sum_{j=k+1}^{\nu-1} \mathcal{N}_{kj}(Dv_j)' + \sum_{j=k+2}^{\nu-1} \mathcal{M}_{kj} v_j + \mathcal{L}_k q \qquad (2.69)$$

for $k = \nu - 1, \ldots, 0$, with $v_k \in C_D^1(\mathcal{J}, \mathbb{R}^m)$ for $k > 0$ and $v_0 \in C(\mathcal{J}, \mathbb{R}^m)$. The coefficients of (2.69) read

$$\mathcal{K}_k = M_k P_{k+1} \cdots P_{\nu-1} G_\nu^{-1} B + M_k (P_{k+1} \cdots P_{\nu-1} - I) D^- (D\Pi_{\nu-1}D^-)'D$$

$$\mathcal{N}_{kj} = M_k P_{k+1} \cdots P_{j-1} Q_j D^-$$

$$\mathcal{M}_{kj} = M_k \left(\sum_{i=k+1}^{j-1} P_{k+1} \cdots P_{i-1}(Q_i - P_i) \right) D^-(DM_j D^-)'D$$

$$\mathcal{L}_k = M_k P_{k+1} \cdots P_{\nu-1} G_\nu^{-1},$$

where the products $P_{k+1} \cdots P_l$ amount to the identity I whenever $l < k+1$.

The particularization of (2.68) and (2.69) to index one, two and three problems, as well as the corresponding expressions for the coefficients of (2.69), can be found in (2.128)-(2.129), (2.132)-(2.134), and (2.137)-(2.139), respectively. It is also worth emphasizing that the terms $\mathcal{N}_{kj}(Dv_j)' + \mathcal{M}_{kj}v_j$ in (2.69) can be joined together into a single one of the form $\mathcal{N}_{kj}^*(Dv_j)'$, as detailed in Remark 2.10 (p. 64).

The importance of this result relies on the fact that solutions of (2.3) are characterized in terms of the original x-variables via the projections $\Pi_{\nu-1}x$ (or, more precisely, $u = D\Pi_{\nu-1}x$) and $v_k = M_k x$, $k = \nu - 1, \ldots, 0$. This is performed by means of the inherent ODE (2.68); note that this equation (and hence u) lies on $\mathbb{R}^n$, although the contribution of u is "driven back" to $\mathbb{R}^m$ via the D^- coefficient in (2.67). The explicit algebraic relations depicted in (2.69) yield the v_k components, which are successively computed, for $k = \nu - 1, \ldots, 0$, from u, q and (when $k \leq \nu - 2$) the v_j components with $j > k$. In higher ($\nu \geq 2$) index cases, the variables v_k involve, for $0 \leq k \leq \nu - 2$, certain derivatives of the excitation $q(t)$ via the terms $\mathcal{N}_{kj}(Dv_j)'$. Mind also that the terms $\mathcal{M}_{kj}v_j$ are only present in problems with index $\nu \geq 3$.

Remark 2.8. There exist special admissible projectors yielding a *fine* decoupling in which the coefficients $\mathcal{K}_1, \ldots, \mathcal{K}_{\nu-1}$ above vanish. If $\mathcal{K}_0$ also

2.2. Properly stated linear time-varying DAEs

vanishes then the decoupling is said to be a *complete* one; note that in this case the u variable is not at all involved in (2.69). For projectors yielding a complete decoupling, then $\Pi_{\nu-1} = P_0 \cdots P_{\nu-1}$ becomes uniquely defined. See [192] for details in this regard.

Remark 2.9. The existence of a fine decoupling allows for the precise statement of smoothness requirements on q. The continuous excitations $q(t)$ leading to a solvable (in C_D^1) index ν DAE are defined, provided that the coefficients $\mathcal{N}_{kj}$, $\mathcal{M}_{kj}$ and $\mathcal{L}_k$ come from a fine decoupling, by the conditions [191]

$$s_{\nu-1} = D\mathcal{L}_{\nu-1}q \in C^1(\mathcal{J}, \mathbb{R}^n), \quad \text{if } \nu \geq 2,$$
$$s_{\nu-2} = D\mathcal{N}_{\nu-2\,\nu-1}s'_{\nu-1} + D\mathcal{L}_{\nu-2}q \in C^1(\mathcal{J}, \mathbb{R}^n), \quad \text{if } \nu \geq 3,$$

and

$$s_k = \sum_{j=k+1}^{\nu-1} D\mathcal{N}_{kj}s'_j + \sum_{j=k+2}^{\nu-1} D\mathcal{M}_{kj}D^- s_j + D\mathcal{L}_k q \in C^1(\mathcal{J}, \mathbb{R}^n)$$

for $k = \nu - 3, \ldots, 1$, if $\nu \geq 4$.

Theorem 2.3 was proved, with slight differences, by Roswitha März in a remarkable *tour de force* [189, 191]. It involves many technicalities which can be understood as the price for its broad generality. In the remainder of this subsection, we detail a proof of this theorem which is essentially based on the one in [191] (see specifically Section 3 and Appendix B there). The use of Π_i and M_i projectors somehow clarifies the proof and allows for some simplification; compare e.g. the expressions for $\mathcal{K}_k$ and specially for $\mathcal{M}_{kj}$ above with the ones given in [191]. See also Remark 2.10 on page 64.

Outline of the proof of Theorem 2.3. The proof of Theorem 2.3 will be detailed after Lemmas 2.1 and 2.2. A brief outline may however be of help at this point. The semistate vector x will be decomposed as

$$x = \Pi_{\nu-1}x + M_{\nu-1}x + \ldots + M_1 x + M_0 x. \tag{2.70}$$

This is allowed by the relations $M_0 = I - \Pi_0$ and $M_i = \Pi_{i-1} - \Pi_i$ (cf. (2.49)), which yield $M_0 + \ldots + M_{\nu-1} = I - \Pi_{\nu-1}$. It is worth emphasizing that, due to (2.59a), all the components $v_k = M_k x$ will lie on the characteristic space $N_0 \oplus \ldots \oplus N_{\nu-1}$. In the P-framework, (2.70) corresponds to the decomposition

$$x = P_0 \cdots P_{\nu-1}x + P_0 \cdots P_{\nu-2}Q_{\nu-1}x + \ldots + P_0 Q_1 x + Q_0 x.$$

The decoupling of the DAE can be roughly summarized as follows. Rescaling the DAE by G_ν^{-1} will lead to equation (2.84); the projection of (2.84) onto $\operatorname{im} D\Pi_{\nu-1}$ will then yield the inherent ODE (2.68). Additionally, premultiplication of (2.84) by $V_k = Q_k P_{k+1} \cdots P_{\nu-1}$ would project the equation onto the N_k spaces (cf. Lemma 2.2); in order to obtain the decoupling in terms of $v_k = M_k x$, as depicted in (2.69), we would further premultiply by Π_{k-1}. Certainly, this amounts to premultiplying (2.84) directly by $U_k = \Pi_{k-1} V_k = P_0 \cdots P_{k-1} Q_k P_{k+1} \cdots P_{\nu-1}$ and we will proceed this way in the proof.

Lemma 2.1 below shows that the premultiplication of the DAE (as will be reformulated in (2.83) below) by G_ν^{-1} drives certain key components of the equation to the space $N_0 \oplus \ldots \oplus N_{\nu-1}$; note that Q_i projects onto N_i for $0 \leq i \leq \nu - 1$. This result extends to DAEs with arbitrary index the properties stated in (2.30) for index one problems.

Lemma 2.1. *Let the DAE (2.3) be regular with tractability index ν on $\mathcal{J}$. Then*

$$G_\nu^{-1} G_0 = I - Q_{\nu-1} - Q_{\nu-2} - \ldots - Q_0 \qquad (2.71)$$

and, for $0 \leq j \leq \nu - 1$,

$$G_\nu^{-1} B M_j = Q_j + \sum_{i=1}^{j} (I - Q_{\nu-1} - \ldots - Q_i) D^-(D\Pi_i D^-)' DM_j. \qquad (2.72)$$

Proof. Since $G_\nu = G_{\nu-1} + B_{\nu-1} M_{\nu-1}$, $\ker M_{\nu-1} = \ker Q_{\nu-1} = \operatorname{im} P_{\nu-1}$ (see (2.59b) and (2.65)), and $P_{\nu-1}$ projects along $\ker G_\nu$, we have

$$G_\nu P_{\nu-1} = G_{\nu-1} P_{\nu-1} = G_{\nu-1}.$$

Similarly, $G_{\nu-1} P_{\nu-2} = G_{\nu-2} P_{\nu-2} = G_{\nu-2}$. Iteratively, this leads to the relation $G_\nu P_{\nu-1} \cdots P_0 = G_0$, that is,

$$G_\nu^{-1} G_0 = P_{\nu-1} \cdots P_0. \qquad (2.73)$$

On the other hand, from (2.23) one gets

$$P_{\nu-1} \cdots P_0 = P_{\nu-1} - Q_{\nu-2} - \ldots - Q_0. \qquad (2.74)$$

Equations (2.73) and (2.74) yield (2.71).

In order to prove the relation displayed in (2.72), write $B_j M_j = BM_j - \sum_{i=1}^{j} G_i D^-(D\Pi_i D^-)' DM_j$. Letting $G_i = G_\nu P_{\nu-1} \cdots P_i$ as above,

2.2. Properly stated linear time-varying DAEs

premultiplying by G_ν^{-1} and using the relation $G_\nu^{-1} B_j M_j = Q_j$ depicted in (2.55), we get

$$G_\nu^{-1} B M_j = Q_j + \sum_{i=1}^{j} P_{\nu-1} \cdots P_i D^- (D\Pi_i D^-)' D M_j. \qquad (2.75)$$

Proceeding as before with the products $P_{\nu-1} \cdots P_i$ we finally obtain (2.72). Note that, in particular, $G_\nu^{-1} B M_0 = G_\nu^{-1} B Q_0 = Q_0$. □

Lemma 2.2. *The products*

$$V_k = Q_k P_{k+1} \cdots P_{\nu-1} \qquad (2.76)$$
$$U_k = M_k P_{k+1} \cdots P_{\nu-1} = P_0 \cdots P_{k-1} Q_k P_{k+1} \cdots P_{\nu-1} \qquad (2.77)$$

are projectors, for which

$$\operatorname{im} V_k = N_k \qquad (2.78a)$$
$$\ker V_k = N_0 \oplus \ldots \oplus N_{k-1} \oplus N_{k+1} \oplus \ldots \oplus N_{\nu-1} \oplus \operatorname{im} \Pi_{\nu-1} \qquad (2.78b)$$

and

$$\operatorname{im} U_k = \operatorname{im} \Pi_{k-1} \cap (N_0 \oplus \ldots \oplus N_k) \qquad (2.79a)$$
$$\ker U_k = N_0 \oplus \ldots \oplus N_{k-1} \oplus N_{k+1} \oplus \ldots \oplus N_{\nu-1} \oplus \operatorname{im} \Pi_{\nu-1}. \qquad (2.79b)$$

Additionally,

$$U_k P_i = M_k P_{k+1} \cdots P_{\nu-1} P_i = M_k P_{k+1} \cdots P_{\nu-1} \quad \text{if } i \neq k \qquad (2.80)$$
$$U_k P_k = M_k P_{k+1} \cdots P_{\nu-1} P_k = M_k (P_{k+1} \cdots P_{\nu-1} - I), \qquad (2.81)$$

and

$$U_k \Pi_i = M_k P_{k+1} \cdots P_{\nu-1} \Pi_i = \begin{cases} M_k P_{k+1} \cdots P_{\nu-1} & \text{if } i < k \\ M_k (P_{k+1} \cdots P_{\nu-1} - I) & \text{if } i = k \\ M_k (P_{k+1} \cdots P_{\nu-1} - P_{k+1} \cdots P_i) & \text{if } i > k. \end{cases} \qquad (2.82)$$

Proof. The statements in (2.78a), (2.78b) for $V_k = Q_k P_{k+1} \cdots P_{\nu-1}$ follow from item (ii) in Proposition 2.5. Consider first the product $Q_k P_{k+1}$. Due to (2.51) we have $\operatorname{im} Q_k = N_k \subseteq \operatorname{im} P_{k+1} = N_0 \oplus \ldots \oplus N_k \oplus \operatorname{im} \Pi_{k+1}$, and therefore $Q_k P_{k+1}$ is a projector onto $\operatorname{im} Q_k = N_k$ along the space $\ker P_{k+1} \oplus (\ker Q_k \cap \operatorname{im} P_{k+1})$, i.e.

$$N_{k+1} \oplus ([N_0 \oplus \ldots \oplus N_{k-1} \oplus \operatorname{im} \Pi_k] \cap [N_0 \oplus \ldots \oplus N_{k-1} \oplus N_k \oplus \operatorname{im} \Pi_{k+1}]).$$

Using the properties
$$[N_0 \oplus \ldots \oplus N_{k-1} \oplus \operatorname{im} \Pi_k] \cap N_k = \emptyset$$
$$N_0 \oplus \ldots \oplus N_{k-1} \oplus \operatorname{im} \Pi_{k+1} \subseteq N_0 \oplus \ldots \oplus N_{k-1} \oplus \operatorname{im} \Pi_k,$$
the expression depicted above for $\ker P_{k+1} \oplus (\ker Q_k \cap \operatorname{im} P_{k+1})$ amounts to
$$N_0 \oplus \ldots \oplus N_{k-1} \oplus N_{k+1} \oplus \operatorname{im} \Pi_{k+1}.$$

Inductively, assuming that $Q_k P_{k+1} \cdots P_i$ is a projector onto N_k along $N_0 \oplus \ldots \oplus N_{k-1} \oplus N_{k+1} \oplus \ldots \oplus N_i \oplus \operatorname{im} \Pi_i$ for a given i with $k+1 \leq i \leq \nu - 2$, the product $Q_k P_{k+1} \cdots P_{i+1} = (Q_k P_{k+1} \cdots P_i) P_{i+1}$ can be checked, proceeding as above, to be a projector onto N_k along
$$N_0 \oplus \ldots \oplus N_{k-1} \oplus N_{k+1} \ldots \oplus N_{i+1} \oplus \operatorname{im} \Pi_{i+1}.$$
The case $i = \nu - 2$ yields (2.78a) and (2.78b).

The claims for $U_k = M_k P_{k+1} \cdots P_{\nu-1} = P_0 \cdots P_{k-1} Q_k P_{k+1} \cdots P_{\nu-1}$ follow in a simpler way from the identity $U_k = \Pi_{k-1} V_k$ together with item (iv) in Proposition 2.5. Indeed, $\ker \Pi_{k-1} = N_0 \oplus \ldots \oplus N_{k-1} \subseteq \ker V_k = N_0 \oplus \ldots \oplus N_{k-1} \oplus N_{k+1} \oplus \ldots \oplus N_{\nu-1} \oplus \operatorname{im} \Pi_{\nu-1}$, according to (2.78b). Therefore, $U_k = \Pi_{k-1} V_k$ is a projector along $\ker V_k$ (the expression depicted in (2.78b) thus proving (2.79b)) onto $\operatorname{im} \Pi_{k-1} \cap (\ker \Pi_{k-1} \oplus \operatorname{im} V_k)$, that is,
$$\operatorname{im} \Pi_{k-1} \cap (N_0 \oplus \ldots \oplus N_{k-1} \oplus N_k),$$
as claimed in (2.79a).

Finally, (2.80), (2.81) and (2.82) are obtained from the multiplication of (2.24), (2.25) and (2.26), respectively, by $\Pi_{k-1} = P_0 \cdots P_{k-1}$. $\square$

Proof of Theorem 2.3. Assume first that a given map $x \in C_D^1$ solves (2.3), and write
$$x = D^- D\Pi_{\nu-1} x + M_{\nu-1} x + \ldots + M_1 x + M_0 x,$$
where we have used $D^- D\Pi_{\nu-1} = P_0 \Pi_{\nu-1} = \Pi_{\nu-1}$, based on the relations $\operatorname{im} \Pi_{\nu-1} \subseteq \operatorname{im} \Pi_0 = \operatorname{im} P_0$ and item (ii) of Proposition 2.5. We show below that $u = D\Pi_{\nu-1} x$ and $v_k = M_k x$ (for $k = \nu-1, \ldots, 0$) satisfy the relations (2.68) and (2.69), respectively. For the sake of clarity we proceed via several steps which are labeled with uppercase Roman numerals.

I. *Reformulation.* Since $R = DD^-$ is a projector along $\ker A$, the leading matrix A in (2.3) can be written as $AR = ADD^- = G_0 D^-$. This makes it possible to recast the DAE as
$$G_0 D^- (Dx)' + Bx = q(t). \qquad (2.83)$$

2.2. Properly stated linear time-varying DAEs

Premultiplying (2.83) by G_ν^{-1} we get

$$G_\nu^{-1}G_0D^-(Dx)' + G_\nu^{-1}Bx = G_\nu^{-1}q(t), \qquad (2.84)$$

which reads, according to (2.71) in Lemma 2.1,

$$(I - Q_{\nu-1} - Q_{\nu-2} - \ldots - Q_0)D^-(Dx)' + G_\nu^{-1}Bx = G_\nu^{-1}q(t). \qquad (2.85)$$

II. The inherent ODE (2.68). Since $\Pi_{\nu-1}$ projects along $N_0 \oplus \ldots \oplus N_{\nu-1}$ and $N_i = \operatorname{im} Q_i$, premultiplying (2.85) by $D\Pi_{\nu-1}$ we obtain

$$D\Pi_{\nu-1}D^-(Dx)' + D\Pi_{\nu-1}G_\nu^{-1}Bx = D\Pi_{\nu-1}G_\nu^{-1}q(t). \qquad (2.86)$$

On the other hand, using item (iii) of Proposition 2.5 one can check that $\Pi_{\nu-1}D^-D = \Pi_{\nu-1}P_0 = \Pi_{\nu-1}$. From this relation we can write $D\Pi_{\nu-1}x$ as $D\Pi_{\nu-1}D^-Dx$, where the factors $D\Pi_{\nu-1}D^-$ and Dx are in C^1, and differentiating this product we are allowed to write

$$D\Pi_{\nu-1}D^-(Dx)' = (D\Pi_{\nu-1}x)' - (D\Pi_{\nu-1}D^-)'Dx$$

to convert (2.86) into

$$(D\Pi_{\nu-1}x)' - (D\Pi_{\nu-1}D^-)'Dx + D\Pi_{\nu-1}G_\nu^{-1}Bx = D\Pi_{\nu-1}G_\nu^{-1}q(t). \qquad (2.87)$$

In order to simplify the term $D\Pi_{\nu-1}G_\nu^{-1}Bx$, since $\Pi_{\nu-1}$ projects along $N_0 \oplus \ldots \oplus N_{\nu-1}$ we get from (2.72)

$$D\Pi_{\nu-1}G_\nu^{-1}BM_j = D\Pi_{\nu-1}D^- \sum_{i=1}^{j}(D\Pi_iD^-)'DM_j. \qquad (2.88)$$

Additionally $\Pi_{\nu-1}\Pi_i = \Pi_{\nu-1}$ (cf. (2.61)) yields

$$D\Pi_{\nu-1}D^-(D\Pi_iD^-)' = (D\Pi_{\nu-1}D^-)' - (D\Pi_{\nu-1}D^-)'D\Pi_iD^-$$

and because of the relations $\Pi_i M_j = M_j$ if $i < j$ and $\Pi_i M_j = 0$ if $i = j$ stated in (2.62), equation (2.88) amounts to

$$\sum_{i=1}^{j}(D\Pi_{\nu-1}D^-)'DM_j - \sum_{i=1}^{j}(D\Pi_{\nu-1}D^-)'D\Pi_iD^-DM_j = (D\Pi_{\nu-1}D^-)'DM_j,$$

that is,

$$D\Pi_{\nu-1}G_\nu^{-1}BM_j = (D\Pi_{\nu-1}D^-)'DM_j. \qquad (2.89)$$

Since $\sum_{j=0}^{\nu-1} M_j = I - \Pi_{\nu-1}$, from (2.89) we get

$$D\Pi_{\nu-1}G_\nu^{-1}B(I - \Pi_{\nu-1}) = (D\Pi_{\nu-1}D^-)'D(I - \Pi_{\nu-1}). \qquad (2.90)$$

Decomposing $x = \Pi_{\nu-1}x + (I - \Pi_{\nu-1})x$, the relation depicted in (2.90) allows us to rewrite (2.87) in the form
$$(D\Pi_{\nu-1}x)' - (D\Pi_{\nu-1}D^-)'D\Pi_{\nu-1}x+ \\ +D\Pi_{\nu-1}G_\nu^{-1}BD^-D\Pi_{\nu-1}x = D\Pi_{\nu-1}G_\nu^{-1}q(t). \quad (2.91)$$
With $u = D\Pi_{\nu-1}x$, equation (2.91) yields the inherent ODE (2.68).

The space $\operatorname{im} D\Pi_{\nu-1}D^- = \operatorname{im} D\Pi_{\nu-1}$ is invariant for this ODE since $y = (I - D\Pi_{\nu-1}D^-)u$ can be checked to satisfy the homogeneous equation $y' + (D\Pi_{\nu-1}D^-)'y = 0$ on $\mathcal{J}$; the initial condition $y_0 = 0$ for y, which captures the relation $u_0 = D\Pi_{\nu-1}D^-u(0) \in \operatorname{im} D\Pi_{\nu-1}D^-$ on an initial condition for u, yields as unique solution the trivial one $y \equiv 0$ on $\mathcal{J}$, and then $u(t) \in \operatorname{im} D\Pi_{\nu-1}D^-$ for all $t \in \mathcal{J}$.

III. The algebraic components v_k in (2.69). The relations depicted in (2.69) for $v_k = M_k x$, $k = \nu - 1, \ldots, 0$, are obtained after multiplying (2.84) by $U_k = M_k P_{k+1} \cdots P_{\nu-1}$ (cf. Lemma 2.2). This yields
$$U_k G_\nu^{-1} G_0 D^-(Dx)' + U_k G_\nu^{-1} Bx = U_k G_\nu^{-1} q(t). \quad (2.92)$$
The case $k = \nu - 1$ is somehow simpler than the remaining ones and is discussed in IIIa. For the cases $k = \nu - 2, \ldots, 0$ addressed in IIIb-IIId below, the different terms arising in the computations will be arranged according to the following scheme:
$$\underbrace{U_k G_\nu^{-1} G_0 D^-(Dx)'}_{-\sum_j \mathcal{N}_{kj}(Dv_j)' + \tilde{\mathcal{K}}_k D^- u - \sum_j \tilde{\mathcal{M}}_{kj} v_j} + \underbrace{U_k G_\nu^{-1} Bx}_{v_k + \hat{\mathcal{K}}_k D^- u - \sum_j \hat{\mathcal{M}}_{kj} v_j} = \underbrace{U_k G_\nu^{-1} q(t)}_{\mathcal{L}_k q(t)}. \quad (2.93)$$

IIIa. The $v_{\nu-1}$ component. The case $k = \nu - 1$ proceeds by premultiplying (2.84) by $U_{\nu-1} = M_{\nu-1}$. From (2.71) we have $M_{\nu-1} G_\nu^{-1} G_0 = 0$, since $M_{\nu-1} Q_{\nu-1} = M_{\nu-1}$ and $M_{\nu-1} Q_j = 0$ if $j < \nu - 1$ (see (2.64)). Similarly, from (2.72) it follows that $M_{\nu-1} G_\nu^{-1} B M_{\nu-1} x = M_{\nu-1} x$ and $M_{\nu-1} G_\nu^{-1} B M_j x = 0$ for $0 \leq j \leq \nu - 2$.

Thereby, premultiplying equation (2.84) by $M_{\nu-1}$ we get, using $x = M_0 x + \ldots + M_{\nu-1} x + \Pi_{\nu-1} x$ and $\Pi_{\nu-1} x = D^- D\Pi_{\nu-1} x$,
$$M_{\nu-1} x + M_{\nu-1} G_\nu^{-1} B D^- D\Pi_{\nu-1} x = M_{\nu-1} G_\nu^{-1} q(t),$$
that is,
$$v_{\nu-1} + \mathcal{K}_{\nu-1} D^- u = \mathcal{L}_{\nu-1} q(t), \quad (2.94)$$
with $v_{\nu-1} = M_{\nu-1} x$ and $u = D\Pi_{\nu-1} x$. Note that in this case the terms with coefficients $\mathcal{N}_{kj}$ and $\mathcal{M}_{kj}$ are not present in (2.69), and the coefficient $\mathcal{K}_{\nu-1}$ amounts to $M_{\nu-1} G_\nu^{-1} B$.

2.2. Properly stated linear time-varying DAEs

IIIb. Terms coming from $U_k G_\nu^{-1} G_0 D^-(Dx)'$ in (2.93). For $k = \nu - 2, \ldots, 0$ we make use of the fact that $U_k = M_k P_{k+1} \cdots P_{\nu-1}$ projects along

$$N_0 \oplus \ldots \oplus N_{k-1} \oplus N_{k+1} \oplus \ldots \oplus N_{\nu-1} \oplus \operatorname{im} \Pi_{\nu-1},$$

as stated in (2.79b). This means that $U_k Q_i = 0$ if $k \neq i$. Using (2.71) we then get

$$\begin{aligned} U_k G_\nu^{-1} G_0 D^-(Dx)' &= M_k P_{k+1} \cdots P_{\nu-1}(I - Q_k) D^-(Dx)' \\ &= M_k (P_{k+1} \cdots P_{\nu-1} - I) D^-(Dx)', \end{aligned} \quad (2.95)$$

where we have made use of (2.81).

Now, write

$$P_{k+1} \cdots P_{\nu-1} - I = -Q_{k+1} - P_{k+1} Q_{k+2} - \ldots - P_{k+1} \cdots P_{\nu-2} Q_{\nu-1}. \quad (2.96)$$

Since $Q_j M_j = Q_j$, as displayed in (2.66), (2.96) can be restated as

$$-Q_{k+1} M_{k+1} - P_{k+1} Q_{k+2} M_{k+2} - \ldots - P_{k+1} \cdots P_{\nu-2} Q_{\nu-1} M_{\nu-1},$$

which makes it possible to recast (2.95) in the form

$$-M_k [Q_{k+1} M_{k+1} + \ldots + P_{k+1} \cdots P_{\nu-2} Q_{\nu-1} M_{\nu-1}] D^-(Dx)'. \quad (2.97)$$

For $j > 0$, differentiating the product $DM_j x = DM_j D^- Dx$ we have

$$DM_j D^-(Dx)' = (DM_j x)' - (DM_j D^-)' Dx$$

and, since $M_j = P_0 M_j = D^- DM_j$ for $j > 0$, the expression depicted in (2.97) reads

$$\begin{aligned} &- M_k [Q_{k+1} D^-(DM_{k+1} x)' + \ldots + P_{k+1} \cdots P_{\nu-2} Q_{\nu-1} D^-(DM_{\nu-1} x)'] \\ &+ M_k [Q_{k+1} D^-(DM_{k+1} D^-)' + \ldots + P_{k+1} \cdots P_{\nu-2} Q_{\nu-1} D^-(DM_{\nu-1} D^-)'] Dx. \end{aligned} \quad (2.98)$$

The first line of (2.98) yields the terms

$$-\mathcal{N}_{k\,k+1}(Dv_{k+1})' - \ldots - \mathcal{N}_{k\,\nu-1}(Dv_{\nu-1})' = -\sum_{j=k+1}^{\nu-1} \mathcal{N}_{kj}(Dv_j)', \quad (2.99)$$

with

$$\mathcal{N}_{k\,k+1} = M_k Q_{k+1} D^-, \quad \mathcal{N}_{kj} = M_k P_{k+1} \cdots P_{j-1} Q_j D^- \quad \text{for } j \geq k+2. \quad (2.100)$$

For the terms coming from the second line of (2.98), we decompose $Dx = D\Pi_{\nu-1} x + D \sum_{j=1}^{\nu-1} M_j x$. Note that $DM_0 x = 0$ since $M_0 = Q_0$ projects onto $N_0 = \ker D$.

The term in $D\Pi_{\nu-1}x$ reads
$$M_k[Q_{k+1}D^-(DM_{k+1}D^-)' + P_{k+1}Q_{k+2}D^-(DM_{k+2}D^-)' + \ldots \\ + P_{k+1}\cdots P_{\nu-2}Q_{\nu-1}D^-(DM_{\nu-1}D^-)']D\Pi_{\nu-1}x. \quad (2.101)$$

The relation $M_jD^-D\Pi_{\nu-1} = M_j\Pi_{\nu-1} = 0$ (cf. (2.63)) yields
$$(DM_jD^-)'D\Pi_{\nu-1}D^- = -DM_jD^-(D\Pi_{\nu-1}D^-)',$$
which makes it possible to rewrite (2.101), using (2.96), as
$$M_k(P_{k+1}\cdots P_{\nu-1} - I)D^-(D\Pi_{\nu-1}D^-)'D\Pi_{\nu-1}x = \tilde{\mathcal{K}}_kD^-u \quad (2.102)$$
with
$$\tilde{\mathcal{K}}_k = M_k(P_{k+1}\cdots P_{\nu-1} - I)D^-(D\Pi_{\nu-1}D^-)'D. \quad (2.103)$$

The remaining terms from the second line of (2.98) are
$$M_k[Q_{k+1}D^-(DM_{k+1}D^-)' + P_{k+1}Q_{k+2}D^-(DM_{k+2}D^-)' \\ + \ldots + P_{k+1}\cdots P_{\nu-2}Q_{\nu-1}D^-(DM_{\nu-1}D^-)']D\sum_{j=1}^{\nu-1}M_jx \\ = -\sum_{j=1}^{\nu-1}\tilde{\mathcal{M}}_{kj}M_jx, \quad (2.104)$$
where, for later convenience, $\tilde{\mathcal{M}}_{kj}$ will be written as
$$\tilde{\mathcal{M}}_{kj} = -M_k[Q_{k+1}D^-(DM_{k+1}D^-)' + P_{k+1}Q_{k+2}D^-(DM_{k+2}D^-)' \\ + \ldots + P_{k+1}\cdots P_{\nu-2}Q_{\nu-1}D^-(DM_{\nu-1}D^-)']DM_jD^-D.$$

Using
$$(DM_jD^-)'DM_jD^- = (DM_jD^-)' - DM_jD^-(DM_jD^-)' \\ (DM_iD^-)'DM_jD^- = -DM_iD^-(DM_jD^-)' \quad \text{if } i \neq j$$
from (2.60), and proceeding as above, the coefficient $\tilde{\mathcal{M}}_{kj}$ amounts to
$$\tilde{\mathcal{M}}_{kj} = M_k(I - P_{k+1}\cdots P_{\nu-1} - P_{k+1}\cdots P_{j-1}Q_j)D^-(DM_jD^-)'D \quad (2.105)$$
if $j > k$, and to
$$\tilde{\mathcal{M}}_{kj} = M_k(I - P_{k+1}\cdots P_{\nu-1})D^-(DM_jD^-)'D \quad (2.106)$$
if $j \leq k$.

Altogether, via (2.98) and subsequently (2.99), (2.102) and (2.104), we have restated $U_kG_\nu^{-1}G_0D^-(Dx)'$ in (2.93) as
$$-\sum_{j=k+1}^{\nu-1}\mathcal{N}_{kj}(Dv_j)' + \tilde{\mathcal{K}}_kD^-u - \sum_{j=1}^{\nu-1}\tilde{\mathcal{M}}_{kj}v_j \quad (2.107)$$

2.2. Properly stated linear time-varying DAEs

with $u = D\Pi_{\nu-1}x$, $v_j = M_j x$, and coefficients $\mathcal{N}_{kj}$, $\tilde{\mathcal{K}}_k$ and $\tilde{\mathcal{M}}_{kj}$ given by (2.100), (2.103) and (2.105)-(2.106), respectively.

IIIc. Terms coming from $U_k G_\nu^{-1} Bx$ in (2.93). Via the decomposition

$$x = \Pi_{\nu-1}x + \sum_{j=0}^{\nu-1} M_j x,$$

write $U_k G_\nu^{-1} Bx = M_k P_{k+1} \cdots P_{\nu-1} G_\nu^{-1} Bx$ in (2.93) as

$$M_k P_{k+1} \cdots P_{\nu-1} G_\nu^{-1} B(\Pi_{\nu-1}x + \sum_{j=0}^{\nu-1} M_j x). \tag{2.108}$$

Using $\Pi_{\nu-1} = D^- D\Pi_{\nu-1}$, the first term gives

$$M_k P_{k+1} \cdots P_{\nu-1} G_\nu^{-1} B D^- D\Pi_{\nu-1} x = \hat{\mathcal{K}}_k D^- u, \tag{2.109}$$

where

$$\hat{\mathcal{K}}_k = M_k P_{k+1} \cdots P_{\nu-1} G_\nu^{-1} B. \tag{2.110}$$

The additional terms coming from (2.108) are

$$\sum_{j=0}^{\nu-1} M_k P_{k+1} \cdots P_{\nu-1} G_\nu^{-1} B M_j x$$

which, using (2.72), (2.79b) and (2.81), can be written as

$$M_k x - \sum_{j=1}^{\nu-1} \hat{\mathcal{M}}_{kj} M_j x \tag{2.111}$$

with

$$\hat{\mathcal{M}}_{kj} = -M_k \left\{ \sum_{i=1}^{k} (P_{k+1} \cdots P_{\nu-1} - I) D^- (D\Pi_i D^-)' \right.$$
$$\left. + \sum_{i=k+1}^{j} P_{k+1} \cdots P_{\nu-1} D^- (D\Pi_i D^-)' \right\} DM_j D^- D \text{ if } j > k \tag{2.112}$$

and

$$\hat{\mathcal{M}}_{kj} = -M_k \sum_{i=1}^{j} (P_{k+1} \cdots P_{\nu-1} - I) D^- (D\Pi_i D^-)' DM_j D^- D \text{ if } j \leq k. \tag{2.113}$$

To simplify the expressions depicted in (2.112)-(2.113) for $\hat{\mathcal{M}}_{kj}$ we will make use of the identities

$$D^-(D\Pi_i D^-)' DM_j D^- = (I - \Pi_i) D^- (DM_j D^-)', \; i < j$$
$$D^-(D\Pi_j D^-)' DM_j D^- = -\Pi_j D^- (DM_j D^-)',$$

which follow from $\Pi_i M_j = M_j$ if $i < j$ and $\Pi_j M_j = 0$, as depicted in (2.62). For the case $j > k$ in (2.112), the reader can check from (2.82) that

$$-M_k(P_{k+1}\cdots P_{\nu-1} - I)(I - \Pi_i) = 0$$

if $1 \leq i \leq k$, whereas

$$-M_k P_{k+1}\cdots P_{\nu-1}(I - \Pi_i) = -M_k P_{k+1}\cdots P_i$$

for the cases in which $k+1 \leq i < j$, and

$$M_k P_{k+1}\cdots P_{\nu-1}\Pi_j = M_k(P_{k+1}\cdots P_{\nu-1} - P_{k+1}\cdots P_j)$$

since $j > k$. These expressions yield

$$\hat{\mathcal{M}}_{kj} = M_k\left(-\sum_{i=k+1}^{j} P_{k+1}\cdots P_i + P_{k+1}\cdots P_{\nu-1}\right) D^-(DM_j D^-)'D \quad (2.114)$$

if $j > k$.

Similarly, for the case $j \leq k$ in (2.113) we use the properties

$$-M_k(P_{k+1}\cdots P_{\nu-1} - I)(I - \Pi_i) = 0$$

for $i < j \leq k$, and

$$M_k(P_{k+1}\cdots P_{\nu-1} - I)\Pi_j = M_k(P_{k+1}\cdots P_{\nu-1} - I)$$

for $i = j \leq k$. These relations follow also from (2.82) and yield

$$\hat{\mathcal{M}}_{kj} = M_k(P_{k+1}\cdots P_{\nu-1} - I)D^-(DM_j D^-)'D \quad \text{if } j \leq k. \quad (2.115)$$

This way, splitting (2.108) via (2.109) and (2.111), we have rewritten the term $U_k G_\nu^{-1} Bx$ in (2.93) as

$$\hat{\mathcal{K}}_k D^- u + v_k - \sum_{j=1}^{\nu-1} \hat{\mathcal{M}}_{kj} v_j \quad (2.116)$$

with $u = D\Pi_{\nu-1}x$, $v_k = M_k x$, $v_j = M_j x$, and coefficients $\hat{\mathcal{K}}_k$ and $\hat{\mathcal{M}}_{kj}$ given by (2.110) and (2.114)-(2.115), respectively.

IIId. The v_k components, $k = \nu - 2, \ldots, 0$. According to IIIb and IIIc, the expressions depicted in (2.107) and (2.116) transform (2.93) into the equivalent form

$$v_k = -(\tilde{\mathcal{K}}_k + \hat{\mathcal{K}}_k)D^- u + \sum_{j=k+1}^{\nu-1} \mathcal{N}_{kj}(Dv_j)' + \sum_{j=1}^{\nu-1}(\tilde{\mathcal{M}}_{kj} + \hat{\mathcal{M}}_{kj})v_j + \mathcal{L}_k q.$$

Setting $\mathcal{K}_k = \hat{\mathcal{K}}_k + \tilde{\mathcal{K}}_k$, from (2.103) and (2.110) we get

2.2. Properly stated linear time-varying DAEs

$$\mathcal{K}_k = M_k P_{k+1} \cdots P_{\nu-1} G_\nu^{-1} B + M_k (P_{k+1} \cdots P_{\nu-1} - I) D^- (D\Pi_{\nu-1} D^-)' D.$$

In turn, if $j > k$ then $\mathcal{M}_{kj} = \tilde{\mathcal{M}}_{kj} + \hat{\mathcal{M}}_{kj}$ reads, from (2.105) and (2.114),

$$\mathcal{M}_{kj} = M_k (I - P_{k+1} \cdots P_{j-1} Q_j - \sum_{i=k+1}^{j} P_{k+1} \cdots P_i) D^- (DM_j D^-)' D \quad (2.117)$$

which, using $P_{k+1} \cdots P_{j-1} Q_j + P_{k+1} \cdots P_j = P_{k+1} \cdots P_{j-1}$ together with $I - P_{k+1} \cdots P_{j-1} = \sum_{i=k+1}^{j-1} P_{k+1} \cdots P_{i-1} Q_i$, yields

$$\mathcal{M}_{kj} = M_k \left(\sum_{i=k+1}^{j-1} P_{k+1} \cdots P_{i-1}(Q_i - P_i) \right) D^-(DM_j D^-)' D \quad \text{if } j \geq k+2$$

and $\mathcal{M}_{kj} = 0$ if $j = k+1$.

On the other hand, (2.106) and (2.115) lead to $\mathcal{M}_{kj} = 0$ if $j \leq k$.

These expressions make it possible then to write the algebraic component v_k, for $k = \nu - 2, \ldots, 0$ as

$$v_k = -\mathcal{K}_k D^- u + \sum_{j=k+1}^{\nu-1} \mathcal{N}_{kj}(Dv_j)' + \sum_{j=k+2}^{\nu-1} \mathcal{M}_{kj} v_j + \mathcal{L}_k q,$$

as stated in (2.69). Note that this expression also accommodates the case $k = \nu - 1$ addressed in IIIa since the $\mathcal{N}_{kj}(Dv_j)'$ and $\mathcal{M}_{kj} v_j$ terms are not present in this situation, consistently with (2.94).

Remark finally that $Dv_k = DM_k x$ is in C^1 for $k \geq 1$ due to the relation $DM_k x = DM_k D^- Dx$ and the fact that both $DM_k D^-$ (which equals $D\Pi_{k-1} D^- - D\Pi_k D^-$) and Dx are in C^1.

The proof of the converse assertion, stating that u, v_k from (2.68)-(2.69) yield a solution of (2.3) via the decomposition $x = D^- u + v_{\nu-1} + \ldots + v_1 + v_0$ stated in (2.67), is simpler; we just need to show that the above process is reversible. Note first that the identity $Dv_0 = DM_0 v_0 = 0$ implies that $Dx = DD^- u + Dv_{\nu-1} + \ldots + Dv_1 = u + Dv_{\nu-1} + \ldots + Dv_1$ is in $C^1(\mathcal{J}, \mathbb{R}^{n \times n})$; we have made use of the identity $u = DD^- u$ which owes to $u \in \operatorname{im} D\Pi_{\nu-1} D^-$.

Additionally, due to the expressions displayed for the coefficients $\mathcal{K}_k$, $\mathcal{N}_{kj}$, $\mathcal{M}_{kj}$ and $\mathcal{L}_k$ on p. 52, we have $v_k = M_k v_k$ and this means that the relations $D\Pi_{\nu-1} x = u$, $M_k x = v_k$ hold since $\Pi_{\nu-1} M_k = 0$, $M_k \Pi_{\nu-1} = 0$ and $M_k M_j = 0$ if $k \neq j$ (cf. (2.62), (2.63) and (2.60), respectively).

With these relations in mind, it only remains to note that the inherent ODE (2.68) is a restatement of (2.84) premultiplied by $D\Pi_{\nu-1}$, i.e.

$$D\Pi_{\nu-1}G_{\nu}^{-1}A(Dx)' + D\Pi_{\nu-1}G_{\nu}^{-1}Bx = D\Pi_{\nu-1}G_{\nu}^{-1}q(t), \qquad (2.118)$$

which premultiplied in turn by D^- yields, using $D^- D\Pi_{\nu-1} = \Pi_{\nu-1}$,

$$\Pi_{\nu-1}G_{\nu}^{-1}A(Dx)' + \Pi_{\nu-1}G_{\nu}^{-1}Bx = \Pi_{\nu-1}G_{\nu}^{-1}q(t), \qquad (2.119)$$

whereas the algebraic relations (2.69) stand for

$$U_k G_{\nu}^{-1}A(Dx)' + U_k G_{\nu}^{-1}Bx = U_k G_{\nu}^{-1}q(t), \; k = \nu - 1, \ldots, 0 \qquad (2.120)$$

(cf. (2.92)). Now, the expressions for V_k, U_k depicted in (2.76), (2.77) together with the identity $Q_k M_k = Q_k$ from (2.66), imply that $Q_k U_k = V_k$. Hence, premultiplying (2.120) by Q_k we obtain

$$V_k G_{\nu}^{-1}A(Dx)' + V_k G_{\nu}^{-1}Bx = V_k G_{\nu}^{-1}q(t), \; k = \nu - 1, \ldots, 0. \qquad (2.121)$$

The identities $\Pi_{\nu-1} = P_0 \cdots P_{\nu-1}$ and $V_k = Q_k P_{k+1} \cdots P_{\nu-1}$ mean that $\Pi_{\nu-1} + \sum_{k=0}^{\nu-1} V_k = I$. Therefore, adding (2.119) and (2.121) for the values $k = \nu - 1, \ldots, 0$, and multiplying by G_{ν}, we finally obtain that

$$A(Dx)' + Bx = q(t)$$

holds, as we aimed to show. □

Remark 2.10. As detailed below, the terms $\mathcal{N}_{kj}(Dv_j)' + \mathcal{M}_{kj}v_j$ in (2.69) can be joined in a single one of the form $\mathcal{N}^*_{kj}(Dv_j)'$. This makes it possible to write (2.69) in the somewhat simpler form

$$v_k = -\mathcal{K}_k D^- u + \sum_{j=k+1}^{\nu-1} \mathcal{N}^*_{kj}(Dv_j)' + \mathcal{L}_k q. \qquad (2.122)$$

The chance to do this stems from the factor $D^-(DM_j D^-)'D$ at the end of $\mathcal{M}_{kj}$ for $j \geq k+2$. Indeed, using $v_j = M_j v_j$, in the $\mathcal{M}_{kj}v_j$ term within (2.69) we may write $(DM_j D^-)'Dv_j = (Dv_j)' - DM_j D^-(Dv_j)'$ and then

$$D^-(DM_j D^-)'Dv_j = D^-(Dv_j)' - M_j D^-(Dv_j)'. \qquad (2.123)$$

Now, rewrite the term in front of $D^-(DM_j D^-)'D$ in the expression depicted for $\mathcal{M}_{kj}$ in (2.117) as

$$M_k(I - P_{k+1} - P_{k+1}P_{k+2} - \ldots - P_{k+1}\cdots P_{j-2} - 2P_{k+1}\cdots P_{j-1}).$$

Use, for $i \leq j-1$, the identities $\operatorname{im} M_j \subseteq \operatorname{im} \Pi_{j-1} \subseteq \operatorname{im} \Pi_i \subseteq \operatorname{im} P_i$ following from (2.48), (2.51) and (2.59a), together with item (i) of Proposition 2.5, to show that $P_i M_j = M_j$ for $i < j$ and then

$$M_k(I - P_{k+1} - P_{k+1}P_{k+2} - \ldots - P_{k+1}\cdots P_{j-2} - 2P_{k+1}\cdots P_{j-1})M_j = 0$$

2.2. Properly stated linear time-varying DAEs

in the light of (2.60).

This means that in $\mathcal{M}_{kj}v_j$ there is actually no contribution coming from the term $-M_j D^-(Dv_j)'$ in (2.123), so that $\mathcal{M}_{kj}v_j$ can be rewritten as $\mathcal{M}_{kj}^*(Dv_j)'$, with

$$\mathcal{M}_{kj}^* = M_k \left(I - P_{k+1}\cdots P_{j-1}Q_j - \sum_{i=k+1}^{j} P_{k+1}\cdots P_i \right) D^-$$

$$= M_k \left(\sum_{i=k+1}^{j-1} P_{k+1}\cdots P_{i-1}(Q_i - P_i) \right) D^-,$$

the coefficient $\mathcal{N}_{kj}^* = \mathcal{N}_{kj} + \mathcal{M}_{kj}^*$ in (2.122) thus having the expression

$$\mathcal{N}_{kj}^* = M_k \left(I - \sum_{i=k+1}^{j} P_{k+1}\cdots P_i \right) D^-$$

for $j \geq k+1$.

2.2.6 A tutorial example

The linear, time-varying analog of Chua's circuit [69] with current-controlled resistors depicted in Figure 2.1 was proposed in [195] to illustrate the computation of the tractability index. Some singularities of this system have been analyzed in [196]. We summarize below the index analysis of [195] and then use the DAE modeling this circuit to illustrate the decoupling presented in Theorem 2.3. The interest of this problem relies on the fact that, being simple enough to keep computations at a minimum, displays however a rich variety of indices depending on device characteristics and has therefore a value for illustrative purposes.

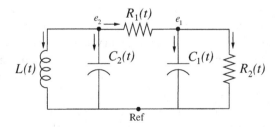

Fig. 2.1 Linear time-varying Chua's circuit with current-controlled resistors.

The reader is here referred to Chapter 5 for details concerning the form of the following model for the dynamics of this circuit:

$$(C_1(t)e_1)' - i_{r_1} + i_{r_2} = 0 \tag{2.124a}$$
$$(C_2(t)e_2)' + i_l + i_{r_1} = 0 \tag{2.124b}$$
$$(L(t)i_l)' - e_2 = 0 \tag{2.124c}$$
$$e_2 - e_1 - R_1(t)i_{r_1} = 0 \tag{2.124d}$$
$$e_1 - R_2(t)i_{r_2} = 0. \tag{2.124e}$$

Here, e_1 and e_2 represent node potentials (cf. Figure 2.1), whereas i_l, i_{r_1} and i_{r_2} stand for the currents through the inductor and the resistors, respectively. These variables define the semistate vector $x = (e_1, e_2, i_l, i_{r_1}, i_{r_2})$.

All devices are linear and time-dependent, and their characteristics are assumed to be defined on some working interval $\mathcal{J} \subseteq \mathbb{R}$. Both resistors are supposed to be current-controlled via the relations $v_{r_1} = R_1(t)i_{r_1}$ and $v_{r_2} = R_2(t)i_{r_2}$. If both $R_1(t)$ and $R_2(t)$ do not vanish on $\mathcal{J}$, then i_{r_1} and i_{r_2} can be written in terms of e_1 and e_2 using (2.124d) and (2.124e). Inserting these relations into (2.124a) and (2.124b) one gets the linear time-varying analog of Chua's equation [69]. If, additionally, the reactances are in C^1 and do not vanish, then the system can be rewritten as an explicit state equation on e_1, e_2, i_l; this case is framed below as an index one problem. By contrast, if we assume that the resistances $R_1(t)$ or $R_2(t)$ may vanish then the attention must be driven to DAE models such as (2.124).

We remove hereafter explicit dependences on t. The differential-algebraic model (2.124) can be written in the form $A(Dx)' + Bx = 0$ with

$$A = \begin{pmatrix} 1 & 0 & 0 \\ 0 & 1 & 0 \\ 0 & 0 & 1 \\ 0 & 0 & 0 \\ 0 & 0 & 0 \end{pmatrix}, \; D = \begin{pmatrix} C_1 & 0 & 0 & 0 & 0 \\ 0 & C_2 & 0 & 0 & 0 \\ 0 & 0 & L & 0 & 0 \end{pmatrix}, \; B = \begin{pmatrix} 0 & 0 & 0 & -1 & 1 \\ 0 & 0 & 1 & 1 & 0 \\ 0 & -1 & 0 & 0 & 0 \\ -1 & 1 & 0 & -R_1 & 0 \\ 1 & 0 & 0 & 0 & -R_2 \end{pmatrix}.$$

In the sequel we analyze the index of this DAE under different assumptions on the device characteristics, also showing how the decoupling of Theorem 2.3 unravels the circuit dynamics. Contrary to [195], we work with the Π-framework of subsection 2.2.4. The reader should have in mind that the goal is not just to provide the system solutions, but to illustrate how the projector framework drives the analysis of the DAE to an explicit ODE on the variable u, with explicit algebraic expressions for the v_k components of the solution (cf. (2.67)); the key aspect here is that exactly the same arrangements can be used in the analysis of high-scale problems.

2.2.6.1 Index one

Assume that the resistances $R_1(t)$, $R_2(t)$, the capacitances $C_1(t)$, $C_2(t)$, and the inductance $L(t)$ do not meet any zero on $\mathcal{J}$. These conditions will define an index one configuration which, as indicated above, amounts to the standard state space setting for Chua's circuit.

Indeed, from the non-vanishing of C_1, C_2 and L it follows that the leading term is properly stated and we may take

$$D^- = \begin{pmatrix} 1/C_1 & 0 & 0 \\ 0 & 1/C_2 & 0 \\ 0 & 0 & 1/L \\ 0 & 0 & 0 \\ 0 & 0 & 0 \end{pmatrix}.$$

Note that $R = DD^- = I$. The matrix $G_0 = AD$ reads

$$G_0 = \begin{pmatrix} C_1 & 0 & 0 & 0 & 0 \\ 0 & C_2 & 0 & 0 & 0 \\ 0 & 0 & L & 0 & 0 \\ 0 & 0 & 0 & 0 & 0 \\ 0 & 0 & 0 & 0 & 0 \end{pmatrix},$$

and

$$N_0(t) = \ker G_0(t) = \ker D(t) = \mathrm{span}\left[\begin{pmatrix} 0 \\ 0 \\ 0 \\ 1 \\ 0 \end{pmatrix}, \begin{pmatrix} 0 \\ 0 \\ 0 \\ 0 \\ 1 \end{pmatrix} \right]. \qquad (2.125)$$

Take

$$\Pi_0 = \begin{pmatrix} 1 & 0 & 0 & 0 & 0 \\ 0 & 1 & 0 & 0 & 0 \\ 0 & 0 & 1 & 0 & 0 \\ 0 & 0 & 0 & 0 & 0 \\ 0 & 0 & 0 & 0 & 0 \end{pmatrix}, \quad M_0 = I - \Pi_0 = \begin{pmatrix} 0 & 0 & 0 & 0 & 0 \\ 0 & 0 & 0 & 0 & 0 \\ 0 & 0 & 0 & 0 & 0 \\ 0 & 0 & 0 & 1 & 0 \\ 0 & 0 & 0 & 0 & 1 \end{pmatrix}, \qquad (2.126)$$

which, together with $B_0 = B$, yield

$$G_1 = G_0 + B_0 M_0 = \begin{pmatrix} C_1 & 0 & 0 & -1 & 1 \\ 0 & C_2 & 0 & 1 & 0 \\ 0 & 0 & L & 0 & 0 \\ 0 & 0 & 0 & -R_1 & 0 \\ 0 & 0 & 0 & 0 & -R_2 \end{pmatrix}. \qquad (2.127)$$

The working assumptions $R_1 \neq 0 \neq R_2$, $C_1 \neq 0 \neq C_2$, $L \neq 0$ make G_1 nonsingular and the problem index one.

The decoupling (2.68)-(2.69) reads, in an index one context,

$$u' + (-R' + DG_1^{-1}BD^-)u = DG_1^{-1}q, \qquad (2.128a)$$
$$v_0 = -\mathcal{K}_0 D^- u + \mathcal{L}_0 q, \qquad (2.128b)$$

having used the identities $D\Pi_0 D^- = DD^- = R$ and $D\Pi_0 = D$. The coefficients in (2.128b) read

$$\mathcal{K}_0 = M_0 G_1^{-1} B, \ \mathcal{L}_0 = M_0 G_1^{-1}. \qquad (2.129)$$

Since the circuit has no excitation we have $q = 0$, and then the terms $DG_1^{-1}q$ and $\mathcal{L}_0 q$ vanish. Additionally, $R = I$ makes $R' = 0$, so that the only non-trivial term in (2.128a) comes from the product $DG_1^{-1}BD^-$. A straightforward computation yields for the inherent ODE (2.128a) the expression

$$u' + \begin{pmatrix} \frac{R_1+R_2}{R_1 R_2 C_1} & \frac{-1}{R_1 C_2} & 0 \\ \frac{-1}{R_1 C_1} & \frac{1}{R_1 C_2} & \frac{1}{L} \\ 0 & \frac{-1}{C_2} & 0 \end{pmatrix} u = 0, \qquad (2.130)$$

whereas the product $-\mathcal{K}_0 D^- = -M_0 G_1^{-1} B D^-$ in (2.128b) reads

$$\begin{pmatrix} 0 & 0 & 0 \\ 0 & 0 & 0 \\ 0 & 0 & 0 \\ \frac{-1}{R_1 C_1} & \frac{1}{R_1 C_2} & 0 \\ \frac{1}{R_2 C_1} & 0 & 0 \end{pmatrix}.$$

Solutions of the problem are then given by $x = D^- u + v_0$, where the vector $u = (u_1, u_2, u_3)$ solves (2.130) in the invariant space im $D\Pi_0 D^- = \mathbb{R}^3$. This yields the solutions of the DAE (2.124) in the form

$$x = \begin{pmatrix} e_1 \\ e_2 \\ i_l \\ i_{r_1} \\ i_{r_2} \end{pmatrix} = \begin{pmatrix} \frac{u_1}{C_1} \\ \frac{u_2}{C_2} \\ \frac{u_3}{L} \\ 0 \\ 0 \end{pmatrix} + \begin{pmatrix} 0 \\ 0 \\ 0 \\ \frac{-u_1}{R_1 C_1} + \frac{u_2}{R_1 C_2} \\ \frac{u_1}{R_2 C_1} \end{pmatrix}.$$

2.2.6.2 Index two

Let us now assume that $R_1(t) = 0$ on the working interval $\mathcal{J}$, and that the remaining device parameters do not vanish on $\mathcal{J}$, namely, that $R_2(t) \neq 0$, $C_1(t) \neq 0 \neq C_2(t)$, and $L(t) \neq 0$ for all $t \in \mathcal{J}$. Suppose additionally that $C_1(t) + C_2(t) \neq 0$ on $\mathcal{J}$. As detailed below, these conditions make the DAE (2.124) index two.

In this situation, the matrix G_1 depicted in (2.127) has constant rank $r_1 = 4$. The kernel N_1 is defined by

$$N_1(t) = \ker G_1(t) = \mathrm{span}\left[\begin{pmatrix} 1/C_1 \\ -1/C_2 \\ 0 \\ 1 \\ 0 \end{pmatrix}\right]$$

and satisfies the condition $N_0 \cap N_1 = \{0\}$, with N_0 given in (2.125). Additionally,

$$N_0(t) \oplus N_1(t) = \mathrm{span}\left[\begin{pmatrix} 0 \\ 0 \\ 0 \\ 1 \\ 0 \end{pmatrix}, \begin{pmatrix} 0 \\ 0 \\ 0 \\ 0 \\ 1 \end{pmatrix}, \begin{pmatrix} 1/C_1 \\ -1/C_2 \\ 0 \\ 0 \\ 0 \end{pmatrix}\right].$$

We may then take the projector Π_1 along $N_0 \oplus N_1$ as

$$\Pi_1 = \begin{pmatrix} 0 & 0 & 0 & 0 & 0 \\ C_1/C_2 & 1 & 0 & 0 & 0 \\ 0 & 0 & 1 & 0 & 0 \\ 0 & 0 & 0 & 0 & 0 \\ 0 & 0 & 0 & 0 & 0 \end{pmatrix},$$

which satisfies the admissibility condition $\mathrm{im}\,\Pi_1 \subseteq \mathrm{im}\,\Pi_0$, with Π_0 defined in (2.126). Some simple computations yield

$$B_1 = \begin{pmatrix} 0 & 0 & 0 & 0 & 0 \\ 0 & 0 & 1 & 0 & 0 \\ 0 & -1 & 0 & 0 & 0 \\ -1 & 1 & 0 & 0 & 0 \\ 1 & 0 & 0 & 0 & 0 \end{pmatrix}, \quad G_2 = \begin{pmatrix} C_1 & 0 & 0 & -1 & 1 \\ 0 & C_2 & 0 & 1 & 0 \\ C_1/C_2 & 0 & L & 0 & 0 \\ -1 - C_1/C_2 & 0 & 0 & 0 & 0 \\ 1 & 0 & 0 & 0 & -R_2 \end{pmatrix}. \quad (2.131)$$

Since C_2, L, and R_2 do not vanish, we conclude that the matrix G_2 is nonsingular if and only if $1 + C_1/C_2 \neq 0$, condition which holds indeed because of the assumption $C_1 + C_2 \neq 0$. The DAE (2.124) is therefore regular with index two under these assumptions on the devices.

For index two DAEs, the decoupling (2.68)-(2.69) is defined by

$$u' + (-(D\Pi_1 D^-)' + D\Pi_1 G_2^{-1} BD^-)u = D\Pi_1 G_2^{-1} q, \qquad (2.132)$$

together with

$$v_1 = -\mathcal{K}_1 D^- u + \mathcal{L}_1 q, \qquad (2.133a)$$

$$v_0 = -\mathcal{K}_0 D^- u + \mathcal{N}_{01}(Dv_1)' + \mathcal{L}_0 q. \qquad (2.133b)$$

The matrix mappings in (2.133) have the expressions

$$\mathcal{K}_1 = M_1 G_2^{-1} B, \quad \mathcal{L}_1 = M_1 G_2^{-1}, \qquad (2.134a)$$

$$\mathcal{K}_0 = M_0(P_1 G_2^{-1} B - Q_1 D^-(D\Pi_1 D^-)'D), \qquad (2.134b)$$

$$\mathcal{N}_{01} = M_0 Q_1 D^-, \quad \mathcal{L}_0 = M_0 P_1 G_2^{-1}. \qquad (2.134c)$$

The inherent ODE (2.132) reads, for this configuration,

$$u' + \begin{pmatrix} 0 & 0 & 0 \\ \frac{1}{R_2(C_1+C_2)} & \frac{1}{R_2(C_1+C_2)} & \frac{1}{L} \\ \frac{-1}{C_1+C_2} & \frac{-1}{C_1+C_2} & 0 \end{pmatrix} u = 0, \qquad (2.135)$$

with invariant space

$$\operatorname{im} D\Pi_1 D^- = \operatorname{im} \begin{pmatrix} 0 & 0 & 0 \\ 1 & 1 & 0 \\ 0 & 0 & 1 \end{pmatrix} = \operatorname{span} \left[\begin{pmatrix} 0 \\ 1 \\ 0 \end{pmatrix}, \begin{pmatrix} 0 \\ 0 \\ 1 \end{pmatrix} \right].$$

Within this space, the inherent ODE (2.135) can be described in terms of the components u_2, u_3 of u as

$$u_2' + \frac{1}{R_2(C_1 + C_2)} u_2 + \frac{1}{L} u_3 = 0 \qquad (2.136a)$$

$$u_3' - \frac{1}{C_1 + C_2} u_2 = 0. \qquad (2.136b)$$

The coefficient $-\mathcal{K}_1 D^- = -M_1 G_2^{-1} BD^-$ in (2.133a) is given by

$$-\mathcal{K}_1 D^- = \begin{pmatrix} \frac{-C_2}{C_1(C_1+C_2)} & \frac{1}{C_1+C_2} & 0 \\ \frac{1}{C_1+C_2} & \frac{-C_1}{C_2(C_1+C_2)} & 0 \\ 0 & 0 & 0 \\ 0 & 0 & 0 \\ 0 & 0 & 0 \end{pmatrix},$$

2.2. Properly stated linear time-varying DAEs

whereas $-\mathcal{K}_0 D^-$ and $\mathcal{N}_{01}$ in (2.133b) read

$$-\mathcal{K}_0 D^- = \begin{pmatrix} 0 & 0 & 0 \\ 0 & 0 & 0 \\ 0 & 0 & 0 \\ \frac{1}{R_2(C_1+C_2)} & \frac{1}{R_2(C_1+C_2)} & 0 \\ \frac{1}{R_2(C_1+C_2)} & \frac{1}{R_2(C_1+C_2)} & 0 \end{pmatrix}, \quad \mathcal{N}_{01} = \begin{pmatrix} 0 & 0 & 0 \\ 0 & 0 & 0 \\ 0 & 0 & 0 \\ 1 & 0 & 0 \\ 0 & 0 & 0 \end{pmatrix}.$$

Solutions $x = D^- u + v_1 + v_0$ are then defined by

$$x = \begin{pmatrix} e_1 \\ e_2 \\ i_l \\ i_{r_1} \\ i_{r_2} \end{pmatrix} = \begin{pmatrix} 0 \\ \frac{u_2}{C_2} \\ \frac{u_3}{L} \\ 0 \\ 0 \end{pmatrix} + \begin{pmatrix} \frac{u_2}{C_1+C_2} \\ \frac{-C_1 u_2}{C_2(C_1+C_2)} \\ 0 \\ 0 \\ 0 \end{pmatrix} + \begin{pmatrix} 0 \\ 0 \\ 0 \\ \frac{u_2}{R_2(C_1+C_2)} + \left(\frac{C_1 u_2}{C_1+C_2}\right)' \\ \frac{u_2}{R_2(C_1+C_2)} \end{pmatrix},$$

where u_2 and u_3 are given by (2.136). Mind the term $\left(\frac{C_1 u_2}{C_1+C_2}\right)'$ coming from $\mathcal{N}_{01}(Dv_1)'$ in v_0.

2.2.6.3 Index three

Finally, consider the setting of 2.2.6.2 with $C_1(t) + C_2(t) = 0$, i.e., assume that $R_1(t) = 0$, $R_2(t) \neq 0$, $L(t) \neq 0$ and $C_1(t) = -C_2(t) \neq 0$ for all $t \in \mathcal{J}$. The computations of 2.2.6.2 are still valid, but in this situation

$$G_2 = \begin{pmatrix} C_1 & 0 & 0 & -1 & 1 \\ 0 & -C_1 & 0 & 1 & 0 \\ -1 & 0 & L & 0 & 0 \\ 0 & 0 & 0 & 0 & 0 \\ 1 & 0 & 0 & 0 & -R_2 \end{pmatrix},$$

is a singular matrix with $\operatorname{rk} G_2 = r_2 = 4$. Now

$$N_2(t) = \ker G_2(t) = \operatorname{span}\left[\begin{pmatrix} 1 \\ 1 + \frac{1}{R_2 C_1} \\ \frac{1}{L} \\ C_1 + \frac{1}{R_2} \\ \frac{1}{R_2} \end{pmatrix}\right]$$

which satisfies $(N_0 \oplus N_1) \cap N_2 = \{0\}$. Additionally,

$$N_0(t) \oplus N_1(t) \oplus N_2(t) = \mathrm{span}\left[\begin{pmatrix}0\\0\\0\\1\\0\end{pmatrix}, \begin{pmatrix}0\\0\\0\\0\\1\end{pmatrix}, \begin{pmatrix}1\\1\\0\\0\\0\end{pmatrix}, \begin{pmatrix}0\\ \frac{L}{R_2C_1}\\1\\0\\0\end{pmatrix}\right].$$

An admissible choice for the projector Π_2 along $N_0 \oplus N_1 \oplus N_2$ is

$$\Pi_2 = \begin{pmatrix}0 & 0 & 0 & 0 & 0\\ -1 & 1 & \frac{-L}{R_2C_1} & 0 & 0\\ 0 & 0 & 0 & 0 & 0\\ 0 & 0 & 0 & 0 & 0\\ 0 & 0 & 0 & 0 & 0\end{pmatrix},$$

which yields, after some computations,

$$B_2 = \begin{pmatrix}0 & 0 & 0 & 0 & 0\\ 0 & 0 & 1-L\left(\frac{1}{R_2}\right)' & 0 & 0\\ 1 & -1 & 0 & 0 & 0\\ -1 & 1 & 0 & 0 & 0\\ 0 & 0 & 0 & 0 & 0\end{pmatrix}, \quad G_3 = \begin{pmatrix}C_1 & 0 & 0 & -1 & 1\\ 0 & -C_1 & 1-L\left(\frac{1}{R_2}\right)' & 1 & 0\\ -1 & 0 & L-\frac{L}{R_2C_1} & 0 & 0\\ 0 & 0 & \frac{L}{R_2C_1} & 0 & 0\\ 1 & 0 & 0 & 0 & -R_2\end{pmatrix}.$$

The matrix G_3 is nonsingular and then (2.124) is index three in this setting.

In an index three context, the decoupling (2.68)-(2.69) is given by

$$u' + (-(D\Pi_2 D^-)' + D\Pi_2 G_3^{-1} B D^-)u = D\Pi_2 G_3^{-1} q \qquad (2.137)$$

and

$$v_2 = -\mathcal{K}_2 D^- u + \mathcal{L}_2 q, \qquad (2.138a)$$
$$v_1 = -\mathcal{K}_1 D^- u + \mathcal{N}_{12}(Dv_2)' + \mathcal{L}_1 q \qquad (2.138b)$$
$$v_0 = -\mathcal{K}_0 D^- u + \mathcal{N}_{01}(Dv_1)' + \mathcal{N}_{02}(Dv_2)' + \mathcal{M}_{02} v_2 + \mathcal{L}_0 q, \qquad (2.138c)$$

with coefficients

$$\mathcal{K}_2 = M_2 G_3^{-1} B, \quad \mathcal{L}_2 = M_2 G_3^{-1}, \qquad (2.139a)$$
$$\mathcal{K}_1 = M_1(P_2 G_3^{-1} B - Q_2 D^-(D\Pi_2 D^-)'D), \qquad (2.139b)$$
$$\mathcal{N}_{12} = M_1 Q_2 D^-, \quad \mathcal{L}_1 = M_1 P_2 G_3^{-1} \qquad (2.139c)$$
$$\mathcal{K}_0 = M_0(P_1 P_2 G_3^{-1} B + (P_1 P_2 - I) D^-(D\Pi_2 D^-)'D), \qquad (2.139d)$$
$$\mathcal{N}_{01} = M_0 Q_1 D^-, \quad \mathcal{N}_{02} = M_0 P_1 Q_2 D^-, \qquad (2.139e)$$
$$\mathcal{M}_{02} = M_0(Q_1 - P_1)D^-(DM_2 D^-)'D, \quad \mathcal{L}_0 = M_0 P_1 P_2 G_3^{-1}. \qquad (2.139f)$$

2.2. Properly stated linear time-varying DAEs

The inherent ODE (2.137) reads for this index three configuration

$$u' + \begin{pmatrix} 0 & 0 & 0 \\ \frac{R_2}{L}+\frac{R'_2}{R_2} & \frac{R_2}{L}+\frac{R'_2}{R_2} & \frac{R_2}{L}+\frac{R'_2}{R_2} \\ 0 & 0 & 0 \end{pmatrix} u = 0, \qquad (2.140)$$

with invariant space

$$\operatorname{im} D\Pi_1 D^- = \operatorname{im} \begin{pmatrix} 0 & 0 & 0 \\ 1 & 1 & \frac{1}{R_2} \\ 0 & 0 & 0 \end{pmatrix} = \operatorname{span}\left[\begin{pmatrix} 0 \\ 1 \\ 0 \end{pmatrix}\right],$$

where (2.140) can be described in terms of the component u_2 as

$$u'_2 + \left(\frac{R_2}{L} + \frac{R'_2}{R_2}\right) u_2 = 0. \qquad (2.141)$$

The coefficients in (2.138) read

$$-\mathcal{K}_2 D^- = \begin{pmatrix} 0 & 0 & 0 \\ \frac{1}{C_1} & \frac{1}{C_1} & 0 \\ \frac{R_2}{L} & \frac{R_2}{L} & 0 \\ 0 & 0 & 0 \\ 0 & 0 & 0 \end{pmatrix}, \quad -\mathcal{K}_1 D^- = \begin{pmatrix} \frac{-1}{C_1} & 0 & 0 \\ \frac{-1}{C_1} & 0 & 0 \\ 0 & 0 & 0 \\ 0 & 0 & 0 \\ 0 & 0 & 0 \end{pmatrix},$$

$$\mathcal{N}_{12} = \begin{pmatrix} 0 & 0 & 1 \\ 0 & 0 & 1 \\ 0 & 0 & 0 \\ 0 & 0 & 0 \\ 0 & 0 & 0 \end{pmatrix}, \quad \mathcal{N}_{01} = \begin{pmatrix} 0 & 0 & 0 \\ 0 & 0 & 0 \\ 0 & 0 & 0 \\ 1 & 0 & 0 \\ 0 & 0 & 0 \end{pmatrix}, \quad \mathcal{N}_{02} = \begin{pmatrix} 0 & 0 & 0 \\ 0 & 0 & 0 \\ 0 & 0 & 0 \\ 0 & 0 & \frac{1}{R_2} \\ 0 & 0 & \frac{1}{R_2} \end{pmatrix},$$

whereas $-\mathcal{K}_0 D^- = \mathcal{M}_{02} = 0$.

Solutions $x = D^- u + v_2 + v_1 + v_0$ are finally defined by

$$x = \begin{pmatrix} e_1 \\ e_2 \\ i_l \\ i_{r_1} \\ i_{r_2} \end{pmatrix} = \begin{pmatrix} 0 \\ \frac{-u_2}{C_1} \\ 0 \\ 0 \\ 0 \end{pmatrix} + \begin{pmatrix} 0 \\ \frac{u_2}{C_1} \\ \frac{R_2 u_2}{L} \\ 0 \\ 0 \end{pmatrix} + \begin{pmatrix} (R_2 u_2)' \\ (R_2 u_2)' \\ 0 \\ 0 \\ 0 \end{pmatrix} + \begin{pmatrix} 0 \\ 0 \\ 0 \\ (C_1(R_2 u_2)')' + \frac{(R_2 u_2)'}{R_2} \\ \frac{(R_2 u_2)'}{R_2} \end{pmatrix}$$

with u_2 given by (2.141).

2.2.7 Regular points

The tractability index framework as discussed so far is directed to the linear time-varying DAE (2.3) "globally" on the working interval $\mathcal{J}$. We reformulate here this approach in a local manner, introducing the notion of a *regular point*. This will be done in a way such that if all points are regular in a given interval, then the DAE itself is regular on the whole interval, that is, in a somehow uniform manner. The ultimate aim is to provide the background for the analysis of singularities carried out in Chapter 4.

Definition 2.7. A point $t^* \in \mathcal{J}$ is said to be *algebraically nice* at level $k \geq 1$, *nice* at level $k \geq 0$, or *regular* for (2.3) if the DAE itself is so, according to Definitions 2.4 and 2.5, on some open interval $\mathcal{I} \subseteq \mathcal{J}$ with $t^* \in \mathcal{I}$. The set of regular points within $\mathcal{J}$ will be denoted by $\mathcal{J}_{\text{reg}}$.

If a point t^* is regular, we call $\mathcal{I}$ a *regularity interval* for t^*. The tractability index of a regular point t^* can be defined as the tractability index of the DAE in $\mathcal{I}$; the same happens with the characteristic values defined by the ranks r_i. Trivially, if the DAE (2.3) is regular with tractability index ν on $\mathcal{J}$, then all points in this interval are regular, that is, $\mathcal{J}_{\text{reg}} = \mathcal{J}$. The same assertion holds for algebraically nice or nice at level k points. Furthermore, all points inherit the index ν and the characteristic values $r_0, \ldots, r_\nu$ (resp. $r_0, \ldots, r_k$) of the DAE on $\mathcal{J}$.

The converse is also true, as stated below and proved in [195] (see Propositions 5 and 6 there). For simplicity we restrict the statement to regular points, although the obvious analogs for algebraically nice and nice points hold as well. Note that this result guarantees in particular that the choice of the regularity interval results in no ambiguity concerning the index and the characteristic values of a given regular point.

Proposition 2.6. *Let t_1, $t_2 \in \mathcal{J}$ be two different regular points of DAE (2.3) with regularity intervals $\mathcal{I}_1$, $\mathcal{I}_2$, indices ν_1, ν_2, and characteristic values $r_{1,0}, \ldots, r_{1,\nu_1}$ and $r_{2,0}, \ldots, r_{2,\nu_2}$, respectively. If $\mathcal{I}_1 \cap \mathcal{I}_2 \neq \emptyset$, then*

(i) the indices and the characteristic values coincide, namely $\nu_1 = \nu_2$ and $r_{1,i} = r_{2,i}$ for $i = 0, \ldots, \nu_1$, and

(ii) the DAE is regular on $\mathcal{I}_1 \cup \mathcal{I}_2$ with the same index and characteristic values.

Moreover, if all points of an open interval $\mathcal{I} \subseteq \mathcal{J}$ are regular, then they have uniform index and characteristic values $\nu, r_0, \ldots, r_\nu$, and the DAE is regular on $\mathcal{I}$ with this index and these characteristic values.

The set $\mathcal{J}_{\text{reg}}$ is open in $\mathcal{J}$, and may hence be described as the union of a (possibly infinite) number of disjoint open subintervals. Following Proposition 2.10 of [189] (see Remark 2.3 on p. 43 above) this is independent of the actual choice of admissible projectors. Additionally, Proposition 2.6 shows that the DAE has a uniform structure in each of these subintervals. Points in $\mathcal{J} - \mathcal{J}_{\text{reg}}$ are called *singular* and will be analyzed in Section 4.2.

2.3 Standard form linear DAEs

Projector-based methods were originally directed to the standard form DAE (2.2) [107, 108, 125, 179–182, 185]. We can now profit from the framework developed for the properly stated form (2.3), compiled in Section 2.2, to improve on the analysis of standard form problems presented in [182]. Indeed, the tractability index of (2.2) has a nice characterization (Theorem 2.4) in which all the requirements, including smoothness ones, can be directed to the matrices G_i and the projector-independent spaces $N_0 \oplus \ldots \oplus N_i$. Moreover, problems with arbitrary index can be fully characterized by means of the decoupling presented in Theorem 2.5, which is a restatement of the one discussed in Theorem 2.3 for properly stated DAEs.

2.3.1 *The tractability index of standard form DAEs*

Let us then drive our attention back to the standard form linear DAE (2.2). The results of Section 2.2 apply to these problems, provided that $\ker A(t)$ is a C^1-space (cf. Remark 2.1 on p. 36), via the reformulation

$$A(t)(P(t)x(t))' + B(t)x(t) = q(t), \quad t \in \mathcal{J}, \qquad (2.142)$$

with $B(t) = E(t) - A(t)P'(t)$. Here $P(t)$ is a C^1 projector along $\ker A(t)$. Note that, according to Definition 2.2, the proper statement of (2.142) follows from the C^1 nature of the projector $P(t)$.

Definition 2.8. The standard form linear DAE (2.2) is regular with tractability index ν on $\mathcal{J}$ if $\ker A(t)$ is a C^1-space on $\mathcal{J}$ and the properly stated reformulation (2.142) is regular with tractability index ν on $\mathcal{J}$, for some (hence any) C^1 projector $P(t)$ along $\ker A(t)$.

In order to apply the framework of Section 2.2 we set $D(t) = P(t)$, so that $R(t) = P(t)$. We also assume that $D^-(t) = P(t)$ (cf. Remark 2.6 in [189]); this will lead to the identities $\Pi_0 = P$ and $P_0 = P$ used below.

Other choices of D^- (or, respectively, of Π_0 and P_0) are however possible; we might e.g. construct the chain in the Π-framework beginning with the orthogonal choice of Π_0, without requiring P to be orthogonal. Nevertheless, the identities $\Pi_0 = P$ and $P_0 = P$ will provide several simplifications in the matrix chain construction and also in the decoupling of the DAE.

Proposition 2.7. *The index notion of Definition 2.8 for a standard form linear DAE (2.2) is independent of the choice of the C^1 projector $P(t)$.*

Proof. This is a consequence of Theorem 4.6 in [189]. Assume that (2.142) is regular with index ν on $\mathcal{J}$ and let $\tilde{P}$ be another C^1 projector along $\ker A(t)$, yielding the reformulation

$$A(t)(\tilde{P}(t)x(t))' + \tilde{B}(t)x(t) = q(t)$$

with $\tilde{B}(t) = E(t) - A(t)\tilde{P}'(t)$.

Since $\ker P = \ker \tilde{P}$, item (iii) in Proposition 2.5 implies that $P\tilde{P} = P$ and $\tilde{P}P = \tilde{P}$. Theorem 4.6 in [189] then applies by setting $H = H^- = \tilde{P}$ within equations (4.17) and (4.18) there. Specifically, the requirements stated in equations (4.19) and (4.20) of [189] are met because of the relations $P\tilde{P}\tilde{P}P = P\tilde{P} = P$ and

$$A = A\tilde{P}, \ \tilde{P} = \tilde{P}P, \ E - A\tilde{P}' = E - AP' + AP'\tilde{P},$$

respectively. In the last identity we have used $P' = (P\tilde{P})' = P'\tilde{P} + P\tilde{P}'$, which yields $AP' = AP'\tilde{P} + A\tilde{P}'$ since $AP = A$. $\square$

The matrix chain construction based on the Π-framework of subsection 2.2.4 will be defined, for the reformulation (2.142) of the standard form DAE (2.2), by $\Pi_0 = P$ and

$$G_0 = A, \ B_0 = B = E - AP', \tag{2.143}$$

together with

$$G_1 = G_0 + B_0 M_0, \tag{2.144}$$

where $M_0 = I - \Pi_0$. Additionally

$$B_i = (B_{i-1} - G_i \Pi_0 \Pi_i')\Pi_{i-1} \tag{2.145a}$$

$$G_{i+1} = G_i + B_i M_i, \tag{2.145b}$$

for $i \geq 1$. Here, Π_i is a C^1 projector along the space $N_0 \oplus \ldots \oplus N_i$ verifying $\operatorname{im} \Pi_i \subseteq \operatorname{im} \Pi_{i-1}$, and $M_i = \Pi_{i-1} - \Pi_i$.

Note that in (2.145a) we have used the identities $D = D^- = \Pi_0$, as well as the properties $(\Pi_0 \Pi_i \Pi_0)' = \Pi_i'$ and $\Pi_0 \Pi_{i-1} = \Pi_{i-1}$ following from (2.61).

2.3. Standard form linear DAEs

In terms of the P-projectors of subsection 2.2.3, the matrix chain is equivalently defined by (2.143), jointly with

$$G_1 = G_0 + B_0 Q_0, \qquad (2.146)$$

where Q_0 is defined from $P_0 = P$ as $Q_0 = I - P_0$, and

$$B_i = B_{i-1} P_{i-1} - G_i P_0 (P_0 \cdots P_i)' P_0 \cdots P_{i-1}, \qquad (2.147a)$$
$$G_{i+1} = G_i + B_i Q_i \qquad (2.147b)$$

for $i \geq 1$. As usual, P_i is a projector along $N_i = \ker G_i$, and $Q_i = I - P_i$. The properties in the P-framework making it possible to write B_i in the form depicted in (2.147a) can be simply derived from the ones within the Π-framework via the identity $\Pi_i = P_0 \cdots P_i$.

In Theorem 2.4 below we state the conditions which support these constructions and thereby characterize the tractability index of the standard form DAE (2.2). Such conditions are of course a particularization of those arising in subsections 2.2.3 and 2.2.4, although the smoothness requirement (c) admits a nice geometrical restatement in this context.

Theorem 2.4. *The standard form linear DAE (2.2) is regular with tractability index zero on $\mathcal{J}$ if and only if $A(t)$ is nonsingular on $\mathcal{J}$.*

Provided that $\ker A(t)$ is a C^1-space on $\mathcal{J}$, the DAE (2.2) is regular with tractability index $\nu \geq 1$ on $\mathcal{J}$ if and only if, for $1 \leq i \leq \nu - 1$

(a) G_i has constant rank $r_i < m$ on $\mathcal{J}$,
(b) $N_i = \ker G_i$ verifies $(N_0 \oplus \ldots \oplus N_{i-1}) \cap N_i = \{0\}$ for all $t \in \mathcal{J}$,
(c) the space $N_0 \oplus \ldots \oplus N_i$ admits a C^1-basis on $\mathcal{J}$,

and additionally G_ν is nonsingular on $\mathcal{J}$.

Here the chain construction can be (equivalently) assumed to be based either on the Π- or the P-frameworks. In the Π-based construction, one can choose Π_i as the *orthogonal* projector along $N_0 \oplus \ldots \oplus N_i$ for $i = 0, \ldots, \nu-1$; the C^1-smoothness of this orthogonal projector is equivalent to that of the space $N_0 \oplus \ldots \oplus N_i$ itself. In terms of P-projectors, we need to require explicitly for $Q_i = I - P_i$ the admissibility condition $N_0 \oplus \ldots \oplus N_{i-1} \subseteq \ker Q_i$ and the C^1-smoothness condition on the resulting product $P_0 \cdots P_i$. Note however that the Π_i-projector captures these properties in a natural way.

We emphasize that conditions (a) and (b) above, as well as the spaces $N_0 \oplus \ldots \oplus N_i$ (and thereby condition (c)) are independent of the actual choice of admissible projectors, according to Theorem 2.3 in [191]; see also Proposition 2 in [195]. It follows that the tractability index of (2.2) is independent of the choice of admissible projectors.

2.3.2 Decoupling

The decoupling discussed in Theorem 2.3 applies to index ν standard form DAEs (2.2) via the reformulation (2.142), with some simplifications which are commented on below. We assume that $q(t)$ meets the smoothness requirements emanating from Remark 2.9 (cf. page 53).

Theorem 2.5. *Assume that the standard form linear DAE (2.2) is regular with tractability index ν on $\mathcal{J}$. Then,*

$$x(t) \in C_P^1(\mathcal{J}, \mathbb{R}^m) = \{x \in C(\mathcal{J}, \mathbb{R}^m) \ / \ Px \in C^1(\mathcal{J}, \mathbb{R}^m)\},$$

solves the reformulation (2.142) if and only if it can be written as

$$x = u + v_{\nu-1} + \ldots + v_1 + v_0, \qquad (2.148)$$

where $u \in C^1(\mathcal{J}, \mathbb{R}^m)$ is a solution of the inherent ODE

$$u' - \Pi'_{\nu-1} u + \Pi_{\nu-1} G_\nu^{-1} B u = \Pi_{\nu-1} G_\nu^{-1} q \qquad (2.149)$$

lying on the invariant space $\operatorname{im} \Pi_{\nu-1}$, whereas

$$v_k = -\mathcal{K}_k u + \sum_{j=k+1}^{\nu-1} \mathcal{N}_{kj} v'_j + \sum_{j=k+2}^{\nu-1} \mathcal{M}_{kj} v_j + \mathcal{L}_k q \qquad (2.150)$$

for $k = \nu - 1, \ldots, 0$, with $v_k \in C^1(\mathcal{J}, \mathbb{R}^m)$ for $k > 0$ and $v_0 \in C(\mathcal{J}, \mathbb{R}^m)$. The coefficients of (2.150) read

$$\mathcal{K}_k = M_k P_{k+1} \cdots P_{\nu-1} G_\nu^{-1} B + M_k (P_{k+1} \cdots P_{\nu-1} - I) \Pi'_{\nu-1}$$

$$\mathcal{N}_{kj} = M_k P_{k+1} \cdots P_{j-1} Q_j$$

$$\mathcal{M}_{kj} = M_k \left(\sum_{i=k+1}^{j-1} P_{k+1} \cdots P_{i-1} (Q_i - P_i) \right) \Pi_0 M'_j$$

$$\mathcal{L}_k = M_k P_{k+1} \cdots P_{\nu-1} G_\nu^{-1},$$

where the products $P_{k+1} \cdots P_l$ amount to the identity I whenever $l < k+1$.

Theorem 2.5 is just a restatement of Theorem 2.3. Note that we have used $D = D^- = \Pi_0$ and $\Pi_0 \Pi_i \Pi_0 = \Pi_0 \Pi_i = \Pi_i$. The product $D^- u$ has been written as u since solutions of the inherent ODE contribute to the DAE solutions only when lying on the space $\operatorname{im} D \Pi_{\nu-1} D^- = \operatorname{im} \Pi_{\nu-1}$, and then we can use $D^- \Pi_{\nu-1} = \Pi_{\nu-1}$. Similarly, $v_j = M_j v_j$ holds because of (2.150) and the leading factor M_k in all the coefficients $\mathcal{K}_k$, $\mathcal{L}_k$, $\mathcal{N}_{kj}$ and $\mathcal{M}_{kj}$; since $DM_j = \Pi_0 M_j = M_j$ by (2.62) (note that $j \geq 1$ always) we have $Dv_j = v_j$. This explains in particular the C^1 condition on v_k for $k > 0$.

2.3. Standard form linear DAEs

The only additional aspects which need to be clarified are the following. Regarding the D^- factor within the $\mathcal{K}_k$ coefficient in Theorem 2.3, we have used $D^- = \Pi_0$ and $M_k(P_{k+1}\cdots P_{\nu-1} - I)\Pi_0 = M_k(P_{k+1}\cdots P_{\nu-1} - I)$, which owes to (2.82). The D term at the end of $\mathcal{K}_k$ and $\mathcal{M}_{kj}$ within the general statement in Theorem 2.3 can be now removed since $D = \Pi_0$ will multiply $D^-u = \Pi_{\nu-1}u$ and $v_j = M_j v_j$, respectively; use then the identities $\Pi_0 \Pi_{\nu-1} = \Pi_{\nu-1}$ and $\Pi_0 M_j = M_j$. Finally, the D^- factor at the end of $\mathcal{N}_{kj}$ in Theorem 2.3 is not present here because $Q_j \Pi_0 = Q_j$ if $j > 0$, which in turn is due to the relation $\ker \Pi_0 = N_0 \subseteq \ker Q_j$ (cf. (2.65)).

Remark 2.11. The reader can easily check that the decoupling defined by (2.33) and (2.34) for standard form DAEs with index one discussed in subsection 2.2.1 is a particular instance of (2.149) and (2.150) above. Note that in an index one problem we have $\mathcal{K}_0 = M_0 G_1^{-1} B$ and $\mathcal{L}_0 = M_0 G_1^{-1}$, and that Π_0, M_0, G_1 in the current setting stand for P, Q, A_1 in 2.2.1.

It is worth indicating that if a solution of (2.142) given by Theorem 2.5 belongs to $C^1(\mathcal{J}, \mathbb{R}^m)$, then it can be easily checked to solve the original standard form equation (2.2). Note finally that the ideas discussed in Remark 2.10 (p. 64) and in subsection 2.2.7 apply, with straightforward modifications, in the standard form setting of (2.2).

2.3.3 Time-invariant problems revisited

The results above apply in particular to the linear time-invariant DAE (2.1) considered in Section 2.1, provided that the pencil $\{A, E\}$ is regular. Indeed, in this situation all projectors can be chosen to be constant, so that all time derivatives of projector products vanish. This idea was already present in the original chain construction (see [179] and [108]). The construction of the matrix chain for (2.1) via (2.143), (2.146) and (2.147) yields

$$G_0 = A, \ B_0 = E \qquad (2.151)$$

and

$$G_{i+1} = G_i + B_i Q_i, \ B_{i+1} = B_i P_i, \qquad (2.152)$$

where Q_i is a constant projector onto $N_i = \ker G_i$, and $P_i = I - Q_i$. This construction is identical to the one displayed in (2.21). Using Theorem 2.1, this means that in constant coefficient cases the tractability index notion depicted in Definition 2.8 amounts to the Kronecker index of the pencil.

The Π-framework provides the following simple characterization of a regular pencil and its Kronecker index [251].

Proposition 2.8. *Consider the matrix pencil $\{A, E\}$ with $A, E \in \mathbb{R}^{m \times m}$. Write $G_0 = A$ and, for $i \geq 0$,*

$$G_{i+1} = A + ET_i,$$

where T_i is the orthogonal projector onto $N_0 + \ldots + N_i$, with $N_i = \ker G_i$. Then the matrix pencil $\{A, E\}$ is regular with Kronecker index ν if and only if G_i is singular for $i < \nu$ and G_ν is nonsingular.

Proof. Use the chain construction based on (2.56), (2.57) and (2.58) with the orthogonal choice of Π_i. Since the derivatives $(D\Pi_i D^-)'$ in (2.57) vanish, we have $H_i = H_0 = E$ for all $i \geq 0$ and then

$$G_{i+1} = G_i + EM_i = G_i + E(\Pi_{i-1} - \Pi_i) = G_{i-1} + E(\Pi_{i-2} - \Pi_{i-1})$$
$$+ E(\Pi_{i-1} - \Pi_i) = G_{i-1} + E(\Pi_{i-2} - \Pi_i) = \ldots = A + E(I - \Pi_i).$$

The result then follows from the equivalence of this construction with the one based on (2.152) (cf. Theorem 2.2), together with Theorem 2.1. □

Note that there is no need here to assume that $N_0 + \ldots + N_i$ can be written as a direct sum. This is due to the fact that any non-trivial intersection $(N_0 + \ldots + N_{i-1}) \cap N_i$ can be shown to yield a singular matrix G_k for any k; in this situation the pencil is necessarily a singular one [189].

The decoupling presented in Theorem 2.5 applies, with straightforward simplifications, in the constant coefficient setting. If the pencil $\{A, E\}$ is regular, $x \in C_P^1(\mathcal{J}, \mathbb{R}^m)$ solves (2.1) if and only if it can be written as

$$x = u + v_{\nu-1} + \ldots + v_1 + v_0,$$

where $u \in C^1(\mathcal{J}, \mathbb{R}^m)$ is a solution of the inherent ODE

$$u' + \Pi_{\nu-1} G_\nu^{-1} E u = \Pi_{\nu-1} G_\nu^{-1} q \qquad (2.153)$$

lying on the invariant space $\operatorname{im} \Pi_{\nu-1}$, and the components v_k verify

$$v_k = -\mathcal{K}_k u + \sum_{j=k+1}^{\nu-1} \mathcal{N}_{kj} v_j' + \mathcal{L}_k q \qquad (2.154)$$

for $k = \nu - 1, \ldots, 0$, with $v_k \in C^1(\mathcal{J}, \mathbb{R}^m)$ for $k > 0$ and $v_0 \in C(\mathcal{J}, \mathbb{R}^m)$. The coefficients of (2.154) read, in this constant coefficient context,

$$\mathcal{K}_k = M_k P_{k+1} \cdots P_{\nu-1} G_\nu^{-1} E$$

$$\mathcal{N}_{kj} = M_k P_{k+1} \cdots P_{j-1} Q_j$$

$$\mathcal{L}_k = M_k P_{k+1} \cdots P_{\nu-1} G_\nu^{-1}.$$

2.4. Other approaches for linear DAEs: Reduction techniques

Mind the fact that, within these coefficients, all factors involving time derivatives vanish in the time-invariant setting; this is in particular the case for $\mathcal{M}_{kj}$.

Remark 2.12. The reader can particularize (2.153) and (2.154) to derive (2.15) and (2.16), respectively, in the index one setting considered in 2.1.3.2. Note that in index one problems the terms $\mathcal{N}_{kj} v'_j$ are not present.

2.4 Other approaches for linear DAEs: Reduction techniques

The reader is referred to [30, 42, 54] and [147, 148, 151], respectively, for comprehensive discussions of the differentiation and strangeness index frameworks for the analysis of linear time-varying DAEs. The differentiation index and derivative arrays will be briefly examined, in a nonlinear context, in Section 3.7. We sketch below the reduction framework of Rabier and Rheinboldt for linear DAEs in the standard form (2.2), as presented in [225, 228]; these techniques will be discussed in great detail for nonlinear systems in Chapter 3. Reduction methods accommodating discontinuous inputs and inconsistent initial conditions in linear problems are tackled within a distributional framework in [226, 228]. Some interrelations of reduction techniques and projector-based methods are addressed in [250].

The reduction of (2.2) is supported on the assumption that $\{A(t), E(t)\}$ is a *regular pair* on the interval $\mathcal{J}$, in the sense that, for some fixed $r \leq m$ and all t in $\mathcal{J}$, the conditions

$$\operatorname{rk} A(t) = r \qquad (2.155a)$$

$$\operatorname{rk}(A(t)\ E(t)) = m \qquad (2.155b)$$

hold; here $(A\ E)$ stands for the $m \times 2m$ matrix map constructed by joining the columns of A and E. It is also assumed that both $A(t)$ and $E(t)$ belong to $C^1(\mathcal{J}, \mathbb{R}^{m \times m})$.

If $r < m$, from the above-mentioned requirements it follows that both $\operatorname{im} A(t)$ and $S(t) = \{x \in \mathbb{R}^m : E(t)x \in \operatorname{im} A(t)\}$ are C^1-spaces (cf. Remark 2.1 on page 36). This is due to the existence of a C^1, maximal rank mapping $H(t) : \mathcal{J} \to \mathbb{R}^{(m-r) \times m}$ such that $v \in \operatorname{im} A(t) \Leftrightarrow H(t)v = 0$. The product $H(t)E(t)$ has constant rank $m - r$ due to (2.155b) and the identity $H(t)A(t) = 0$. Hence, the set $S(t)$ may be described as the C^1-space $\ker H(t)E(t)$, which is r-dimensional for every fixed t.

The C^1-structure of $\operatorname{im} A(t)$ and $S(t)$ makes it possible to choose matrix-valued maps $C(t) \in C^1(\mathcal{J}, \mathbb{R}^{m \times r})$, $D(t) \in C^1(\mathcal{J}, \mathbb{R}^{r \times m})$ such that, for every $t \in \mathcal{J}$, $C(t)$ and the restriction of $D(t)$ to im $A(t)$ yield isomorphisms $\mathbb{R}^r \to S(t)$ and im $A(t) \to \mathbb{R}^r$, respectively. The pair

$$A_1 = DAC, \quad E_1 = D(EC + AC'), \qquad (2.156)$$

is then called a *reduction* of $\{A, E\}$.

The regular pair $\{A, E\}$ is said to have (geometric) index one if $A_1(t)$ is nonsingular for all $t \in \mathcal{J}$. This index one notion is proved in [228] independent of the choice of C, D, and equivalent to the tractability index one notion provided that the C^1 smoothness condition required by the reduction framework is met. See [225] for a discussion in the analytic setting.

The analysis of (2.2) in this index one context requires also the introduction of a "reduced" excitation $q_1 = D(q - Eu_0 - Au_0')$, where $Eu_0 - q \in \text{im } A$. This requirement can be met since it holds in particular for $u_0 = E^\mathsf{T}(AA^\mathsf{T} + EE^\mathsf{T})^{-1}q$. Classical solutions of a DAE (2.2) with an index one, C^1 pair $\{A(t), E(t)\}$ and a C^1 excitation $q(t)$ can be then recovered from those of the *reduced ODE*

$$A_1(t)\xi' + E_1(t)\xi = q_1(t), \qquad (2.157)$$

through the relation $x = C\xi + u_0$.

In cases in which the matrix mapping A_1 is singular, under additional regularity and smoothness assumptions the same procedure can be applied to $\{A_1, E_1\}$. If the procedure can be continued indefinitely, it will eventually become stationary, and the number of reduction steps defines the (so-called *geometric*) index: cf. [225, 228]. In these higher index cases the reduced ODE is formulated in terms of the pair $\{A_\nu(t), B_\nu(t)\}$ which results from the iteration of the reduction procedure outlined above.

When compared with projector methods and the tractability index framework, the reduction approach presents some limitations which concern the need to parametrize the set $S(t)$, and also the smoothness requirements on $A(t)$, $E(t)$ and $q(t)$; for instance, the explicit appearance of the derivative u_0' in the expression for q_1 depicted above might lead to the erroneous interpretation that solutions of index one problems actually depend on the first derivative of the excitation, against common understanding in the context of the perturbation index [121, 122]; compare also with the solution characterization (2.33)-(2.34) provided by projector methods, not involving any derivative of $q(t)$. On the other hand, this approach provides a rather direct method for the analysis of DAE, suitable for extension to higher index cases without additional complexity. Reduction techniques define also a nice framework for the analysis of nonlinear DAEs, as detailed in the next Chapter.

Chapter 3

Nonlinear DAEs and reduction methods

The present Chapter addresses nonlinear differential-algebraic equations, with special emphasis on quasilinear problems. Our attention will be mainly centered on *reduction* methods which, roughly speaking, unveil the behavior of a given nonlinear DAE in terms of an explicit ODE defined on a lower dimensional manifold. This approach is particularly well-suited for the analysis of dynamical aspects in nonlinear semistate systems; model reduction aspects and state formulations may also be advantageously tackled within this framework. The focus on reduction methods will provide the reader with a perspective and a set of tools different, and somehow complementary, to the projector-based ones presented in Chapter 2, and also to those supported on the differentiation index and derivative arrays, widely discussed in the DAE literature: see [30, 45, 46, 94] and references therein.

A key reference within the reduction framework is the paper [238] by W. C. Rheinboldt. The geometric spirit of this approach is also present in [239], as well as in the research of S. Reich on this topic [229–231] and in the joint work of P. Rabier and Rheinboldt [220, 221, 228]. See also [106, 284, 285]. The geometric motivation is also strong in other related references [17, 73, 101, 102, 170, 171, 205, 208, 261–263], whose origins can be traced back to the work of Dirac on generalized Hamiltonian dynamics [79–81]. An excellent compilation of the framework of Rabier and Rheinboldt can be found in Chapter 4 of [228], where the focus on quasilinear DAEs provides a significant simplification of the fully nonlinear setting of [221].

The reduction approach will be motivated via some structured problems (semiexplicit index one and Hessenberg DAEs) in Sections 3.1 and 3.2. They are intended to provide a friendly introduction to these methods for readers without a background on this topic. The somewhat simple structure of these systems allows also for a discussion of some relations and differences with the framework based on the differentiation index.

From the relatively simple form of semiexplicit and Hessenberg systems and the easy derivation of a reduced ODE for regular instances of them, the reader should not conclude that state reductions are straightforward. Reduction methods become highly non-trivial when applied to the analysis of general quasilinear DAEs and specially in the presence of singularities (cf. Chapter 4). Section 3.3 compiles some elementary prerequisites from differential geometry aimed at the study of quasilinear problems.

The reduction of quasilinear DAEs of the form (1.20), discussed in Section 3.4, is the main topic of this Chapter. Our point of view will be a *local* one. The global framework of Rabier and Rheinboldt, summarized following [228] in 3.4.1, requires of course global conditions which often fail to hold in practical cases. Their approach will be rephrased in a local manner by means of the geometric index notion for a *regular point*, around which solutions of the DAE can be locally characterized. This will yield an exact description of the so-called *regular manifold* where the reduction is actually feasible. This is detailed in subsections 3.4.2-3.4.4.

The local setting allows naturally for descriptions of the (local) dynamics via explicit ODEs defined on open subsets of $\mathbb{R}^r$, in terms of a subset of the original problem variables. This will be important in circuit modeling, where state models for higher index configurations can be formulated in terms of variables linked with reactive elements, with a well-defined and clear physical meaning. As a byproduct, this makes the reduction approach easier for readers not so familiar with differential geometry.

Different properties concerning local equivalence of quasilinear DAEs will be considered in 3.4.5. The local equivalence relation introduced there will turn out to be intimately linked with the reduction procedure, and naturally yields the invariance of notions such as the index as well as their independence of the choice of reduction operators. This discussion includes as well certain results of independent interest, in particular the one stating that any two reductions of locally equivalent DAEs are themselves locally equivalent, which leads to the conclusion that any two state space local descriptions of a DAE around a regular point are locally conjugate.

Several examples illustrating the reduction procedure for quasilinear DAEs are presented in 3.4.6. Section 3.4 ends with a discussion of nonautonomous quasilinear problems in 3.4.7. Additional results concerning projector-based methods for quasilinear DAEs arising in electrical circuit theory can be found in Chapter 5.

The suitability of reduction methods for the study of local dynamic aspects of DAEs will be illustrated in Section 3.5 by the analysis of qualitative properties of equilibria in quasilinear DAEs. Section 3.6 briefly examines geometric reduction methods for fully nonlinear DAEs; the reasons for this shorter treatment are their rare appearance in practical applications and the fact that fully nonlinear problems can be naturally rewritten in quasilinear form by taking the time derivative of the semistate vector as an additional variable, which renders the system semiexplicit (cf. (1.7)). Finally, the differentiation index and derivative arrays are discussed in Section 3.7.

Due to the intrinsic complexity of nonlinear problems, we will try to keep the terms of the discussion as simple as possible in order to improve readability. In this direction, we restrict the attention to standard form DAEs, instead of using proper formulations [157, 190, 193, 293] which may be less familiar for many readers. Also for the sake of simplicity, from Section 3.3 on all operators will be assumed to be in C^∞, although this requirement can be easily relaxed along the lines indicated in Sections 3.1 and 3.2. Technically, we avoid stating local results in terms of germs (see e.g. [7, 99, 100]), which would yield much more synthetic statements but at the same time would significantly obscure the discussion for many readers.

3.1 Semiexplicit index one DAEs

Consider the autonomous semiexplicit DAE

$$y' = h(y, z) \tag{3.1a}$$
$$0 = g(y, z), \tag{3.1b}$$

and assume that $h \in C^k(W_0, \mathbb{R}^r)$ and $g \in C^k(W_0, \mathbb{R}^p)$ for some $k \geq 1$, the set W_0 being open in $\mathbb{R}^{r+p}$.

Let $(y^*, z^*) \in W_0$ satisfy $g(y^*, z^*) = 0$. If the derivative $g_z(y^*, z^*)$ defines an invertible matrix, then g_z is nonsingular on some open neighborhood U of (y^*, z^*) and the DAE (3.1) is said to have *differentiation index one* on U. We will often say the index to be one *around* the point (y^*, z^*), without explicit mention of U. This notion (cf. [30, 45, 46] and Section 3.7 below) is supported on the fact that one differentiation in (3.1b) suffices to obtain, locally around (y^*, z^*), an explicit *underlying ODE*

$$y' = h(y, z) \tag{3.2a}$$
$$z' = -g_z^{-1}(y, z) g_y(y, z) h(y, z), \tag{3.2b}$$

for which $g(y,z) = 0$ is an invariant comprising the solutions of the original DAE.

The right-hand side of (3.2) defines, at least locally around (y^*, z^*), a C^{k-1} vector field $v = (v_1, v_2)$ given by

$$\begin{pmatrix} v_1(y,z) \\ v_2(y,z) \end{pmatrix} = \begin{pmatrix} h(y,z) \\ -g_z^{-1}(y,z) g_y(y,z) h(y,z) \end{pmatrix}. \qquad (3.3)$$

Actually $v(y, z)$ is well-defined on the set of points $(y, z) \in W_0$ where $g_z(y, z)$ is nonsingular. The invariance of the set $g = 0$ for the underlying equation (3.2) relies on the fact that the derivative of g along any integral curve of the vector field (3.3) reads $g_y v_1 + g_z v_2 = g_y h + g_z(-g_z^{-1} g_y h) = 0$. Note also that the vector field $v(y, z)$ is in C^k if the smoothness requirements $h \in C^k(W_0, \mathbb{R}^r)$, $g \in C^{k+1}(W_0, \mathbb{R}^p)$ hold for the mappings in (3.1).

Reduction methods will unveil the behavior of (3.1) in a different manner. The nonsingularity of $g_z(y^*, z^*)$ implies that the set

$$W_1 = \{(y, z) \in W_0 \;/\; g(y,z) = 0\} \subset W_0, \qquad (3.4)$$

which accommodates the solutions of the DAE, has an r-dimensional manifold structure locally around (y^*, z^*) (cf. Section 3.3). Moreover, if the nonsingularity condition on $g_z(y, z)$ holds at every $(y, z) \in W_1$, then the whole set W_1 will be an r-dimensional C^k-manifold. Under this global assumption, system (3.1) will be said to be a *regular DAE* with geometric index one; throughout this Chapter we will often omit the label 'geometric' to refer to the index. In this index one setting, a C^k-vector field is globally defined on the manifold W_1, and its integral curves give the solutions of the DAE. This vector field can be shown to be the restriction of (3.3) to the tangent bundle (see again Section 3.3) TW_1. The set W_1 may in this situation be called the *solution manifold* of the semiexplicit DAE (3.1).

From a local point of view, this behavior can be described without resorting to these geometric concepts. Provided that $g(y^*, z^*) = 0$, the hypothesis that $g_z(y^*, z^*)$ is nonsingular makes it possible to apply the implicit function theorem to describe the set $g(y, z) = 0$ locally as $z = \psi(y)$ with $\psi \in C^k(\Omega_1, \mathbb{R}^p)$, Ω_1 being an open neighborhood of y^* in $\mathbb{R}^r$. In terms of the y-variables the local behavior can be described via the *reduced ODE*

$$y' = h(y, \psi(y)), \qquad (3.5)$$

defined on $\Omega_1 \subseteq \mathbb{R}^r$. Now, $h(y, \psi(y))$ can be seen as a C^k-vector field locally defined on an open subset of $\mathbb{R}^r$, and (3.5) can be understood as a local state space description of the problem. Solutions of the DAE (3.1) near

3.1. Semiexplicit index one DAEs

(y^*, z^*) are defined from those of the explicit ODE (3.5) by using additionally $z = \psi(y)$. The vanishing of $g(y^*, z^*)$ together with the nonsingularity of $g_z(y^*, z^*)$ will define (y^*, z^*) as a *regular point with geometric index one* or simply as an *index one point*.

The connection between the approaches sketched in the last two paragraphs comes from looking at the mapping $y \to (y, \psi(y))$ as a local parametrization (cf. Section 3.3) of the manifold W_1. The right-hand side of the state space description (3.5) can be understood as a local coordinate representation of the above-defined vector field on W_1. It is important to emphasize that other parametrizations of W_1 would lead to different state space descriptions of the DAE. Nevertheless, all state representations will be equivalent in a sense made precise in subsection 3.4.5 (see specifically Theorem 3.4 there). In this case, the y-variables are privileged by the semiexplicit form of the system, which makes it possible to distinguish these "dynamic" variables from the "algebraic" ones, namely, z. But a feature that will be shared with general quasilinear DAEs with a well-defined geometric index is that a local state space representation in terms of some of the original problem variables will always be available, actually coming out in a natural way from the reduction procedure.

At this point it is worth clarifying the distinction between two concepts which, though elementary, are sometimes confused in the literature. We mean the notion of an underlying ODE such as (3.2), defined on the whole semistate space W_0 (which is open in $\mathbb{R}^{r+p}$) and for which the r-dimensional manifold W_1 is an invariant, and that of a reduced ODE, which is defined only on the lower-dimensional manifold W_1 and, in local coordinates, yields an equation such as (3.5) on some $\Omega_1 \subseteq \mathbb{R}^r$. The same distinction can be made in terms of vector fields or flows, and also holds in higher index and unstructured problems.

Note finally that initial value problems for (3.1) are only well-posed for initial conditions lying on W_1. Specially within the numerical literature, points on the solution manifold W_1 are called *consistent initial conditions* for an index one DAE. Characterizing the solution manifold or, in these terms, the set of consistent initial conditions for higher index DAEs is a more involved problem, as discussed below.

3.2 Hessenberg systems

The ideas presented above need to be refined in order to apply to higher index equations. Consider the autonomous Hessenberg DAE of size two

$$y' = h(y, z) \tag{3.6a}$$
$$0 = g(y), \tag{3.6b}$$

where $h \in C^k(W_0, \mathbb{R}^r)$, $g \in C^{k+1}(\hat{W}_0, \mathbb{R}^p)$, $k \geq 1$ and $r \geq p$. Here W_0 and $\hat{W}_0$ are open subsets of $\mathbb{R}^{r+p}$ and $\mathbb{R}^r$, respectively, and $\hat{W}_0$ is assumed to include the y-projection of W_0.

Taking the time derivative of g along trajectories of (3.6) we can check that solutions of the DAE (3.6) must satisfy the *hidden constraint*

$$0 = g_y(y)h(y, z). \tag{3.7}$$

Assume that a given (y^*, z^*) satisfies $g(y^*) = 0$, $g_y(y^*)h(y^*, z^*) = 0$. If the product $g_y h_z$ is invertible at (y^*, z^*), then it is so on a neighborhood U of this point and (3.6) is said to have differentiation index two on U or, simply, around (y^*, z^*). Indeed, differentiating (3.7) we get

$$y' = h(y, z) \tag{3.8a}$$
$$z' = -(g_y h_z)^{-1}(y, z)(g_y h)_y(y, z)h(y, z), \tag{3.8b}$$

which is an underlying ODE for (3.6). Analogously to the index one case, the right-hand side of (3.8) is a vector field defined on points of W_0 with a nonsingular $g_y h_z$, and the equations $g(y) = 0$, $g_y(y)h(y, z) = 0$ can be checked to define an invariant set comprising the DAE solutions.

From the point of view of reduction methods, the nonsingularity requirement on $g_y h_z$ implies in particular that g_y has maximal rank p. Globally, if $\operatorname{rk} g_y(y) = p$ whenever $g(y) = 0$, then the set

$$W_1 = \{(y, z) \in W_0 \ / \ g(y) = 0\} \tag{3.9}$$

naturally admits an r-dimensional C^{k+1}-manifold structure. Furthermore, if we define

$$W_2 = \{(y, z) \in W_0 \ / \ g(y) = 0, \ g_y(y)h(y, z) = 0\} \subset W_1 \tag{3.10}$$

then the assumption $\operatorname{rk} g_y(y)h_z(y, z) = p$ for all $(y, z) \in W_2$ makes this set an $(r - p)$-dimensional C^k-manifold. This will be due (see Section 3.3) to the maximal rank condition

$$\operatorname{rk} \begin{pmatrix} g_y & 0 \\ (g_y h)_y & g_y h_z \end{pmatrix} = 2p \tag{3.11}$$

3.2. Hessenberg systems

which follows from $\operatorname{rk} g_y(y)h_z(y,z) = p$. Note in particular that the identity $\operatorname{rk} g_y(y)h_z(y,z) = p$ comprises the "intermediate" requirement $\operatorname{rk} g_y(y) = p$.

The global assumption $\operatorname{rk} g_y(y)h_z(y,z) = p$ for all $(y,z) \in W_2$ yields a well-defined vector field and a flow on W_2. This manifold is filled by the solutions of (3.6), which are integral curves of this vector field, and therefore W_2 will be termed the *solution manifold* of the problem. The DAE (3.6) will be said to be a *regular DAE* with geometric index two. Points in W_2 are sometimes called *consistent initial values* for index two DAEs.

Locally, from the condition $\operatorname{rk} g_y(y^*) = p$ it follows that y can be split into $r - p$ variables u (which for simplicity can be assumed w.l.o.g. to be $y_1, \ldots, y_{r-p}$), and p variables w (corresponding to $y_{r-p+1}, \ldots, y_r$) in a way such that $\operatorname{rk} g_w(y^*) = p$. Hence, by the implicit function theorem, $g = 0$ amounts to $w = \psi(u)$ for some locally defined C^{k+1}-map ψ. Additionally, using the condition $\operatorname{rk} g_y(y^*)h_z(y^*,z^*) = p$ and the splitting $y = (u,w)$, we can describe locally the set $g_y h = 0$ by a C^k-relation of the form $z = \eta(u,w)$. On the intersection with $g = 0$ we get $z = \eta(u, \psi(u))$.

Using these coordinates and splitting $h = (h_1, h_2)$, where h_1 (resp. h_2) denotes the first $r - p$ (resp. last p) components of h, we get from (3.6a) the reduced explicit ODE

$$u' = h_1((u, \psi(u)), \eta(u, \psi(u))). \tag{3.12}$$

In cases in which the u-variables qualifying for the description $w = \psi(u)$ via the implicit function theorem are not the first $r-p$ ones of y, we simply have to partition h taking the indices which define h_1 and h_2 from the corresponding splitting of y into u and w.

The state equation (3.12), together with $w = \psi(u)$ (or, equivalently, $y = (u, \psi(u))$) and $z = \eta(u, \psi(u))$, yields a complete description of the local solution behavior. Here u lies on an open set $\Omega_2 \subseteq \mathbb{R}^{r-p}$. The three conditions $g(y^*) = 0$, $g_y(y^*)h(y^*,z^*) = 0$ and $\operatorname{rk} g_y(y^*)h_z(y^*,z^*) = p$ will define (y^*, z^*) as a *regular point with geometric index two* or an *index two point* for the Hessenberg DAE (3.6).

In terms of the above-defined manifolds W_1 and W_2, the relation $w = \psi(u)$ leads to a local parametrization $(u, z) \to ((u, \psi(u)), z)$ of W_1, with $(u, z) \in \Omega_1 \subseteq \mathbb{R}^r$ for some open set Ω_1. In turn, $z = \eta(u, w)$ defines, via $(u, w) \to ((u, w), \eta(u, w))$, a parametrization of the "hidden" manifold $g_y h = 0$. Finally, the mapping $z = \eta(u, \psi(u))$ yields the local parametrization $u \to ((u, \psi(u)), \eta(u, \psi(u)))$ of the solution manifold W_2.

These ideas also apply, *mutatis mutandis*, to the Hessenberg DAE of arbitrarily large size (1.18), under the assumption that the product in (1.19) is nonsingular. For instance, solutions of an autonomous Hessenberg DAE of size three

$$y'_1 = h_1(y_1, y_2, z)$$
$$y'_2 = h_2(y_1, y_2)$$
$$0 = g(y_2)$$

lie on the manifold W_3 defined by the identities $g = 0$, $g_{y_2} h_2 = 0$ and $g_{y_2} h_{2y_1} h_1 + (g_{y_2} h_2)_{y_2} h_2 = 0$, provided that the product $g_{y_2} h_{2y_1} h_{1z}$ is nonsingular.

The notions here presented can be extended in a non-trivial way to general quasilinear DAEs, as detailed in Section 3.4, using some elementary concepts from differential geometry which are previously compiled in Section 3.3. Hessenberg DAEs will be revisited in 3.4.6 in order to illustrate the discussion performed in subsections 3.4.2 - 3.4.5. Nonautonomous analogs of the problems above will be considered in 3.4.7.

3.3 Some notions from differential geometry

The reader is referred to [1, 27, 31] for extensive introductions to differential geometry and, in particular, for a more detailed discussion of the notions compiled here for later use. Henceforth we will restrict the attention to smooth (this term being used to mean C^∞ in the remainder of the Chapter) problems, although virtually all results can be restated in a C^k context with finite k.

We will assume throughout that $\mathbb{R}^m$ is furnished with the usual topology, and restrict the attention to subsets of $\mathbb{R}^m$ equipped with the relative topology. In this context, an r-dimensional *local parametrization* of a nonempty set $W \subseteq \mathbb{R}^m$ is a pair (Ω, φ), with Ω and $\varphi(\Omega)$ nonempty and open in $\mathbb{R}^r$ and W, respectively, such that the C^∞-mapping $\varphi : \Omega \to \mathbb{R}^m$ verifies $\mathrm{rk}\, \varphi'(\xi) = r$ for all $\xi \in \Omega$ and yields a homeomorphism $\Omega \to \varphi(\Omega) \subseteq W$. With terminological abuse, sometimes we will use 'parametrization' to mean the mapping φ, regardless of the domain Ω. Hereafter we will refer to the inverse mapping $\varphi^{-1} : \varphi(\Omega) \to \Omega \subseteq \mathbb{R}^r$ as a *local coordinate system* or a *chart*, which is of course defined only on the image of φ.

Now, a nonempty set $W \subseteq \mathbb{R}^m$ is said to be a *smooth manifold* if for every nonisolated point $x \in W$ there exists a local parametrization (Ω, φ) with $x \in \varphi(\Omega)$; if this parametrization is r-dimensional, r is termed the *local dimension* of W at x. As it is done e.g. in [1, 6], we do not

3.3. Some notions from differential geometry

require the domains of all parametrizations of W to be open subsets of the same space $\mathbb{R}^r$; should all parametrizations be r-dimensional for some constant r, the manifold is said to be r-dimensional. We also assume that any set composed of isolated points is a 0-dimensional manifold, and allow isolated points to be 0-dimensional components of any manifold; actually, accommodating these 0-dimensional components will be only necessary for the sake of completeness in 3.4.4, and throughout the present Section we exclude isolated points without explicit mention in situations in which they become meaningless. Note that $\mathbb{R}^m$ itself, as well as any open subset of $\mathbb{R}^m$, can be seen as an m-dimensional smooth manifold by using the identity as a parametrization. Hence, a smooth manifold in $\mathbb{R}^m$ may have different components whose dimensions range from 0 to m.

Roughly speaking, a *submanifold* of a given smooth manifold $W \subseteq \mathbb{R}^m$ is a subset of W which has the relative topology and is itself a manifold. Formally, given a manifold W, a nonempty subset $S \subseteq W$ is said to be a smooth submanifold of W either if S is open in W (being then an *open submanifold* of W), or if for every $x \in S$ there exists a local parametrization $(\Omega_1 \times \Omega_2, \varphi)$ of W in which Ω_1 and Ω_2 are open neighborhoods of the origin in $\mathbb{R}^{d_1}$ and $\mathbb{R}^{d_2}$, respectively, $x = \varphi(0,0)$, and $S \cap \varphi(\Omega_1 \times \Omega_2) = \varphi(\Omega_1 \times \{0\})$) (see e.g. [59]). The corresponding chart φ^{-1} yields a *preferred coordinate system* for W relative to S [27]. In particular, if W is r-dimensional and there exists a constant $d \geq 1$ such that $d_1 = r - d$ and $d_2 = d$ for all local parametrizations of S of the form above, S is said to be a codimension-d submanifold of W; S is then $(r-d)$-dimensional. Note that an r-dimensional manifold $W \subseteq \mathbb{R}^m$ is a submanifold of $\mathbb{R}^m$ with codimension $m - r$.

A mapping $f : W \to \tilde{W}$ is said to be *smooth* if for every $x \in W$ there exist local parametrizations (Ω, φ) and $(\tilde{\Omega}, \tilde{\varphi})$ of W and $\tilde{W}$, with $x \in \varphi(\Omega)$ and $f(\varphi(\Omega)) \subseteq \tilde{\varphi}(\tilde{\Omega})$ (thus in particular $f(x) \in \tilde{\varphi}(\tilde{\Omega})$) such that $\tilde{\varphi}^{-1} \circ f \circ \varphi$ belongs to $C^\infty(\Omega, \tilde{\Omega})$. Since Ω and $\tilde{\Omega}$ are open sets within certain spaces $\mathbb{R}^r$ and $\mathbb{R}^s$, this relies on standard concepts of real differential calculus. It is easy to check that this notion does not depend on the specific choice of the parametrizations. A smooth bijection f of W onto $\tilde{W}$ is called a *diffeomorphism* if the inverse $f^{-1} : \tilde{W} \to W$ is also smooth.

Letting $W \subseteq \mathbb{R}^m$ be a smooth manifold, the set of tangent vectors at $x \in W$ of all smooth curves on W that pass through x is called the *tangent space* to W at x and will be denoted by $T_x W$. Mind that we are restricting the attention to submanifolds of $\mathbb{R}^m$; in abstract settings, this space can be defined as the quotient set under an equivalence relation given by the tangency of smooth curves leaving x, cf. [1]. The tangent space can be shown to be a vector subspace of $\mathbb{R}^m$, its dimension given by the local dimension of W at x. If φ is a local parametrization of W with $x = \varphi(\xi)$, then $\operatorname{im} \varphi'(\xi)$ defines the tangent space to W at x. The disjoint union $\bigcup_{x \in W} T_x W$ is called the *tangent bundle* of W; it can be given a manifold structure, which is $2r$-dimensional if W is r-dimensional. In particular, the tangent space to $\mathbb{R}^m$ (or to an open set $\Omega \subseteq \mathbb{R}^m$) at any point can be canonically identified with $\mathbb{R}^m$ itself, and we may therefore write the corresponding tangent bundles as $\mathbb{R}^m \times \mathbb{R}^m$ and $\Omega \times \mathbb{R}^m$, respectively.

A smooth mapping $f : W \to \tilde{W}$ induces at every $x \in W$ a linear map $df_x : T_x W \to T_{f(x)} \tilde{W}$ called the *differential*. It can be defined as the mapping which carries a vector v tangent at $t = 0$ to a smooth curve $\phi(t)$ on W leaving x (i.e. a smooth curve such that $\phi(t) \in W$ for all t, with $\phi(0) = x$, $\phi'(0) = v$), to the vector $(f \circ \phi)'(0)$, which is tangent to the curve $f(\phi(t))$ at $t = 0$; see e.g. [1, 6]. This notion does not depend on the particular choice of the smooth curve ϕ. The *rank* of a smooth mapping $f : W \to \tilde{W}$ at $x \in W$ is then defined as the rank of the differential df_x, that is, as the dimension of the image space $df_x(T_x W)$. It can be proved to be given by the rank of the Jacobian matrix in any coordinate representation; with the parametrizations used in the above definition of a smooth mapping, the rank of f at $x = \varphi(\xi)$ equals the rank of $(\tilde{\varphi}^{-1} \circ f \circ \varphi)'(\xi)$.

A smooth map $f : W \to \tilde{W}$ is said to be an *immersion at* $x \in W$ if the rank of f at x equals the local dimension of W at x; it is called an *immersion* if it is an immersion at every $x \in W$. An immersion $f : W \to \tilde{W}$ that is a homeomorphism of W onto $f(W)$ is called an *embedding* of W in $\tilde{W}$. In particular, a local parametrization $\varphi : \Omega \to W$ is an embedding of Ω in W. Locally, if $f : W \to \tilde{W}$ is an immersion at x, then there exists a neighborhood U of x such that $f|_U$ is an embedding of U in $\tilde{W}$ and a diffeomorphism of U onto $f(U)$ (see Theorem 3.5.7 in [1] and Theorem III.4.12 in [27]). These notions will be the key to guaranteeing that the reduction procedure discussed below preserves all the local dynamic information of a given quasilinear DAE.

In Sections 3.1 and 3.2 we assumed at several points that the derivative of the map g defining explicit constraints (cf. (3.1) and (3.6)) has maximal rank. A smooth mapping $g : W \to \tilde{W}$ is called a *submersion at* $x \in W$ if the rank of g at x equals the local dimension of $\tilde{W}$ at $g(x)$, and it is called a *submersion* if it is a submersion at every $x \in W$. As it happens in the DAEs (3.1) and (3.6), smooth manifolds will often be implicitly described in terms of submersions. Actually this idea provides an alternative manifold definition (cf. Definition 9.2 in [59]). The key remark in this regard is that a local r-dimensional parametrization of $W \subseteq \mathbb{R}^m$ exists around $x \in W$ if and only if there is an open set $U \subseteq \mathbb{R}^m$ containing x and a submersion $g \in C^\infty(U, \mathbb{R}^{m-r})$ such that $W \cap U = \{x \in U \;/\; g(x) = 0\}$. This type of local description will also be used in the general quasilinear setting of Section 3.4, in particular with $g(x) = H(x)f(x)$ in (3.23); see also Lemma 3.1. In this setting, the tangent space to W at x is defined by $\ker g'(x)$.

The submersion condition on g at a given x^* verifying $g(x^*) = 0$ yields, via the implicit function theorem, a variable splitting $x = (u, w)$ which makes it possible to describe locally the set $g = 0$ as $w = \psi(u)$ for some locally defined smooth map ψ, thereby yielding a local parametrization of the form $u \to (u, \psi(u))$ of $g = 0$ around x^*; here the choice of the variables w is of course defined by a nonsingular derivative $g_w(x^*)$. In particular, the nonsingularity of $g_z(y^*, z^*)$ in the index one setting of Section 3.1 led to the local description $z = \psi(y)$ of the set W_1 in (3.4).

Immersions and submersions are particular instances of a *subimmersion*, which in this finite-dimensional setting can be defined as a smooth map of locally constant rank, cf. Definition 3.5.15 and Proposition 3.5.16(ii) in [1]. The subimmersion theorem, which can be understood as a geometric analog of the *rank theorem* [78], states that if W and $\tilde{W}$ are smooth manifolds with dimensions r and s, and a given smooth mapping $g: W \to \tilde{W}$ has constant rank d on W, then for every y in $g(W)$ the set $g^{-1}(y)$ is a submanifold of W with dimension $r - d$ (see e.g. Th. III.5.8 in [27]). The result also holds if g has constant rank on an open neighborhood of $g^{-1}(y)$ (cf. Th. 3.5.17 in [1]). If g is a submersion (i.e. if $d = s$) then $g^{-1}(y)$ has dimension $r - s$; in this maximal rank case, it is enough to require that the rank of g equals s on the set $g^{-1}(y)$, as stated for instance in Corollary III.5.9 of [27]. We already used this idea in the above-mentioned local implicit description of a manifold as $g^{-1}(0)$, $g: U \to \mathbb{R}^{m-r}$ being a submersion defined on the m-dimensional manifold $U \subseteq \mathbb{R}^m$. This result supports the manifold structure of W_1 in (3.4) under the assumption of nonsingularity on g_z, as well as that of W_2 in (3.10) under the maximal rank requirement (3.11).

3.4 Quasilinear DAEs: The geometric index

As already indicated in subsection 1.4.3, the semiexplicit problems (3.1) and (3.6) considered in Sections 3.1 and 3.2 can be rewritten in the form

$$\begin{pmatrix} I_r & 0 \\ 0 & 0 \end{pmatrix} \begin{pmatrix} y' \\ z' \end{pmatrix} = \begin{pmatrix} h \\ g \end{pmatrix}.$$

Writing $(y, z) = x$ and $(h, g) = f$, these equations can be seen as particular instances of the autonomous quasilinear DAE

$$A(x)x' = f(x), \qquad (3.13)$$

with $A \in C^\infty(W_0, \mathbb{R}^{m \times m})$, $f \in C^\infty(W_0, \mathbb{R}^m)$, W_0 open in $\mathbb{R}^m$. The results in this Section also hold if the smoothness requirement is reduced to C^m and, eventually, to C^ν if the index does not exceed ν. The pair $(A(x), f(x))$ is sometimes called a *generalized vector field* [67, 68, 202, 203], and the quasilinear DAE (3.13) will be often referred to simply as (A, f). We detail in this Section a reduction procedure for general problems of the form (3.13) which extends the ideas introduced in Sections 3.1 and 3.2. The nonautonomous counterpart of (3.13) (namely, the DAE (3.70) on page 123) will be considered in subsection 3.4.7.

3.4.1 The framework of Rabier and Rheinboldt

Geometric methods for the reduction of nonlinear DAEs have been developed in the last two decades mainly by Rheinboldt [238, 239], Reich [229, 230], and Rabier and Rheinboldt [220, 221, 228], among other authors. We summarize in this subsection the global reduction method of Rabier and Rheinboldt for quasilinear problems of the form (3.13), following Chapter 4 of [228].

The attention will be focused on DAEs defined on an open submanifold W_0 of $\mathbb{R}^m$, although the results also hold if the problem domain is a lower-dimensional submanifold of $\mathbb{R}^m$ as it is assumed in [228]. Note that any C^1 solution to (3.13) must obviously lie on the set (cf. in particular (3.4) and (3.9) above)

$$W_1 = \{x \in W_0 \;/\; f(x) \in \operatorname{im} A(x)\}, \qquad (3.14)$$

which by certain global conditions will be guaranteed to be an r-dimensional submanifold of W_0.

With this aim, define the map $F : TW_0 \simeq W_0 \times \mathbb{R}^m \to \mathbb{R}^m$ as

$$F(x, p) = A(x)p - f(x), \qquad (3.15)$$

and let

$$M_0 = F^{-1}(0) = \{(x, p) \in TW_0 \;/\; A(x)p - f(x) = 0\}. \qquad (3.16)$$

Denoting by $\pi : \mathbb{R}^m \times \mathbb{R}^m \to \mathbb{R}^m$ the projection onto the first factor we have $W_1 = \pi(M_0)$. Now, if a given mapping $x(t)$ solves (3.13), then it follows immediately from (3.16) that the pair $(x(t), x'(t))$ must belong to M_0.

3.4. Quasilinear DAEs: The geometric index

The reduction method as discussed by Rabier and Rheinboldt then proceeds through the following two assumptions.

(G1) $A(x)$ has constant rank $r_1 \leq m$ for all $x \in W_1$.
(G2) $F(x,p)$ is a submersion on its zero set M_0.

As explained in Section 3.3, the submersion hypothesis G2 requires the derivative F' to have maximal rank m at every point of M_0. Assumption 22.1 in [228] is a slightly more general condition which, nevertheless, is acknowledged by the authors to be checked in practice via the above-stated submersion assumption.

Under these working conditions, the mapping $\pi|_{M_0} : M_0 \to W_1$ is proved to have constant rank r_1 and also to be an open map, that is, to carry M_0 and open subsets of M_0 to open subsets of W_1. Via the subimmersion theorem, it follows that $W_1 = \pi(M_0)$ is a smooth r_1-dimensional submanifold of W_0; find details in Theorems 21.1 and 22.3 of [228].

A key remark at this stage is the following. Provided that $x(t)$ is a solution to (3.13), thereby lying entirely on W_1, the pair $(x(t), x'(t))$ must belong to the tangent bundle TW_1. This means that $x'(t)$ must be tangent not only to W_0 but also to W_1 itself. Hence, the pair $(x(t), x'(t))$ needs to be in the intersection $M_1 = TW_1 \cap M_0$ and, in particular, $x(t)$ must lie on $W_2 = \pi(M_1)$. Letting $F_1 = F|_{TW_1}$, we can describe $M_1 = TW_1 \cap M_0$ as $F_1^{-1}(0)$, whereas the set W_2 reads

$$W_2 = \{x \in W_1 \ / \ f(x) \in \operatorname{im} A(x)|_{T_x W_1}\}. \tag{3.17}$$

Again, (3.17) can be checked to amount to (3.10) for the Hessenberg problem (3.6) considered in Section 3.2. If the analogs of assumptions G1 and G2 hold when applied to F_1 and $A(x)|_{T_x W_1}$, W_2 will be an r_2-dimensional manifold with $r_2 = \operatorname{rk} A(x)|_{T_x W_1}$, and the same reasoning can be performed one step further.

This way, if the the above-mentioned working assumptions hold at every step, the procedure yields two sequences of smooth manifolds which will eventually stabilize, namely, $M_0 \supset M_1 \supset \ldots \supset M_\nu = M_{\nu+1}$ and

$$W_0 \supset W_1 \supset \ldots \supset \ldots \supset W_\nu \supseteq W_{\nu+1} = W_{\nu+2}. \tag{3.18}$$

The dimensions of the manifolds in (3.18) are given by the rank sequence

$$r_0 > r_1 > r_2 > \ldots > r_\nu = r_{\nu+1} = r_{\nu+2},$$

where r_0 stands for m. Under these conditions the pair (W_0, M_0) is said to be *completely reducible*; Rabier and Rheinboldt then call the smallest integer ν such that either $M_\nu = \emptyset$ or $M_\nu \neq \emptyset$ and $r_\nu = r_{\nu+1}$ the *geometric index* of the quasilinear DAE (3.13).

The procedure involves $\nu+1$ steps, although in the $(\nu+1)$-th one only the constant rank condition $r_{\nu+1} = \operatorname{rk} A(x)|_{T_x W_\nu} = r_\nu$ for $x \in W_{\nu+1}$ needs to be checked, as shown in Theorem 26.4 of [228]. In index ν problems with $M_\nu \neq \emptyset$, the manifold $W_{\nu+1}$ turns out to be open in W_ν. Always under the above-stated working assumptions, this manifold comprises all the smooth solutions of the DAE, which can be described in terms of a vector field uniquely defined on $W_{\nu+1}$. Details can be found in Theorems 23.2 and 24.1 of [228].

Via local parametrizations, solutions of index ν DAEs can also be locally described in terms of reduced equations. As detailed in Sects. 26 and 27 of [228], if U is an open set such that $W_\nu \cap U$ admits a local parametrization $x = \varphi(u)$ with domain $\Omega \subseteq \mathbb{R}^{r_\nu}$, and U is small enough as to guarantee the existence of $P \in \mathbb{R}^{r_\nu \times m}$ yielding an isomorphism from $\operatorname{im} A(x)|_{T_x W_{\nu-1}}$ onto $\mathbb{R}^{r_\nu}$ for all x in $W_\nu \cap U$, then $x(t)$ is proved to be a solution of (3.13) lying on U if and only if $x(t) \in W_{\nu+1} \subseteq W_\nu$ for all t and $u(t) = \varphi^{-1}(x(t))$ is a solution of the reduced equation

$$A_\nu(u)u' = f_\nu(u) \qquad (3.19)$$

on $\Omega \subseteq \mathbb{R}^{r_\nu}$, with

$$A_\nu(u) = PA(\varphi(u))\varphi'(u), \quad f_\nu(u) = Pf(\varphi(u)). \qquad (3.20)$$

The leading matrix A_ν in (3.19) can be proved to be nonsingular along solutions of the DAE, and actually on $\varphi^{-1}(W_{\nu+1} \cap U)$. This makes it possible to describe locally the solutions of the DAE via the state space representation

$$u' = A_\nu^{-1}(u)f_\nu(u), \qquad (3.21)$$

on $\varphi^{-1}(W_{\nu+1} \cap U) \subseteq \Omega$. Equation (3.21) is a generalization of the local state space descriptions (3.5) and (3.12).

Why a local approach?

The above-summarized framework provides a nice approach for the analysis of quasilinear DAEs when the global assumptions G1 and G2 on page 95 hold at every reduction step. But its obvious limitation is the exclusion of DAEs for which these assumptions are not met. Consider for example a situation in which the constant rank assumption in G1 does not hold (not even locally) around a given $x^* \in W_1$ (cf. for instance (4.35) or (4.63) on pages 167 and 186, respectively). In this case W_1 may not admit locally a manifold structure, and thereby $T_{x^*}W_1$ need not be well-defined. In turn this rules out the definition of M_1, W_2, and so on.

This is actually a non-trivial, two-fold problem. First, restating the working conditions and the construction in a local manner will make it possible, by means of the concept of a *regular point*, to characterize the so-called regular set where a complete reduction is actually feasible, possibly with different dimensions and indices in different components. In the terms detailed in subsections 3.4.2 - 3.4.4, and disregarding for the moment minor differences, behind the approach of Rabier and Rheinboldt is the requirement that all points on W_1 are 0-regular (see Definition 3.2 below). In the forthcoming framework, even if W_1 does not have a manifold structure, the regular part W_1^{reg} will be indeed a smooth manifold (cf. Proposition 3.2, p. 109). The same will happen in later steps, thus allowing to rename the above-mentioned regular set the *regular manifold* of the DAE, which is well-defined for any quasilinear problem. The concepts of a *locally regular* DAE and its solution manifold follow naturally, as detailed in subsection 3.4.4.

The second aspect of this problem concerns the points where the (so far vague) regularity notion fails; these singular problems will be tackled in Chapter 4. In the above-sketched situation in which the constant rank assumption fails locally around a given $x^* \in W_1$, this point would be called an *inner singularity*; see the example (4.63) discussed in 4.4.6.3. These singular points cannot be accommodated in the framework of Rabier and Rheinboldt; the singular phenomena discussed in [228] (see specifically p. 353) amount to what may be called "last-step singularities", which do not affect the validity of the reduction procedure. In terms of the reduced equation (3.19), $A_\nu(u)$ will not be invertible on $\Omega - \varphi^{-1}(W_{\nu+1} \cap U)$ and therefore (3.19) will be in general a quasilinear ODE with singularities, but the reduction procedure was assumed to be feasible up to step ν. A more general setting will be discussed in Chapter 4.

3.4.2 Index zero and index one points

The present subsection, together with 3.4.3, introduces a local reduction procedure that overcomes the limitations discussed at the end of 3.4.1. Readers without a previous background on this topic might profit from taking a look at the examples in subsection 3.4.6 before or in parallel to reading this construction.

3.4.2.1 Index zero points

Definition 3.1. A point $x^* \in W_0$ is called *regular with geometric index zero* for the DAE (3.13) if $A(x^*) \in \mathbb{R}^{m \times m}$ is a nonsingular matrix.

The set of regular points with index zero will be denoted by $W^{\text{ind}0}$.

We use hereafter the expression 'index zero point' (and later on 'index ν point') for regular points with index zero (resp. with index ν). Note that we are often omitting the label 'geometric' for the index within this Chapter.

The set $W^{\text{ind}0}$, if nonempty, is immediately seen by the condition $\det A(x) \neq 0$ to be an open submanifold of W_0. The behavior on this submanifold trivially amounts to that of the explicit ODE

$$x' = A^{-1}(x)f(x), \ x \in W^{\text{ind}0}.$$

3.4.2.2 Index one points

As already remarked in 3.4.1, an obvious necessary condition for a C^1 solution of (3.13) to pass through a given point is that it belongs to the set $W_1 = \{x \in W_0 \ / \ f(x) \in \operatorname{im} A(x)\}$ defined in (3.14). This set plays a key role in the analysis of points which are not index zero, as detailed in the sequel. Specifically, the below-defined subset $W_1^{\text{reg}} \subseteq W_1$ of so-called 0-regular points captures the algebraic requirements which yield a smooth manifold structure, supporting not only the local reduction of Theorem 3.1 but also the notion of an index one point (see Definition 3.3 on p. 102) and the analysis of higher index problems undertaken in 3.4.3. Recall that the map $F: W_0 \times \mathbb{R}^m \to \mathbb{R}^m$ was defined in (3.15) as $F(x,p) = A(x)p - f(x)$.

Definition 3.2. A point $x^* \in W_0$ is said to be *0-regular* for the DAE (3.13) if $x^* \in W_1$ and the following two conditions hold:

(R1) $A(x)$ has constant rank $r_1 \leq m$ on some neighborhood of x^*.
(R2) F is a submersion at (x^*, p^*), for some p^* satisfying $A(x^*)p^* = f(x^*)$.

The set of 0-regular points will be denoted by W_1^{reg}.

By construction W_1^{reg} is open in W_1 and therefore the sets W_1^{reg} and W_1 coincide locally around any 0-regular point. More precisely, for all $x^* \in W_1^{\text{reg}}$ there exist a neighborhood U of x^* such that $W_1^{\text{reg}} \cap U = W_1 \cap U$. Henceforth we will abbreviate this kind of relation as

$$W_1^{\text{reg}} \stackrel{\text{loc}}{=} W_1. \tag{3.22}$$

Further relations between W_0, $W^{\text{ind}0}$, W_1 and W_1^{reg} can be found in 3.4.4.

3.4. Quasilinear DAEs: The geometric index

The constant rank condition in R1 above is a local version of G1 on page 95, with the slightly stronger requirement that the rank is constant within a whole neighborhood (say $\tilde{U}_0$) of x^* in W_0. Note that a local rank deficiency in A along W_1 seems to be unlikely to occur except in incidental examples. If this locally constant rank r_1 verifies $r_1 < m$, then there will exist another open neighborhood $\hat{U}_0 \subseteq \tilde{U}_0 \cap U$ of x^* and a smooth matrix-valued map $H \in C^\infty(\hat{U}_0, \mathbb{R}^{(m-r_1)\times m})$ such that

$$\ker H(x) = \operatorname{im} A(x) \quad \forall x \in \hat{U}_0,$$

see e.g. Proposition 3.4.18(ii) in [1] or Lemma 22.1 in [228]. Note that $\ker H(x) = \operatorname{im} A(x)$ yields of course $H(x)A(x) = 0$ but also the maximal rank condition $\operatorname{rk} H(x) = m - r_1$ on $\hat{U}_0$. Now $v \in \operatorname{im} A(x) \Leftrightarrow H(x)v = 0$ for $x \in \hat{U}_0$, allowing for the following local implicit description of $W_1^{\mathrm{reg}} \stackrel{\mathrm{loc}}{=} W_1$:

$$W_1^{\mathrm{reg}} \cap \hat{U}_0 = W_1 \cap \hat{U}_0 = \{x \in \hat{U}_0 \,/\, H(x)f(x) = 0\}. \tag{3.23}$$

The submersion condition R2 requires $\operatorname{rk} F'(x^*, p^*) = m$. This is a key hypothesis because it characterizes the situations in which the product $H(x)f(x)$ is a submersion, as shown in Lemma 3.1 below; in this setting the implicit description (3.23) will lead to a local parametrization of the set $W_1^{\mathrm{reg}} \stackrel{\mathrm{loc}}{=} W_1$, guaranteeing a local smooth structure on it and thereby paving the way for the local reduction discussed in Theorem 3.1. Lemma 3.1 also shows that, under assumption R1, the submersion condition in R2 does not depend on the specific choice of p^* such that $A(x^*)p^* = f(x^*)$; this means that R2 is an exact local version of G2 on page 95.

Lemma 3.1. *Let $x^* \in W_1$. Assume that $A(x)$ has constant rank r_1, with $0 < r_1 < m$, on some open neighborhood $\tilde{U}_0$ of x^*, and let the matrix-valued map $H \in C^\infty(\hat{U}_0, \mathbb{R}^{(m-r_1)\times m})$ verify $\ker H(x) = \operatorname{im} A(x) \; \forall x \in \hat{U}_0 \subseteq \tilde{U}_0$. Then the following two statements are equivalent:*

(i) $H(x)f(x)$ *is a submersion at* x^*;
(ii) $F(x, p)$ *is a submersion at* (x^*, p^*) *for some (hence any) p^* satisfying* $A(x^*)p^* = f(x^*)$.

Proof. We will denote $(Hf)'(x^*)\cdot = (H'(x^*)\cdot)f(x^*) + H(x^*)f'(x^*)\cdot$ and, with notational abuse, the Jacobian matrix $F'(x^*, p^*)$ will be written as

$$((A'(x^*)\cdot)p^* - f'(x^*)\cdot \quad A(x^*)),$$

in order to denote $F'(x^*, p^*)(v_1, v_2) = (A'(x^*)v_1)p^* - f'(x^*)v_1 + A(x^*)v_2$.

Assume that (i) holds, that is, $\operatorname{rk}(Hf)'(x^*) = m - r_1$, and suppose that for some p^* satisfying $A(x^*)p^* = f(x^*)$ the submersion condition on

$F(x^*, p^*)$ fails, i.e., that $\operatorname{rk} F'(x^*, p^*) < m$. This would mean that there exists a non-vanishing $v \in \mathbb{R}^m \simeq \mathbb{R}^{m \times 1}$ such that $v^\mathsf{T} F'(x^*, p^*) = 0$, namely

$$v^\mathsf{T}[(A'(x^*)\cdot)p^* - f'(x^*)\cdot] = 0 \qquad (3.24\text{a})$$
$$v^\mathsf{T} A(x^*) = 0. \qquad (3.24\text{b})$$

From the latter and the definition of H, v^T can be written as $w^\mathsf{T} H(x^*)$ for some non-vanishing $w \in \mathbb{R}^{m-r_1}$. Equation (3.24a) then reads

$$w^\mathsf{T} H(x^*)[(A'(x^*)\cdot)p^* - f'(x^*)\cdot] = 0. \qquad (3.25)$$

Due to the relation $H(x^*)A'(x^*)\cdot = -(H'(x^*)\cdot)A(x^*)$ following from $H(x)A(x) = 0$, (3.25) reads $w^\mathsf{T}[-(H'(x^*)\cdot)A(x^*)p^* - H(x^*)f'(x^*)\cdot] = 0$; in turn, this can be written as $-w^\mathsf{T}[(H'(x^*)\cdot)f(x^*) + H(x^*)f'(x^*)\cdot] = 0$, against the maximal rank condition on $(Hf)'(x^*)$. This shows that, if (i) is met, then (ii) holds for any p^* as long as it satisfies $A(x^*)p^* = f(x^*)$.

Let now the statement (ii) hold for some p^* satisfying $A(x^*)p^* = f(x^*)$. The product $H(x^*)F'(x^*, p^*)$ must then have maximal rank, that is, $\operatorname{rk}(\, H(x^*)((A'(x^*)\cdot)p^* - f'(x^*)\cdot) \quad 0\,) = m - r_1$, and then

$$\operatorname{rk}(-(H'(x^*)\cdot)A(x^*)p^* - H(x^*)f'(x^*)\cdot)$$
$$= \operatorname{rk}(-(H'(x^*)\cdot)f(x^*) - H(x^*)f'(x^*)\cdot) = m - r_1,$$

so that $\operatorname{rk}(Hf)'(x^*) = m - r_1$, which is the submersion condition in (i).

Note finally that, as shown above, when (i) holds, (ii) holds for any p^* verifying $A(x^*)p^* = f(x^*)$. This proves that the submersion condition on F at (x^*, p^*) does not depend on the specific choice of p^*. □

Theorem 3.1. *Let $x^* \in W_0$ be a 0-regular point for (3.13), and denote by r_1 the locally constant rank of $A(x)$ around x^*. If $r_1 > 0$, then there exists an open neighborhood $U_0 \subseteq W_0 \subseteq \mathbb{R}^m$ of x^* such that*

(i) *$W_1^{\text{reg}} \cap U_0 = W_1 \cap U_0$ admits a smooth r_1-dimensional parametrization $x = \varphi_1(\xi)$ with surjective $\varphi_1 : \Omega_1 \to W_1^{\text{reg}} \cap U_0$;*
(ii) *there exists a C^∞ matrix-valued mapping $P_1 : U_0 \to \mathbb{R}^{r_1 \times m}$ verifying that $P_1(x)|_{\operatorname{im} A(x)}$ is an isomorphism $\operatorname{im} A(x) \to \mathbb{R}^{r_1}$ for all $x \in U_0$.*

For any such φ_1, P_1, $x(t)$ is a solution of (3.13) within U_0 if and only if $x(t) \in W_1^{\text{reg}} \stackrel{\text{loc}}{=} W_1$ for all t and $\xi(t) = \varphi_1^{-1}(x(t))$ is a solution of

$$A_1(\xi)\xi' = f_1(\xi), \quad \xi \in \Omega_1 \subseteq \mathbb{R}^{r_1} \qquad (3.26)$$

with

$$A_1(\xi) = P_1(\varphi_1(\xi))A(\varphi_1(\xi))\varphi_1'(\xi) \qquad (3.27\text{a})$$
$$f_1(\xi) = P_1(\varphi_1(\xi))f(\varphi_1(\xi)). \qquad (3.27\text{b})$$

3.4. Quasilinear DAEs: The geometric index

Proof. Choose U_0 small enough as to guarantee the coincidence on it of W_1 and W_1^{reg} pointed out in (3.22), as well as the existence of the parametrization φ_1 in (i) which follows from the local conditions supporting the notion of a 0-regular point together with Lemma 3.1, and the existence of $P_1(x)$ in (ii) which is due to the vector bundle structure locally associated with im $A(x)$ (cf. Proposition 3.4.18(ii) in [1] and Lemma 22.1 in [228]).

Now, assume that $x(t)$ solves (3.13) with $x(t) \in U_0$ for all t. Since $x'(t)$ solves pointwise the linear system $A(x(t))p = f(x(t))$, it must be $x(t) \in W_1 \stackrel{\text{loc}}{=} W_1^{\text{reg}}$ for all t. Therefore $\xi(t) = \varphi_1^{-1}(x(t))$ is well-defined. Writing $x(t) = \varphi_1(\xi(t))$, $x'(t) = \varphi_1'(\xi(t))\xi'(t)$, and premultiplying (3.13) by $P_1(\varphi_1(\xi(t)))$ we get that (3.26) holds.

Conversely, the assumption that $\xi(t)$ solves (3.26) reads, in the light of (3.27),

$$P_1(\varphi_1(\xi(t)))A(\varphi_1(\xi(t)))\varphi_1'(\xi(t))\xi'(t) = P_1(\varphi_1(\xi(t)))f(\varphi_1(\xi(t))). \quad (3.28)$$

Note that $\varphi_1(\xi(t)) = x(t) \in W_1$ and then $f(\varphi_1(\xi(t))) \in \text{im} A(\varphi_1(\xi(t)))$. Since P_1 yields an isomorphism when restricted to im A, (3.28) leads to

$$A(\varphi_1(\xi(t)))\varphi_1'(\xi(t))\xi'(t) = f(\varphi_1(\xi(t))),$$

showing that $x(t) = \varphi_1(\xi(t))$ solves (3.13); obviously $x(t) \in U_0$ since φ_1 maps onto $W_1^{\text{reg}} \cap U_0$. $\square$

Theorem 3.1 holds trivially if $r_1 = m$, that is, in an index zero setting, although in this case the result is only needed for later consistency; in this situation, (3.27) actually yields a form of *local equivalence* which will examined in detail in subsection 3.4.5.

Equation (3.26), as well as the pair (A_1, f_1), will be called a *one-step local reduction* or, sometimes, simply a *local reduction* of (3.13), whereas (P_1, φ_1) will be termed a *reduction pair*. Allowing the mapping P_1 to be non-constant (in contrast to [228]) may simplify the reduced equations; see for instance example (4.61) on p. 185. Additional remarks on the reduction operator P_1 can be found in 3.4.5.3 below.

Just for the sake of completeness, if x^* is 0-regular with $r_1 = 0$ we may think of a local reduction simply as the algebraic problem $f(\varphi_1(\xi)) = 0$ defined from a trivial map $\varphi_1 : \{\xi^*\} \to \{x^*\}$ (cf. Remark 3.3).

Remark 3.1. According to Lemma 3.1, the "reduced" variables ξ in item (i) of Theorem 3.1 can be taken as a subset of the original variables x, by a straightforward application of the implicit function theorem to $H(x)f(x) = 0$. This particular choice is not necessary for the validity of the reduction, but will be important from the modeling point of view e.g. in Chapter 6.

Definition 3.3. A point $x^* \in W_0$ is called *regular with geometric index one* for (3.13)

(a) either if it is 0-regular with $m > r_1 > 0$ and $A_1(\xi^*)$ is nonsingular for some (hence any) reduction pair (P_1, φ_1) satisfying $x^* = \varphi_1(\xi^*)$;
(b) or if it is 0-regular with $r_1 = 0$.

The set of index one points will be denoted by $W^{\text{ind}1}$.

The independence of the specific choice of the reduction pair will be proved in broader generality in Theorem 3.3 (cf. subsection 3.4.5, p. 112).

In the index one setting described by item (a) of Definition 3.3, the reduction (3.26) can be rewritten as a explicit ODE on some neighborhood of ξ^*, possibly smaller than Ω_1. This will be stated in more generality in Theorem 3.2.

Remark 3.2. If x^* is 0-regular with $m > r_1 > 0$, the index one condition may be equivalently formulated as the requirement that $\xi^* = \varphi^{-1}(x^*)$ is an index zero point for the local reduction (3.26).

Remark 3.3. The cases accommodated in item (b), in which the rank r_1 of $A(x)$ verifies $r_1 = 0$ locally around x^*, correspond to DAEs in which $A(x)$ vanishes on some neighborhood of x^*. The problem then amounts locally to the algebraic one $f(x) = 0$. The 0-regularity requirement yields in this situation the conditions $f(x^*) = 0$, $\operatorname{rk} f'(x^*) = m$; these index one points are therefore isolated, and can be understood as equilibrium solutions of the DAE. Most problems of interest correspond of course to cases with $r_1 > 0$.

The index one condition may be checked without actually computing a parametrization of W_1, as detailed below.

Proposition 3.1. *A point $x^* \in W_1$ is regular with index one for (3.13) if $A(x)$ has constant rank r_1 around x^*, $0 < r_1 < m$, and the matrix*

$$\begin{pmatrix} P_1(x^*)A(x^*) \\ (Hf)'(x^*) \end{pmatrix} \tag{3.29}$$

is nonsingular, with P_1 and H given by Theorem 3.1 and Lemma 3.1, respectively.

Proof. Note first that x^* is 0-regular because of Lemma 3.1 and the maximal rank condition on $(Hf)'(x^*)$ which follows from the nonsingularity of (3.29). The latter is equivalent to $\ker P_1(x^*)A(x^*) \cap \ker (Hf)'(x^*) = \{0\}$. In addition, $\ker (Hf)'(x^*) = T_{x^*} W_1^{\text{reg}}$ because $W_1^{\text{reg}} \stackrel{\text{loc}}{=} W_1$ is locally defined by the condition $H(x)f(x) = 0$, as depicted in (3.23). We may hence restate the nonsingularity of (3.29) as $\ker P_1(x^*)A(x^*) \cap T_{x^*}W_1^{\text{reg}} = \{0\}$. This implies that

$$\operatorname{rk} P_1(x^*)A(x^*)|_{T_{x^*}W_1^{\text{reg}}} = r_1.$$

The result then follows from the fact that this maximal rank condition is equivalent to the nonsingularity of the $r_1 \times r_1$ matrix $A_1(\xi^*) = P_1(\varphi_1(\xi^*))A(\varphi_1(\xi^*))\varphi_1'(\xi^*)$ since, for any local parametrization $\varphi_1(\xi)$ of $W_1^{\text{reg}} \stackrel{\text{loc}}{=} W_1$ with $x^* = \varphi_1(\xi^*)$, it is $\operatorname{im}\varphi_1'(\xi^*) = T_{x^*}W_1^{\text{reg}}$. □

3.4.3 Higher index points

3.4.3.1 Index two points

The expert reader may directly jump to 3.4.3.2, which comprises as a particular case the index two setting here considered. The present discussion is intended to glance at the recursion process which supports the geometric index concept. Being deliberately vague, our aim is to introduce an index two notion in a way such that the point x^* is index two for (3.13) if $\xi^* = \varphi_1^{-1}(x^*)$ is index one for the one-step reduction (3.26).

In order to make this idea precise, let $x^* \in W_0$ be a 0-regular point with $m > r_1 > 0$, assume for the moment $A_1(\xi^*)$ in (3.26) to be singular, and define

$$V_2 = \{\xi \in \Omega_1 \;/\; f_1(\xi) \in \operatorname{im} A_1(\xi)\} \subseteq \Omega_1 \subseteq \mathbb{R}^{r_1}. \tag{3.30}$$

This set can be seen as a local coordinate description of the below-defined set W_2. Let us elaborate. In the setting of Rabier and Rheinboldt discussed in subsection 3.4.1, the manifold structure on W_1 given by assumptions G1 and G2 made the tangent space T_xW_1 (and thereby W_2 in (3.17)) well-defined. Instead, we now make use of the manifold structure on the set of 0-regular points which follows from item (i) of Theorem 3.1 (see Proposition 3.2 in subsection 3.4.4 for a more detailed discussion) to define

$$W_2 = \{x \in W_1^{\text{reg}} \;/\; f(x) \in \operatorname{im} A(x)|_{T_xW_1^{\text{reg}}}\} \subseteq W_1^{\text{reg}} \subseteq W_1, \tag{3.31}$$

where it is worth emphasizing the identity $T_xW_1^{\text{reg}} = T_xW_1$ following at 0-regular points from (3.22). The precise relation between V_2 and W_2 is

$$W_2 \cap U_0 = \varphi_1(V_2), \tag{3.32}$$

where U_0 is defined in Theorem 3.1. Indeed, since φ_1 is a local parametrization of the set $W_1^{\text{reg}} \stackrel{\text{loc}}{=} W_1$ and P_1 is an isomorphism when restricted to im A, the condition $f_1(\xi) \in \operatorname{im} A_1(\xi)$ is easily proved to be equivalent to $f(x) \in \operatorname{im} A(x)|_{T_x W_1^{\text{reg}}}$ for $x = \varphi_1(\xi)$, and (3.32) follows.

The set V_2 and the reduced equation (3.26) make it possible to extend the above-introduced notions in a straightforward manner, via the replacement of W_0 and W_1 in 3.4.2.2 by Ω_1 and V_2, respectively. Indeed, if ξ^* is a 0-regular point for (3.26), then $x^* = \varphi_1(\xi^*)$ will be called a *1-regular* point of the original DAE (3.13); denote by W_2^{reg} the set of 1-regular points. This requirement expresses not only that $\xi^* \in V_2$ (and then $x^* \in W_2$), but also that the analogs of the regularity conditions R1 and R2 in Definition 3.2 hold in the second reduction step, namely:

(R1') $A_1(\xi)$ has constant rank $r_2 \leq r_1$ around ξ^*; and
(R2') $F_1(\xi, p) = A_1(\xi)p - f_1(\xi)$ is a submersion at (ξ^*, p^*) for some p^* satisfying $A_1(\xi^*)p^* - f_1(\xi^*) = 0$.

Note that, if $r_2 = r_1$, $A_1(\xi^*)$ would be nonsingular; the point x^* would be index one and the discussion makes sense but becomes largely unnecessary. In any case, if $r_2 > 0$ there will exist an open neighborhood $U_1 \subseteq \Omega_1$ of ξ^* accommodating both an r_2-dimensional local parametrization $\varphi_2 : \Omega_2 \to V_2 \cap U_1$ of V_2 and a matrix-valued map $P_2 \in C^\infty(U_1, \mathbb{R}^{r_2 \times r_1})$ which makes $P_2(\xi)|_{\operatorname{im} A_1(\xi)}$ an isomorphism $\operatorname{im} A_1(\xi) \to \mathbb{R}^{r_2}$ for all $\xi \in U_1$. The reader can also check that $\varphi_1 \circ \varphi_2$ defines a local parametrization of $W_2^{\text{reg}} \stackrel{\text{loc}}{=} W_2$ near x^*.

This way we arrive at a reduction of (3.26) defined by

$$A_2(\gamma)\gamma' = f_2(\gamma), \ \gamma \in \Omega_2 \subseteq \mathbb{R}^{r_2}, \tag{3.33}$$

where γ stands for the variables within Ω_2, so that $\xi = \varphi_2(\gamma)$, and

$$A_2(\gamma) = P_2(\varphi_2(\gamma))A_1(\varphi_2(\gamma))\varphi_2'(\gamma) \tag{3.34a}$$
$$f_2(\gamma) = P_2(\varphi_2(\gamma))f_1(\varphi_2(\gamma)). \tag{3.34b}$$

Equation (3.33) can be seen as a two-step local reduction of (3.13). Now, setting $\gamma^* = (\varphi_1 \circ \varphi_2)^{-1}(x^*)$, x^* is said to be *regular with geometric index two* if $m > r_1 > r_2 > 0$ and $A_2(\gamma^*)$ is nonsingular, or else if $m > r_1 > r_2 = 0$. Note that this is equivalent to requiring that $\xi^* = \varphi_1^{-1}(x^*)$ be index one for the first reduction $A_1(\xi)\xi' = f_1(\xi)$ derived in (3.26), provided that x^* is 0-regular with $m > r_1 > 0$. If x^* is not index two (neither one nor zero), then the procedure is suitable for assessment one step further. The general case is described below.

3.4.3.2 Index ν points

The idea above can be generalized as follows in order to yield a recursive definition of a k-regular point and thereby support a general index notion. To begin with, assume that we are given the set W_k^{reg}, open in W_k, for some $k \geq 1$: see specifically (3.14) and Definition 3.2 for $k = 1$. Let $(P_1, \varphi_1), \ldots, (P_k, \varphi_k)$ be a sequence of reduction pairs in which, for $i \geq 3$, φ_i and P_i extend in the natural way the cases $i = 1, 2$ detailed in Theorem 3.1 and p. 104, respectively. Points in W_k^{reg} are $(k-1)$-regular and, via the sequence above, admit a local reduction of the form

$$A_k(\zeta)\zeta' = f_k(\zeta), \quad \zeta \in \Omega_k \subseteq \mathbb{R}^{r_k}, \tag{3.35}$$

cf. (3.26) and (3.33) for $k = 1$ and $k = 2$. Here we are implicitly assuming that the locally constant rank r_k of A_{k-1} (or that of A if $k = 1$) is not null; item (b) in Definition 3.4 below accommodates the case $r_k = 0$. Define

$$W_{k+1} = \{x \in W_k^{\text{reg}} \ / \ f(x) \in \text{im}\, A(x)|_{T_x W_k^{\text{reg}}}\} \subseteq W_k^{\text{reg}} \subseteq W_k \tag{3.36}$$

or, in local coordinates,

$$V_{k+1} = \{\zeta \in \Omega_k \ / \ f_k(\zeta) \in \text{im}\, A_k(\zeta)\} \subseteq \Omega_k \subseteq \mathbb{R}^{r_k}, \tag{3.37}$$

which yields a local description of W_{k+1} as $\varphi_1 \circ \cdots \circ \varphi_k(V_{k+1})$.

From the notion of a 0-regular point introduced in Definition 3.2, we may then define recursively a k-regular point as follows.

Definition 3.4. A point $x^* \in W_0$ is said to be k-*regular*, $k \geq 1$, for the DAE (3.13)

a) either if it is $(k-1)$-regular with $r_k > 0$ and $\zeta^* = (\varphi_1 \circ \cdots \circ \varphi_k)^{-1}(x^*)$ is 0-regular for the k-th reduction (3.35);

b) or if it is $(k-1)$-regular with $r_k = 0$.

The set of k-regular points will be denoted by W_{k+1}^{reg}.

If $r_k > 0$, the 0-regularity of ζ^* implies that $\zeta^* \in V_{k+1}$ by (3.37) and then $x^* \in W_{k+1}$, and also that the analogs of R1' and R2' stated for $k = 1$ on page 104 hold; in particular, $A_k(\zeta)$ will have constant rank r_{k+1} around ζ^*, with $r_k \geq r_{k+1} \geq 0$. It is worth mentioning at this point that the rank r_{k+1} of $A_k(\zeta)$ can be checked by construction to equal that of $A(x)|_{T_x W_k^{\text{reg}}}$, with $x = \varphi_1 \circ \cdots \circ \varphi_k(\zeta)$. Under the condition $r_k = 0$ stated in (b) we set $r_{k+1} = 0$ and understand that the corresponding isolated points of W_k^{reg} automatically belong to W_{k+1} (cf. Remark 3.3).

Locally around a k-regular point for which $r_{k+1} > 0$, a reduction pair (P_{k+1}, φ_{k+1}) is well-defined and yields a reduction

$$A_{k+1}(\eta)\eta' = f_{k+1}(\eta), \ \eta \in \Omega_{k+1} \subseteq \mathbb{R}^{r_{k+1}} \qquad (3.38)$$

with

$$A_{k+1}(\eta) = P_{k+1}(\varphi_{k+1}(\eta))A_k(\varphi_{k+1}(\eta))\varphi'_{k+1}(\eta) \qquad (3.39a)$$
$$f_{k+1}(\eta) = P_{k+1}(\varphi_{k+1}(\eta))f_k(\varphi_{k+1}(\eta)). \qquad (3.39b)$$

The above concept of a k-regular point together with this construction supports the following definition.

Definition 3.5. A point $x^* \in W_0$ is called *regular with geometric index* ν, $\nu \geq 1$, for (3.13)

(a) either if it is $(\nu-1)$-regular with $m > r_1 > r_2 > \ldots > r_\nu > 0$, and the matrix $A_\nu(u^*)$ is nonsingular, for some (hence any) reduction sequence $(P_1, \varphi_1), \ldots, (P_\nu, \varphi_\nu)$ satisfying $x^* = \varphi_1 \circ \cdots \circ \varphi_\nu(u^*)$;
(b) or if it is $(\nu-1)$-regular with $m > r_1 > \ldots > r_\nu = 0$.

The set of index ν points will be denoted by $W^{\text{ind}\nu}$.

A point will be said to be *regular* if it is regular with any index ≥ 0; recall Definition 3.1 on page 98 for the index zero case. It is clear from the decreasing rank sequences in this Definition that, in an m-dimensional problem, the index cannot exceed m.

Theorem 3.3 and Proposition 3.4 will show that the above-introduced concepts are consistently defined; namely, they do not depend on the choice of the reduction operators and are invariant with respect to the local equivalence relation introduced in Definition 3.6.

Remark 3.4. Equivalently, stemming from the index one notion in Definition 3.3, x^* can be recursively defined as index $\nu \geq 2$ if it is 0-regular with $m > r_1 > 0$, and $\xi^* = \varphi_1^{-1}(x^*)$ is index $\nu - 1$ for the one-step reduction $A_1(\xi)\xi' = f_1(\xi)$ depicted in (3.26).

Actually, according to Definition 3.5, x^* is index $\nu \geq 2$ for the original DAE (3.13) if and only if for some (hence any) k verifying $1 \leq k \leq \nu - 1$ the point x^* is $(k-1)$-regular with $r_k > 0$ and $\zeta^* = (\varphi_1 \circ \cdots \circ \varphi_k)^{-1}(x^*)$ is index $\nu - k$ for the k-th reduction $A_k(\zeta)\zeta' = f_k(\zeta)$ in (3.35). This reflects the fact that the DAE reduction conveys an index reduction process.

As a cautionary remark, note that we use the symbols ν and r_i for the geometric index and the ranks of the matrices A_i for the quasilinear

3.4. Quasilinear DAEs: The geometric index

DAE (3.13), as we did in the linear time-varying context of Chapter 2 for the tractability index and the ranks of the G_i matrices, respectively. Mind the fact that both constructions are directed to different types of DAEs and therefore these magnitudes cannot be compared in the current setting; the general characterization of their relation, for instance when applying reduction methods to linear DAEs or projector-based methods to quasilinear problems, is an open problem. With this prevention, we may call the ranks r_i of the matrices A_i the (geometric) *characteristic values* of a regular point of the quasilinear DAE (3.13).

It is also worth emphasizing that Remark 3.3 (p. 102) holds, with the obvious modifications, under item (b) of Definition 3.5, which yields isolated index ν points; this will be for instance the case of the origin for the DAE (3.66) considered in 3.4.6.3, which is an index three isolated point. Excluding these situations, the recursive definition of Remark 3.4 can be easily modified to stem from the index zero notion in Definition 3.1; cf. in this direction Remark 3.2.

Solutions of the original DAE (3.13) near a given $(\nu - 1)$-regular point with $r_\nu > 0$ are mapped bijectively into those of the local reduction $A_\nu(u)u' = f_\nu(u)$, as stated in Theorem 3.2 below. In particular, under an index ν assumption an explicit ODE reduction is possible and thereby local unique solvability properties follow from the corresponding theory for explicit ODEs (see e.g. [3, 110, 123, 127]).

In the statement of the following result, U denotes a neighborhood of x^* for which the identity $W_\nu^{\text{reg}} \cap U = \varphi_1 \circ \cdots \circ \varphi_\nu(\Omega_\nu)$ holds; if $\nu = 1$ we may take $U = U_0$, the latter given by Theorem 3.1.

Theorem 3.2. *Assume that $x^* \in W_0$ is a $(\nu - 1)$-regular point for (3.13) with $r_\nu > 0$, $\nu \geq 1$, and let*

$$A_\nu(u)u' = f_\nu(u), \quad u \in \Omega_\nu \subseteq \mathbb{R}^{r_\nu} \tag{3.40}$$

be a ν-th step reduction of (3.13), given by a sequence of reduction pairs $(P_1, \varphi_1), \ldots, (P_\nu, \varphi_\nu)$, on a neighborhood Ω_ν of $u^* = (\varphi_1 \circ \cdots \circ \varphi_\nu)^{-1}(x^*)$. Then $x(t)$ is a solution of (3.13) within U if and only if $x(t) \in W_\nu^{\text{reg}} \stackrel{\text{loc}}{=} W_\nu$ for all t and $u(t) = (\varphi_1 \circ \cdots \circ \varphi_\nu)^{-1}(x(t))$ solves (3.40).

Moreover, if x^* is index ν, then the matrix $A_\nu(u)$ is nonsingular on some neighborhood of u^* within Ω_ν, and on this neighborhood the reduction (3.40) can be rewritten in the explicit form

$$u' = A_\nu^{-1}(u)f_\nu(u). \tag{3.41}$$

The proof of Theorem 3.2 can be derived in a straightforward manner from the repeated application of Theorem 3.1 and is therefore omitted. For the local coincidence of W_ν^{reg} and W_ν the reader is referred to subsection 3.4.4.

Equation (3.41) is a local state space description of the DAE behavior which generalizes the ones given for semiexplicit index one DAEs and Hessenberg index two problems in (3.5) and (3.12), respectively. Of course, different reduction pairs will yield different state space descriptions, although all of them will be equivalent in the sense specified in Theorem 3.4 (p. 116). As we did in Remark 3.1 we also emphasize that, as a byproduct of the reduction procedure, u in the state space equation (3.41) can be taken as a subset of the original problem variables.

Finally, according to the recursive definition of A_k and f_k (see (3.39)), $A_\nu(u)$ is easily shown to read

$$P_\nu(\varphi_\nu(u)) \cdots P_1(\varphi_1 \circ \cdots \circ \varphi_\nu(u)) A(\varphi_1 \circ \cdots \circ \varphi_\nu(u))(\varphi_1 \circ \cdots \circ \varphi_\nu)'(u)$$

and, similarly,

$$f_\nu(u) = P_\nu(\varphi_\nu(u)) \cdots P_1(\varphi_1 \circ \cdots \circ \varphi_\nu(u)) f(\varphi_1 \circ \cdots \circ \varphi_\nu(u)),$$

where it is worth clarifying the meaning of the involved operators; the composition

$$\varphi = \varphi_1 \circ \cdots \circ \varphi_\nu \qquad (3.42)$$

is actually a local parametrization of $W_\nu^{\text{reg}} \stackrel{\text{loc}}{=} W_\nu$, mapping $\Omega_\nu \subseteq \mathbb{R}^{r_\nu}$ onto $W_\nu^{\text{reg}} \cap U = W_\nu \cap U$, whereas the product

$$P_\nu((\varphi_1 \circ \cdots \circ \varphi_{\nu-1})^{-1}(x)) P_{\nu-1}((\varphi_1 \circ \cdots \circ \varphi_{\nu-2})^{-1}(x)) \cdots P_1(x)$$

can be checked to yield an isomorphism from $\operatorname{im} A(x)|_{T_x W_{\nu-1}^{\text{reg}}}$ onto $\mathbb{R}^{r_\nu}$. This makes it clear that the reduction procedure naturally provides a step-by-step construction of the operators depicted in (3.20). Note that here the matrix-valued maps P_k are allowed to be non-constant.

3.4.4 Manifold sequences and locally regular DAEs

The sets of k-regular points and index ν points define smooth manifolds which are interrelated in the way stated below. For notational consistency, write $A_0 = A$, $W_0^{\text{reg}} = W_0$.

3.4. Quasilinear DAEs: The geometric index

Proposition 3.2. *The sets W_k, W_k^{reg} and $W^{\text{ind}\,k}$ introduced in (3.14), (3.36) and Definitions 3.1 - 3.5 verify the following.*

(i) *If nonempty, the set W_k^{reg} is, for $k \geq 1$, a smooth submanifold of W_0.*
(ii) *The set W_k^{reg} is open in $W_k \subseteq W_{k-1}^{\text{reg}}$, for any $k \geq 1$.*
(iii) *If nonempty, $W^{\text{ind}\,k}$ is an open submanifold of W_k^{reg}, for $0 \leq k \leq m$.*

The manifold $W^{\text{ind}\,0}$ is m-dimensional. For $k \geq 1$, the local dimensions of W_k^{reg} (and thereby of $W^{\text{ind}\,k}$ for $1 \leq k \leq m$) are given by the locally constant values r_k of the rank of $A(x)|_{T_x W_{k-1}^{\text{reg}}}$ or, equivalently, of the matrix A_{k-1} coming, if $k > 1$, from any reduction sequence $(P_1, \varphi_1), \ldots, (P_{k-1}, \varphi_{k-1})$.

Proof. For item (i), it is enough to observe that $\varphi_1 \circ \cdots \circ \varphi_k$ yields a local parametrization of W_k^{reg} with a domain Ω_k which is open in $\mathbb{R}^{r_k}$; this means that W_k^{reg} (and eventually $W^{\text{ind}\,k}$, according to (iii)) has a smooth manifold structure with local dimensions given by the ranks r_k.

In turn, (ii) is an easy consequence of Definitions 3.2 and 3.4, whereas (iii) follows also in a simple manner from Definitions 3.1, 3.3 and 3.5. □

In light of this result the manifold sequence (3.18) can be replaced by

$$W_0 \supseteq W_1^{\text{reg}} \supseteq W_2^{\text{reg}} \supseteq \ldots \supseteq W_m^{\text{reg}}. \tag{3.43}$$

The manifolds in this sequence are well-defined for any quasilinear DAE. Note that W_k^{reg} is actually defined for any $k \in \mathbb{N}$; nevertheless, since the DAE (3.13) is defined on an open submanifold of $\mathbb{R}^m$, one can check that $W_k^{\text{reg}} = W_m^{\text{reg}}$ for any $k > m$. Recall that in an m-dimensional problem the index cannot exceed m.

Near an index ν point x^*, the nonsingularity of $A_\nu(u^*)$ (with the notation of Theorem 3.2) implies that $W_\nu^{\text{reg}} \stackrel{\text{loc}}{=} W_{\nu+1} \stackrel{\text{loc}}{=} W_{\nu+1}^{\text{reg}}$. This means that, locally around an index ν point, the above sequence stabilizes as follows:

$$W_0 \supset W_1^{\text{reg}} \supset \ldots \supset W_\nu^{\text{reg}} \stackrel{\text{loc}}{=} W_{\nu+1}^{\text{reg}} \stackrel{\text{loc}}{=} W^{\text{ind}\,\nu}, \tag{3.44}$$

the local dimensions being given by the corresponding rank sequence

$$r_0 > r_1 > r_2 > \ldots > r_\nu = r_{\nu+1}, \tag{3.45}$$

with $r_0 = m$, $r_1 = \text{rk}\,A(x)$ and $r_k = \text{rk}\,A_{k-1}((\varphi_1 \circ \cdots \circ \varphi_{k-1})^{-1}(x))$ if $k \geq 2$.

For later purposes it will useful to include the sets W_k within the sequence (3.43). This yields

$$W_0 \supseteq W_1 \supseteq W_1^{\text{reg}} \supseteq W_2 \supseteq W_2^{\text{reg}} \supseteq \ldots \supseteq W_m \supseteq W_m^{\text{reg}}, \tag{3.46}$$

which locally stabilizes, around index ν points, in the form

$$W_0 \supset W_1 \supseteq W_1^{\text{reg}} \supset \ldots \supset W_\nu \stackrel{\text{loc}}{=} W_\nu^{\text{reg}} \stackrel{\text{loc}}{=} W_{\nu+1} \stackrel{\text{loc}}{=} W_{\nu+1}^{\text{reg}} \stackrel{\text{loc}}{=} W^{\text{ind}\,\nu}. \tag{3.47}$$

3.4.4.1 Regular manifold, solution manifold, and locally regular DAEs

The notion of a regular point following Definition 3.5 supports defining as the *regular manifold* of (3.13) the set
$$W^{\text{reg}} = W^{\text{ind}0} \cup W^{\text{ind}1} \cup \ldots \cup W^{\text{ind}m}, \tag{3.48}$$
which, by the local stabilization property depicted in (3.47), verifies
$$W^{\text{reg}} = \bigcap_{k \geq 1} W_k^{\text{reg}} = W_m^{\text{reg}}. \tag{3.49}$$
The information about the index is however hidden in (3.49) and therefore we prefer describing W^{reg} in the form (3.48). A vector field and a flow are defined by the DAE on the regular manifold, which is well-defined for any quasilinear problem (although it might be empty, as it happens e.g. for the DAE $0 = 1$). We avoid using the term 'solution manifold' in this general context since certain singular points may accommodate smooth solutions, as discussed in Chapter 4.

Additionally, the DAE (3.13) will be said to be *locally regular* if, for $k = 1, \ldots, m$,

(a) $W_k = W_k^{\text{reg}}$; and

(b) W_k is closed in W_{k-1}.

Under item (a), and in particular for locally regular DAEs, we may properly refer to (3.48) as the *solution manifold*. It may be composed of several components with different dimensions and indices. A locally regular DAE for which all regular points have the same index, that is, for which $W^{\text{reg}} = W^{\text{ind}\nu}$, will be said to be *locally regular with geometric index ν*. Note that even in this context the solution manifold might have different local dimensions, since the rank sequence (3.45) need not be the same at different points. We reserve the term *regular DAE with geometric index ν* for locally regular problems displaying the same rank sequence (3.45) (and hence having the same index ν) at all regular points. The term *index ν DAE* will also be used for the latter, e.g. for semiexplicit index one equations.

The meaning of items (a) and (b) will become clearer in Chapter 4, where they will be shown to rule out *inner* and *boundary* singularities, respectively. For the moment, let us mention that they imply in particular that the regular manifold W^{reg} is closed in W_0, and that the requirement depicted in (b) prevents for instance the DAE
$$xx' = 1, \ x \in W_0 = \mathbb{R},$$
from being labeled as regular, since $W_1 = \mathbb{R} - \{0\}$ is not closed in $W_0 = \mathbb{R}$. Note that (a) holds since $W_1 = W_1^{\text{reg}} = \mathbb{R} - \{0\} = W_k = W_k^{\text{reg}}$ for $k \geq 2$.

3.4. Quasilinear DAEs: The geometric index

3.4.4.2 Manifold sequences within different reduction approaches

Disregarding technical differences, which concern e.g. the nature of constant rank conditions or the P_k operators allowed in the reduction, the DAEs accommodated in the framework or Rabier and Rheinboldt (cf. subsection 3.4.1) are those for which

$$W_0 \supset W_1 = W_1^{\text{reg}} \supset \ldots \supset W_\nu = W_\nu^{\text{reg}} \supseteq W_{\nu+1} = W_{\nu+1}^{\text{reg}} = W_{\nu+2}. \quad (3.50)$$

This sequence displays two major differences with (3.46) and (3.47). The first one stems from the global nature of assumptions G1 and G2 (p. 95) in the framework of Rabier and Rheinboldt, which somehow force all points to stabilize at the same reduction step, in contrast to the local approach supporting (3.47). The second difference relies on the identities $W_k = W_k^{\text{reg}}$ for $k = 1, \ldots, \nu$ in (3.50), which express the requirement that, for any k, all points in W_k must be $(k-1)$-regular. This excludes from the approach of [228] singular points affecting the validity of the reduction procedure, which by contrast can be tackled within the framework here introduced; find details in Chapter 4 (Section 4.4).

Note that the reduction method of Rabier and Rheinboldt do include some singularities, namely, impasse points arising in the last reduction step (cf. again Chapter 4): mind the relation $W_\nu^{\text{reg}} \supseteq W_{\nu+1}$ in (3.50). In this regard it is worth clarifying that, in the working setting of Rabier and Rheinboldt, the geometric index notion introduced in Definition 3.5 above would label as regular points with index ν only those belonging to $W_{\nu+1}$, which is open in W_ν and therefore locally coincident with it. The index concept as defined by Rabier and Rheinboldt applies to the whole of W_ν and therefore includes last-step impasse points, but rules out the existence of other singularities, which in practice may well arise in the last step. Our point of view links the index with the existence of local explicit descriptions such as (3.41) and with the solvability results supported on them.

Other approaches, in particular the one proposed by Reich [229], assume that the index conditions hold globally and exclude singularities at all; in the present terms, this working setting would be characterized by a manifold sequence of the form

$$W_0 \supset W_1 = W_1^{\text{reg}} \supset \ldots \supset W_\nu = W_\nu^{\text{reg}} = W_{\nu+1}. \quad (3.51)$$

3.4.5 Local equivalence

Within the reduction procedure described in subsections 3.4.2-3.4.3, we used at several points the reduction pairs (P_k, φ_k) in order to introduce different notions, notably that of the index in Definition 3.5. We prove here that the index concept, as well as other related ones such as that of a k-regular point in Definition 3.4, are actually independent of the specific choice of reduction operators, thereby supporting the "hence any" assertion in Definitions 3.3 and 3.5.

To achieve this goal, we make use of the local equivalence relation for quasilinear DAEs depicted below [67, 229, 247]. This equivalence notion arises in a natural way from the relation between two one-step local reductions of a given DAE; see specifically the second statement in Proposition 3.3. This connection between the equivalence notion and reduction operators will be further explored at the end of this subsection (cf. 3.4.5.3). As shown in 3.4.5.1, the above-mentioned index and k-regularity notions will turn out to be invariant with respect to this equivalence relation, which will be additionally shown in 3.4.5.2 to amount to C^∞-conjugacy for index zero cases, that is, for explicit ODEs. This will clarify the precise relation between different local state space descriptions of the DAE.

Definition 3.6. Two quasilinear DAEs $A(x)x' = f(x)$, $B(y)y' = g(y)$ defined on W_0^a, W_0^b open in $\mathbb{R}^m$, will be said to be C^∞-equivalent locally around x^*, y^* if there exist open neighborhoods $U_b \subseteq W_0^b$ of y^*, $U_a \subseteq W_0^a$ of x^*, a C^∞-diffeomorphism $\phi : U_b \to U_a$ with $\phi(y^*) = x^*$, and a C^∞ nonsingular matrix-valued mapping $E : U_b \to \mathbb{R}^{m \times m}$, such that

$$B(y) = E(y)A(\phi(y))\phi'(y) \tag{3.52a}$$
$$g(y) = E(y)f(\phi(y)) \tag{3.52b}$$

for all $y \in U_b$.

It is left as an exercise for the reader to check that this is actually an equivalence relation. Equation (3.52b) yields a *contact equivalence* between f and g [100, 200]. Note that, for the equivalence of the quasilinear systems, the pair (E, ϕ) is required to link additionally A and B through (3.52a).

3.4.5.1 *The index: Independence of reduction pairs and invariance*

Theorem 3.3. *The notion of an index ν point introduced in Definition 3.5 does not depend on the choice of the reduction pairs $(P_1, \varphi_1), \ldots, (P_\nu, \varphi_\nu)$.*

Moreover, this notion is invariant with respect to the local equivalence relation of Definition 3.6; namely, x^* is regular with index ν for (A, f) if and only if $y^* = \phi^{-1}(x^*)$ is so for the locally equivalent DAE (B, g).

3.4. Quasilinear DAEs: The geometric index

The proof of this result will be derived from Lemma 3.2 and Propositions 3.3 and 3.4.

In Lemma 3.2 below we denote $W_1^a = \{x \in W_0^a \ / \ f(x) \in \operatorname{im} A(x)\}$ and $W_1^b = \{y \in W_0^b \ / \ g(y) \in \operatorname{im} B(y)\}$ whereas, similarly, $W_1^{a\,\text{reg}}$ and $W_1^{b\,\text{reg}}$ stand for 0-regular points of (A, f) and (B, g), respectively. We additionally set $F_a(x, p) = A(x)p - f(x)$ and $F_b(y, q) = B(y)q - g(y)$.

Lemma 3.2. *Let $A(x)x' = f(x)$, $B(y)y' = g(y)$ be C^∞-equivalent locally around x^*, y^* via the relations depicted in Definition 3.6. Then:*

(i) $W_1^a \cap U_a = \phi(W_1^b \cap U_b)$.
(ii) $\operatorname{rk} B(y) = \operatorname{rk} A(\phi(y))$ *for all* $y \in U_b$.
(iii) *If A, B have constant rank r_1 around $x^* \in W_1^a$, $y^* \in W_1^b$, respectively, and p^*, q^* are such that $A(x^*)p^* - f(x^*) = 0$, $B(y^*)q^* - g(y^*) = 0$, then F_a is a submersion at (x^*, p^*) if and only if so it F_b at (y^*, q^*).*
(iv) $W_1^{a\,\text{reg}} \cap U_a = \phi(W_1^{b\,\text{reg}} \cap U_b)$.

Proof. For item (i), it is enough to prove that $W_1^a \cap U_a \supseteq \phi(W_1^b \cap U_b)$ since the reciprocal inclusion holds automatically working with ϕ^{-1}. Let $x = \phi(y) \in \phi(W_1^b \cap U_b)$. Obviously $x \in U_a$ and we only need to check that $x \in W_1^a$, i.e., that $f(x) \in \operatorname{im} A(x)$. But $y \in W_1^b$ means $g(y) \in \operatorname{im} B(y)$ or, equivalently, $E(y)f(\phi(y)) \in \operatorname{im} E(y)A(\phi(y))\phi'(y)$. Since $E(y)$ and $\phi'(y)$ are isomorphisms, this yields $f(\phi(y)) \in \operatorname{im} A(\phi(y))$, that is, $f(x) \in \operatorname{im} A(x)$.

Item (ii) follows immediately from (3.52a).

For item (iii), the proof of the cases $r_1 = 0$ and $r_1 = m$ is straightforward. If $0 < r_1 < m$ we use the characterization given in Lemma 3.1. Let $H_a(x)$ satisfy $H_a(x)A(x) = 0$ is some neighborhood of x^*, with $\operatorname{rk} H_a(x^*) = \operatorname{rk}(H_a f)'(x^*) = m - r_1$, and define

$$H_b(y) = H_a(\phi(y))E^{-1}(y).$$

We have $H_b(y)B(y) = H_a(\phi(y))E^{-1}(y)E(y)A(\phi(y))\phi'(y) = 0$ due to the identity $H_a(\phi(y))A(\phi(y)) = 0$. Moreover, $\operatorname{rk} H_b(y^*) = \operatorname{rk} H_a(x^*) = m - r_1$. Finally, $H_b(y)g(y) = H_a(\phi(y))E^{-1}(y)E(y)f(\phi(y)) = H_a(\phi(y))f(\phi(y))$, so that

$$(H_b g)'(y^*) = (H_a f)'(\phi(y^*))\phi'(y^*)$$

and therefore $\operatorname{rk}(H_b g)'(y^*) = \operatorname{rk}(H_a f)'(\phi(y^*)) = \operatorname{rk}(H_a f)'(x^*) = m - r_1$.

Finally, (iv) is a straightforward consequence of (i), (ii) and (iii). □

In the next result, by a local reduction we mean the one-step reduction of the form (3.26)-(3.27) arising in Theorem 3.1. The particular case $r_1 = 0$ renders the statement trivial.

Proposition 3.3. *Any two local reductions of locally equivalent quasilinear DAEs around 0-regular points are locally equivalent.*

In particular, any two local reductions of a quasilinear DAE around a 0-regular point are locally equivalent.

Proof. Assume that x^* and y^* are 0-regular points of $A(x)x' = f(x)$ and $B(y)y' = g(y)$, and that these DAEs are C^∞-equivalent locally around x^*, y^* via the relations depicted in Definition 3.6. Let two local reductions of (A, f) and (B, g) be defined by $P_1(x)$, $x = \varphi_1(\xi)$ and $Q_1(y)$, $y = \psi_1(\varrho)$, respectively, yielding $A_1(\xi)\xi' = f_1(\xi)$ and $B_1(\varrho)\varrho' = g_1(\varrho)$. The neighborhoods U_a, U_b arising in Definition 3.6, as well as the parametrization domains Ω_a, Ω_b of φ_1 and ψ_1, are assumed to be shrunk as to guarantee that φ_1 and ψ_1 map Ω_a and Ω_b onto $W_1^a \cap U_a = W_1^{a\,\text{reg}} \cap U_a$ and $W_1^b \cap U_b = W_1^{b\,\text{reg}} \cap U_b$, respectively, and also to meet the below-depicted requirement on $\hat{P}_1(x)$.

We then need to show that there exist a local diffeomorphism $\xi = \phi_1(\varrho)$ and a nonsingular matrix-valued mapping $E_1(\varrho)$ yielding an equivalence between $A_1(\xi)\xi' = f_1(\xi)$ and $B_1(\varrho)\varrho' = g_1(\varrho)$. In order to define ϕ_1, note that $y = \psi_1(\varrho)$ is onto $W_1^b \cap U_b$ and, as shown in item (i) of Lemma 3.2, ϕ maps $W_1^b \cap U_b$ onto $W_1^a \cap U_a$. Since φ_1 is onto $W_1^a \cap U_a$ we may write $\varphi_1^{-1} : W_1^a \cap U_a \to \mathbb{R}^{r_1}$, and then the composition

$$\phi_1 = \varphi_1^{-1} \circ \phi \circ \psi_1 \qquad (3.53)$$

is well-defined and yields a local diffeomorphism of Ω_b onto Ω_a. For later use remark that $\varphi_1 \circ \phi_1 = \phi \circ \psi_1$.

For the definition of $E_1(\varrho)$ we use the fact that the local matrix-valued map $P_1(x)$ yields an isomorphism $P_1(x)|_{\text{im } A(x)} : \text{im } A(x) \to \mathbb{R}^{r_1}$ and, provided that U_a is small enough, there will exist a matrix mapping $\hat{P}_1(x) \in C^\infty(U_a, \mathbb{R}^{m \times r_1})$ such that $\hat{P}_1(x)P_1(x)|_{\text{im } A(x)} = \text{id}|_{\text{im } A(x)}$. Define then

$$E_1(\varrho) = Q_1(\psi_1(\varrho))E(\psi_1(\varrho))\hat{P}_1(\phi(\psi_1(\varrho))). \qquad (3.54)$$

From (3.53) and (3.54), we need to check that $B_1(\varrho) = E_1(\varrho)A_1(\phi_1(\varrho))\phi_1'(\varrho)$ and that $g_1(\varrho) = E_1(\varrho)f_1(\phi_1(\varrho))$.

The product $E_1(\varrho)A_1(\phi_1(\varrho))\phi_1'(\varrho)$ reads

$$Q_1(\psi_1(\varrho))E(\psi_1(\varrho))\hat{P}_1(\phi(\psi_1(\varrho)))P_1(\varphi_1(\phi_1(\varrho)))A(\varphi_1(\phi_1(\varrho)))\varphi_1'(\phi_1(\varrho))\phi_1'(\varrho).$$

By means of $\varphi_1(\phi_1(\varrho)) = \phi(\psi_1(\varrho))$ and $\hat{P}_1(x)P_1(x)|_{\text{im } A(x)} = \text{id}|_{\text{im } A(x)}$, we may simplify $\hat{P}_1(\phi(\psi_1(\varrho)))P_1(\varphi_1(\phi_1(\varrho)))A(\varphi_1(\phi_1(\varrho))) = A(\phi(\psi_1(\varrho)))$. Note additionally that $\varphi_1'(\phi_1(\varrho))\phi_1'(\varrho) = \phi'(\psi_1(\varrho))\psi_1'(\varrho)$. Using these properties, the product $E_1(\varrho)A_1(\phi_1(\varrho))\phi_1'(\varrho)$ amounts to

$$Q_1(\psi_1(\varrho))E(\psi_1(\varrho))A(\phi(\psi_1(\varrho)))\phi'(\psi_1(\varrho))\psi_1'(\varrho). \tag{3.55}$$

Now $E(\psi_1(\varrho))A(\phi(\psi_1(\varrho)))\phi'(\psi_1(\varrho)) = B(\psi_1(\varrho))$ and then the expression displayed in (3.55) reads $Q_1(\psi_1(\varrho))B(\psi_1(\varrho))\psi_1'(\varrho) = B_1(\varrho)$, so that $E_1(\varrho)A_1(\phi_1(\varrho))\phi_1'(\varrho) = B_1(\varrho)$, as we aimed to show.

Finally, write

$$E_1(\varrho)f_1(\phi_1(\varrho)) = Q_1(\psi_1(\varrho))E(\psi_1(\varrho))\hat{P}_1(\phi(\psi_1(\varrho)))P_1(\varphi_1(\phi_1(\varrho)))f(\varphi_1(\phi_1(\varrho))).$$

Since the composition $\phi \circ \psi_1$ maps onto $W_1^a \cap U_a$ we have $\varphi_1(\phi_1(\varrho)) = \phi(\psi_1(\varrho)) \in W_1^a$ and therefore $f(\varphi_1(\phi_1(\varrho))) \in \text{im}\, A(\varphi_1(\phi_1(\varrho)))$. The product $\hat{P}_1(\phi(\psi_1(\varrho)))P_1(\varphi_1(\phi_1(\varrho)))f(\varphi_1(\phi_1(\varrho)))$ then amounts to $f(\phi(\psi_1(\varrho)))$ and the expression depicted above for $E_1(\varrho)f_1(\phi_1(\varrho))$ reads

$$Q_1(\psi_1(\varrho))E(\psi_1(\varrho))f(\phi(\psi_1(\varrho))) = Q_1(\psi_1(\varrho))g(\psi_1(\varrho)) = g_1(\varrho),$$

which shows that indeed $E_1(\varrho)f_1(\phi_1(\varrho)) = g_1(\varrho)$, thus completing the proof. □

Proposition 3.4. *The notion of a k-regular point introduced in Definition 3.4 does not depend on the choice of the sequence of reduction pairs $(P_1, \varphi_1), \ldots, (P_k, \varphi_k)$.*

Additionally, this notion is invariant with respect to the local equivalence relation of Definition 3.6; namely, x^* is k-regular for $A(x)x' = f(x)$ if and only if $y^* = \phi^{-1}(x^*)$ is so for the locally equivalent DAE $B(y)y' = g(y)$.

Proof. Note that there is no reduction involved in the notion of a 0-regular point (cf. Definition 3.2 on page 98), whereas item (iv) of Lemma 3.2 explicitly shows the invariance of 0-regular points with respect to local equivalence. A straightforward inductive reasoning proves, for $k \geq 1$, that k-regular points do not depend on the choice of reduction operators and are invariant, using Proposition 3.3 and item (iv) of Lemma 3.2, or item (ii) in the particular case in which a vanishing rank r_k is met. □

Proposition 3.4 expresses that the sets W_k^{reg} are independent of the choice of reduction operators and invariant with respect to local equivalence. Thereby, from (3.36) it follows that the sets W_k are themselves independent of reduction pairs; these sets are also invariant with respect to local equivalence, following item (i) of Lemma 3.2.

Proof of Theorem 3.3. Recall from Definition 3.5 that a given point x^* is regular with index ν if it is $(\nu - 1)$-regular and $A_\nu(u^*)$ is nonsingular. Proposition 3.4 shows that being $(\nu - 1)$-regular is independent of the reduction sequence and invariant with respect to local equivalence. Additionally, by Proposition 3.3 any two ν-th step reductions of a given DAE, or of two locally equivalent DAEs, are locally equivalent: the result then follows from item (ii) of Lemma 3.2. □

In an analogous way, the reader can easily check that the ranks in (3.45) do not depend on the choice of reduction pairs, and also that they are invariant with respect to local equivalence.

3.4.5.2 C^∞-conjugacy of state space descriptions

Finally, Theorem 3.4 below shows that any two local state space reductions of the form (3.41) are locally equivalent in a strong sense. The key idea is that the local equivalence of Definition 3.6 amounts to a local conjugacy of explicit ODEs when the leading matrices of the quasilinear DAEs are nonsingular. Two explicit ODEs $u' = h_1(u)$ and $v' = h_2(v)$ are said to be C^∞-*conjugate* if there exists a C^∞-diffeomorphism ϕ which carries integral curves of h_2 onto integral curves of h_1 preserving parametrization by time. The conjugacy notion can be reformulated locally in the obvious way, and may be expressed via the relation $\phi'(v)h_2(v) = h_1(\phi(v))$. Note that, in the context of explicit ODEs, the weaker term *equivalence* usually allows for time-reparametrization of trajectories [3, 110].

Theorem 3.4. *Any two local state space descriptions (3.41) of the DAE (3.13) around an index ν point with $r_\nu > 0$ are locally C^∞-conjugate.*

Proof. Using Proposition 3.3 it is straightforward to check that any two ν-th step local reductions, say $A_\nu(u)u' = f_\nu(u)$ and $\tilde{A}_\nu(v)v' = \tilde{f}_\nu(v)$, are locally equivalent in the sense depicted in Definition 3.6; that is, there will exist a local diffeomorphism ϕ_ν and a nonsingular matrix-valued mapping E_ν such that

$$\tilde{A}_\nu(v) = E_\nu(v)A_\nu(\phi_\nu(v))\phi'_\nu(v). \tag{3.56a}$$
$$\tilde{f}(v) = E_\nu(v)f_\nu(\phi_\nu(v)). \tag{3.56b}$$

Due to the index ν notion, by shrinking domains if necessary the matrices $A_\nu(u)$ and $\tilde{A}_\nu(v)$ can be assumed to be nonsingular. Therefore

$$E_\nu(v) = \tilde{A}_\nu(v)(A_\nu(\phi_\nu(v))\phi'_\nu(v))^{-1}$$

and then

$$\tilde{f}(v) = \tilde{A}_\nu(v)(A_\nu(\phi_\nu(v))\phi'_\nu(v))^{-1}f_\nu(\phi_\nu(v)),$$

3.4. Quasilinear DAEs: The geometric index

which obviously can be rewritten as

$$\phi'_\nu(v)\tilde{A}_\nu^{-1}(v)\tilde{f}(v) = A_\nu^{-1}(\phi_\nu(v))f_\nu(\phi_\nu(v)). \tag{3.57}$$

Equation (3.57) expresses that the diffeomorphism ϕ_ν carries solutions of $v' = \tilde{A}_\nu^{-1}(v)\tilde{f}(v)$ onto those of $u' = A_\nu^{-1}(u)f_\nu(u)$, thereby proving that both reductions are locally C^∞-conjugate. □

This result is of course consistent with the fact that state space descriptions of the form (3.41) are local coordinate representations of the differential equation defined by a vector field on the regular manifold. Actually, the strong form of equivalence depicted in Theorem 3.4 somehow reflects that the reduction procedure near regular points preserves all the local dynamic information of the DAE. Some implications of this will be examined in Section 3.5.

Remark 3.5. The reader should be aware of the fact that the equivalence relation of Definition 3.6 requires more than the existence of a local mapping between solutions of both DAEs, and therefore should not be thought as a mere mapping of phase portraits. Note that, even for regular DAEs, the diffeomorphism ϕ is not restricted to the regular manifolds. The invariance of the index proved in Theorem 3.3 suggests that this equivalence, which generalizes the linear one given for matrix pencils in subsection 2.1.1, preserves important structural properties of the DAE. Weaker equivalence notions for DAEs, reflecting only their trajectory behavior (either in a topological or differentiable sense, and maybe allowing for time reparametrization and reversion), may be of interest from a dynamical perspective. In this regard, it is worth noting that DAEs with different indices may depict the same solution behavior, as it is the case in particular for the local reductions at different steps of a given DAE. For instance, the index two system

$$x' = f(x), \ y' = z, \ 0 = y,$$

behaves as the index one problem

$$x' = f(x), \ 0 = z, \ 0 = y,$$

coming from a one-step reduction of the former. Actually the identity map carries the phase portraits of both systems into one another.

3.4.5.3 On the link between local equivalence and reduction operators

It was already mentioned after Theorem 3.1 that allowing for a non-constant operator $P_1(x)$ (vs. the constant P_1 used in the framework of Rabier and

Rheinboldt) may simplify the reduced equations. Nevertheless, the form of this reduction operator admits further consideration, in connection to the local equivalence concept introduced in Definition 3.6. A brief digression in this regard follows, mainly aimed at motivating future study.

From Proposition 3.3 we know that two one-step reductions of locally equivalent DAEs are locally equivalent. The same result holds if we use in particular constant operators P_1 in the reduction. But consider now the following related problem. Assume we are given two locally equivalent quasilinear DAEs (A, f), (B, g), a one-step reduction (A_1, f_1) of the former around a 0-regular point, and (B_1, g_1) which is locally equivalent to (A_1, f_1). Is (B_1, g_1) a one-step local reduction of (B, g)? Note that the notation for (B_1, g_1) is supported on the forthcoming positive answer to this question.

This can be indeed answered in the affirmative, the construction proceeding in parallel to the proof of Proposition 3.3. For the sake of brevity we leave details to the reader: we only sketch that now ψ_1 is constructed as $\phi^{-1} \circ \varphi_1 \circ \phi_1$, whereas $Q_1(y)$ is given by $\tilde{E}_1(y) P_1(\phi(y)) E^{-1}(y)$, $\tilde{E}_1(y)$ being any smooth extension to the whole of U_b of $E_1(\psi_1^{-1}(y))$, which is defined only for $y \in W_1^b \cap U_b$.

The reasoning outlined above is not valid for reductions based on a constant operator P_1, since Q_1 may not turn out to be independent of y. This suggests that there seems to be a natural correspondence between equivalence notions and allowable reduction operators for quasilinear DAEs. In this direction, the reduction technique with constant P_1 proposed by Rabier and Rheinboldt seems to be linked with a *semilinear* equivalence relation for quasilinear DAEs, defined by the requirement that E is constant in Definition 3.6 (cf. [247]).

3.4.6 *Examples*

We illustrate below the ideas introduced in subsections 3.4.2-3.4.5 by means of some examples. In 3.4.6.1 and 3.4.6.2 we recast the results of Sections 3.1 and 3.2 for semiexplicit index one and Hessenberg DAEs in terms of the reduction approach here introduced; nonautonomous analogs of these problems will be considered in 3.4.7. Mind that we frame the discussion in the C^∞ setting just for simplicity, since the smoothness requirements can be weakened along the lines indicated in Sections 3.1 and 3.2. The example in 3.4.6.3 illustrates how a locally regular DAE may display different indices. The reduction approach will be applied to the analysis of electrical circuits in Chapter 6, where additional examples can be found.

3.4.6.1 Semiexplicit index one DAEs

Let us drive our attention back to the autonomous semiexplicit DAE (3.1), namely,

$$y' = h(y, z) \qquad (3.58a)$$
$$0 = g(y, z), \qquad (3.58b)$$

with $h \in C^\infty(W_0, \mathbb{R}^r)$, $g \in C^\infty(W_0, \mathbb{R}^p)$, and W_0 open in $\mathbb{R}^{r+p}$.

The set W_1 is explicitly defined by the condition $g = 0$ in (3.58b). A point $(y, z) \in W_0$ is easily checked to be 0-regular if $g(y, z) = 0$ and

$$\operatorname{rk} g'(y, z) = p, \qquad (3.59)$$

that is, if g is a submersion at (y, z). Therefore, the set W_1^{reg} is defined by the conditions

$$g(y, z) = 0, \ \operatorname{rk} g'(y, z) = p.$$

Note that W_1^{reg} is an r-dimensional smooth manifold but W_1 need not be so, since it may obviously include points where (3.59) does not hold.

Setting $P_1 = (I_r \ 0)$, $H = (0 \ I_p)$, Proposition 3.1 states the index one condition on (y, z) as the nonsingularity of

$$\begin{pmatrix} I_r & 0 \\ g_y(y, z) & g_z(y, z) \end{pmatrix},$$

which obviously amounts to the nonsingularity of $g_z(y, z)$. This yields the result anticipated in Section 3.1 and displayed in Proposition 3.5 below. A different derivation of the corresponding result for nonautonomous problems can be found in 3.4.7.2.

Proposition 3.5. *The point (y, z) is regular with geometric index one for the semiexplicit DAE (3.58) if and only if $g(y, z) = 0$ and $g_z(y, z)$ is nonsingular.*

If the nonsingularity of $g_z(y, z)$ holds whenever $g(y, z) = 0$, then (3.58) is a regular DAE with geometric index one.

Note that the nonsingularity of $g_z(y, z)$ comprises in particular the condition (3.59). The set W^{ind1} is defined by the conditions $g(y, z) = 0$ and $\det g_z(z) \neq 0$, being open in W_1 and an open submanifold of W_1^{reg}.

Locally around any index one point, via the implicit function theorem we may describe the set $g = 0$ as $z = \psi(y)$ for some locally defined map ψ. The local parametrization $\varphi_1(y) = (y, \psi(y))$ yields the reduction (3.26) in the form $y' = h(y, \psi(y))$ depicted in (3.5).

3.4.6.2 Hessenberg DAEs

Consider again the autonomous Hessenberg DAE of size two (3.6)

$$y' = h(y, z) \qquad (3.60a)$$
$$0 = g(y), \qquad (3.60b)$$

where $h \in C^\infty(W_0, \mathbb{R}^r)$, $g \in C^\infty(\hat{W}_0, \mathbb{R}^p)$; the set W_0 is open in $\mathbb{R}^{r+p}$, and $\hat{W}_0$ is open in $\mathbb{R}^r$ and contains the y-projection of W_0.

As in the general semiexplicit case considered in (3.58), the set W_1 reads $\{(y, z) \in W_0 \ / \ g(y) = 0\}$. A point (y, z) is 0-regular if $g(y) = 0$ and the submersion condition (3.59) holds, now reading

$$\operatorname{rk} g_y(y) = p. \qquad (3.61)$$

It is clear from Proposition 3.5 that a 0-regular point (y, z) cannot be index one for (3.60). From (3.61) there must exist a nonsingular, $p \times p$ submatrix g_w of g_y and then, by the implicit function theorem, (3.60b) is equivalent to writing locally the p variables w from within y in terms of the remaining $r - p$ ones of y (to be denoted by u) as $w = \psi(u)$. This yields a one-step reduction of (3.60) of the form

$$u' = h_1((u, \psi(u)), z) \qquad (3.62a)$$
$$\psi'(u)u' = h_2((u, \psi(u)), z) \qquad (3.62b)$$

where, as in Section 3.2, we have assumed w.l.o.g. that u are the first $r - p$ variables of y.

We now make use of the second statement in Theorem 3.3, which guarantees that the index is not affected by premultiplication of nonsingular matrix mappings. Rewrite (3.62) as

$$\begin{pmatrix} I_{r-p} & 0 \\ \psi'(u) & 0 \end{pmatrix} \begin{pmatrix} u' \\ z' \end{pmatrix} = \begin{pmatrix} h_1((u, \psi(u)), z) \\ h_2((u, \psi(u)), z) \end{pmatrix} \qquad (3.63)$$

and premultiply (3.63) by

$$\begin{pmatrix} I_{r-p} & 0 \\ g_u(u, \psi(u)) & g_w(u, \psi(u)) \end{pmatrix}.$$

To avoid a cumbersome notation we have written $g_u(u, \psi(u))$ instead of $g_u((u, \psi(u)))$; the same holds for g_w and (later on) for g_y. We then obtain, via the relation $g_u(u, \psi(u)) + g_w(u, \psi(u))\psi'(u) = 0$ following from the implicit function theorem,

$$u' = h_1((u, \psi(u)), z)$$
$$0 = g_u(u, \psi(u))h_1((u, \psi(u)), z) + g_w(u, \psi(u))h_2((u, \psi(u)), z),$$

3.4. Quasilinear DAEs: The geometric index

that is,

$$u' = h_1((u, \psi(u)), z) \tag{3.64a}$$
$$0 = g_y(u, \psi(u))h((u, \psi(u)), z). \tag{3.64b}$$

Note that (3.64b) describes the set V_2 (cf. (3.30)), which is given by the hidden constraint $g_y h = 0$ in (u, z)-coordinates; this means that, within the original domain W_0, the set W_2 is defined by the two conditions $g(y) = 0$, $g_y(y)h(y, z) = 0$.

Since (3.64) is a semiexplicit DAE, we can use again Proposition 3.5 to conclude that an index one point of (3.64) is characterized by the nonsingularity of $g_y(u, \psi(u))h_z((u, \psi(u)), z)$. We have thus proved the following.

Proposition 3.6. *The point (y, z) is regular with geometric index two for the Hessenberg DAE of size two (3.60) if and only if the identities $g(y) = 0$, $g_y(y)h(y, z) = 0$ hold and $g_y(y)h_z(y, z)$ is nonsingular.*

If the nonsingularity of $g_y(y)h_z(y, z)$ holds whenever the conditions $g(y) = 0$ and $g_y(y)h(y, z) = 0$ are met, then (3.60) is a regular DAE with geometric index two.

Note that around an index two point the above-mentioned nonsingularity of the matrix $g_y(u, \psi(u))h_z((u, \psi(u)), z)$ allows for a local description of $g_y(u, \psi(u))h((u, \psi(u)), z) = 0$ as $z = \zeta(u)$, thereby yielding the local state equation

$$u' = h_1((u, \psi(u)), \zeta(u)), \tag{3.65}$$

which is nothing but (3.12) with $\zeta(u) = \eta(u, \psi(u))$.

Analogous results hold for Hessenberg DAEs of higher size, as indicated in Section 3.2; in particular, the Hessenberg DAE (1.18) is regular with geometric index k if the matrix in (1.19) is nonsingular on the manifold W_k. Details in this regard are left to the reader.

3.4.6.3 A locally regular DAE with different indices

Consider now the Hessenberg DAE

$$x' = z \tag{3.66a}$$
$$y' = x + yz \tag{3.66b}$$
$$0 = y(y - 1) \tag{3.66c}$$

with $(x, y, z) \in \mathbb{R}^3$. The set W_1 is explicitly defined by (3.66c) and is thereby composed of the planes $y = 0$ and $y = 1$. It is straightforward to check that all points within these planes are 0-regular.

On $y = 0$, the DAE admits the parametrization $(x, y, z) = \varphi_1(\xi_1, \xi_2)$ defined by $x = \xi_1$, $y = 0$, $z = \xi_2$. We have

$$\varphi_1'(\xi_1, \xi_2) = \begin{pmatrix} 1 & 0 \\ 0 & 0 \\ 0 & 1 \end{pmatrix}$$

and with $P_1 = \begin{pmatrix} 1 & 0 & 0 \\ 0 & 1 & 0 \end{pmatrix}$ we arrive at

$$A_1 = \begin{pmatrix} 1 & 0 & 0 \\ 0 & 1 & 0 \end{pmatrix} \begin{pmatrix} 1 & 0 & 0 \\ 0 & 1 & 0 \\ 0 & 0 & 0 \end{pmatrix} \begin{pmatrix} 1 & 0 \\ 0 & 0 \\ 0 & 1 \end{pmatrix} = \begin{pmatrix} 1 & 0 \\ 0 & 0 \end{pmatrix}$$

and $f_1(\varphi_1(\xi_1, \xi_2)) = (\xi_2, \xi_1, 0)$. This yields the reduction

$$\xi_1' = \xi_2 \tag{3.67a}$$
$$0 = \xi_1. \tag{3.67b}$$

Although for notational consistency we have detailed here the reduction procedure in terms of $\varphi_1(\xi_1, \xi_2)$ and P_1, in practice it is simpler to retain the original name of the variables, the reduction often following from simple inspection of the DAE. Indeed, (3.67) can be rephrased as

$$x' = z \tag{3.68a}$$
$$0 = x, \tag{3.68b}$$

which, according to 3.4.6.2, is a regular Hessenberg DAE with index two. The hidden constraint reads $z = 0$ and the regular manifold of (3.68) amounts to $(x, z) = (0, 0)$. It then follows that the origin $x = y = z = 0$, which belongs to W_3, is regular with index three for the original DAE (3.66).

On the other hand, on $y = 1$ the DAE (3.66) admits the reduction

$$x' = z \tag{3.69a}$$
$$0 = x + z \tag{3.69b}$$

which is an index one problem with reduced equation $x' = -x$. Therefore, points within the line $y = 1$, $x + z = 0$, belonging to W_2, are regular with index two for the DAE (3.66). Together with the result above, this implies that for this system the regular manifold

$$W^{\mathrm{reg}} = W^{\mathrm{ind}\,2} \cup W^{\mathrm{ind}\,3} = \{(x, y, z) \in \mathbb{R}^3 \;/\; y = 1, x + z = 0\} \cup \{(0, 0, 0)\}$$

is composed of a one-dimensional, index two component and an isolated index three (equilibrium) point. The DAE is a locally regular one and hence the regular manifold may also be properly labeled as the solution manifold.

3.4.7 Nonautonomous problems

We extend below the ideas introduced in subsections 3.4.2 and 3.4.3 to nonautonomous quasilinear DAEs, that is, to problems of the form

$$A(x,t)x' = f(x,t), \qquad (3.70)$$

with $A \in C^\infty(W_0, \mathbb{R}^{m \times m})$ and $f \in C^\infty(W_0, \mathbb{R}^m)$; now W_0 is open in $\mathbb{R}^{m+1}$. We work for simplicity in the C^∞ context but, as in the autonomous setting, this smoothness requirement can be relaxed. The ideas follow closely those of 3.4.2 and 3.4.3 and, for the sake of brevity, we present a summarized discussion in which many details are left to the reader. Note also that the chance to parallelize the autonomous case provides a more efficient way to handle nonautonomous problems than the one resulting from considering t as an additional autonomous variable via $t' = 1$, as it is proposed in [228].

3.4.7.1 Geometric index and reduction in the nonautonomous context

We will say that a point $(x^*, t^*) \in W_0$ is *regular with geometric index zero* or, simply, an *index zero point* for (3.70) if $A(x^*, t^*)$ is a nonsingular matrix.

In order to analyze the behavior at points where this condition is not met, let us define the set W_1 for the nonautonomous DAE (3.70) as

$$W_1 = \{(x,t) \in W_0 \ / \ f(x,t) \in \operatorname{im} A(x,t)\},$$

and the mapping F as $F(t, x, p) = A(x,t)p - f(x,t)$. Inspired in conditions R1 and R2 in Definition 3.2 (page 98), assume that $A(x,t)$ has constant rank r_1 on some neighborhood of $(x^*, t^*) \in W_1$, and that

$$\operatorname{rk} \left(\frac{\partial F}{\partial x} \ \frac{\partial F}{\partial p} \right)\bigg|_{(t^*, x^*, p^*)} = m \qquad (3.71)$$

for some p^* satisfying $A(x^*, t^*)p^* = f(x^*, t^*)$.

Following the ideas introduced in Theorem 3.1, one can show that there exists an open neighborhood $U_0 \subseteq W_0$ of (x^*, t^*) such that $W_1 \cap U_0$ admits an $(r_1 + 1)$-dimensional parametrization $\Psi_1 : \Omega_1 \to W_1 \cap U_0$ of the form $(x,t) = \Psi_1(\xi, t) = (\varphi_1(\xi, t), t)$, and that there exists a C^∞ matrix-valued mapping $P_1 : U_0 \to \mathbb{R}^{r_1 \times m}$ such that $P_1(x,t)|_{\operatorname{im} A(x,t)}$ is an isomorphism $\operatorname{im} A(x,t) \to \mathbb{R}^{r_1}$ for all $(x,t) \in U_0$. Denote by $\Psi_1^{-1}(x,t) = (\theta(x,t), t)$ the inverse map from $W_1 \cap U_0$ onto Ω_1.

For any such φ_1, P_1, $x(t)$ is a solution of (3.13) within U_0 if and only if $(x(t), t) \in W_1$ for all t and $\xi(t) = \theta(x(t), t)$ solves the one-step reduction

$$A_1(\xi, t)\xi' = f_1(\xi, t), \qquad (3.72)$$

with

$$A_1(\xi,t) = P_1(\varphi_1(\xi,t),t)A(\varphi_1(\xi,t),t)\varphi_{1\xi}(\xi,t), \tag{3.73a}$$
$$f_1(\xi,t) = P_1(\varphi_1(\xi,t),t)[f(\varphi_1(\xi,t),t) - A(\varphi_1(\xi,t),t)\varphi_{1t}(\xi,t)]. \tag{3.73b}$$

Setting $\xi^* = \theta(x^*,t^*)$, the requirement that $A_1(\xi^*,t^*)$ is a nonsingular matrix or, equivalently, that $m = r_1$, defines (x^*,t^*) as a *regular point with geometric index one* (or *index one point*).

If this is not the case, iterating the procedure as in subsection 3.4.3 we derive the notion of a *regular point with geometric index ν* (or, again, *index ν point*), which is characterized by a local rank sequence of the form

$$m > r_1 > r_2 > \ldots > r_\nu = r_{\nu+1}.$$

A local solvability result analogous to Theorem 3.2 follows. The cases in which a given rank r_i vanishes can be accommodated as in subsections 3.4.2 and 3.4.3; for instance, around an index one point (x^*,t^*) with $r_1 = 0$ the DAE amounts to an algebraic equation of the form $f(x,t) = 0$, which locally around (x^*,t^*) defines implicitly a curve $x(t)$ because of (3.71).

Additionally, proceeding as in subsection 3.4.5, one can show that this index notion is independent of reduction operators, and that it is invariant with respect to the following equivalence notion: equation (3.70) will be said to be C^∞-equivalent to

$$B(y,t)y' = g(y,t), \; (y,t) \in \tilde{W}_0 \tag{3.74}$$

locally around (x^*,t^*), (y^*,t^*) if there exist open neighborhoods $\tilde{U} \subseteq \tilde{W}_0$ of (y^*,t^*) and $U \subseteq W_0$ of (x^*,t^*), a C^∞-diffeomorphism $\Phi : \tilde{U} \to U$ of the form $\Phi(y,t) = (\phi(y,t),t)$ with $\phi(y^*,t^*) = x^*$, and a C^∞ nonsingular matrix-valued mapping $E : \tilde{U} \to \mathbb{R}^{m \times m}$, such that, for all $(y,t) \in \tilde{U}$,

$$B(y,t) = E(y,t)A(\phi(y,t),t)\phi_y(y,t). \tag{3.75a}$$
$$g(y,t) = E(y,t)[f(\phi(y,t),t) - A(\phi(y,t),t)\phi_t(y,t)]. \tag{3.75b}$$

The reader can check that this is indeed an equivalence relation and that it preserves the geometric index notion.

3.4.7.2 Semiexplicit index one DAEs

In the nonautonomous setting, semiexplicit DAEs read

$$y' = h(y,z,t) \tag{3.76a}$$
$$0 = g(y,z,t), \tag{3.76b}$$

where $h \in C^\infty(W_0, \mathbb{R}^r)$, $g \in C^\infty(W_0, \mathbb{R}^p)$, the set W_0 being open in $\mathbb{R}^{r+p+1}$.

Akin to the autonomous context, the set W_1 is explicitly defined by the condition $g = 0$ depicted in (3.76b). A point $(y, z, t) \in W_1$ satisfies the maximal rank condition (3.71) if and only if rk $(g_y \ g_z) = p$, which implies that the submatrix $(g_{\bar{y}} \ g_{\bar{z}})$ is nonsingular for certain components, grouped in $\bar{y}$ and $\bar{z}$, of the variables y and z, respectively. Write, via the implicit function theorem, $\bar{y} = \psi_1(\hat{y}, \hat{z}, t)$, $\bar{z} = \psi_2(\hat{y}, \hat{z}, t)$, where $\hat{y}$ and $\hat{z}$ stand for the remaining components of y and z. Denoting by $\hat{r}$ and $\bar{r}$ the dimensions of $\hat{y}$ and $\bar{y}$, respectively, and using $P_1 = (I_r \ 0)$, the index one condition can be checked to amount to the nonsingularity of the matrix

$$\begin{pmatrix} I_{\hat{r}} & 0 \\ \psi_{1\hat{y}} & \psi_{1\hat{z}} \end{pmatrix}.$$

Via the implicit function theorem, this matrix can be seen to be the Schur complement (cf. Lemma 3.3 below) of the matrix $(g_{\bar{y}} \ g_{\bar{z}})$ in

$$\begin{pmatrix} I_{\hat{r}} & 0 & 0 & 0 \\ 0 & 0 & I_{\bar{r}} & 0 \\ g_{\hat{y}} & g_{\hat{z}} & g_{\bar{y}} & g_{\bar{z}} \end{pmatrix},$$

which is nonsingular if and only if so it is $g_z(y, z, t)$. This means that the point (y, z, t) is regular with geometric index one for the semiexplicit DAE (3.76) if and only if $g(y, z, t) = 0$ and $g_z(y, z, t)$ is nonsingular. If the nonsingularity of $g_z(y, z, t)$ holds whenever $g(y, z, t) = 0$, then (3.76) will be said to be an index one DAE.

From the nonsingularity of g_z defining index one points, we may describe the set $g = 0$ locally around an index one point (y^*, z^*, t^*) as $z = \psi(y, t)$, for some locally defined mapping ψ. This corresponds to the case $\hat{y} = y$, $\bar{z} = z$ above, and yields a local reduction of the form

$$y' = h(y, \psi(y, t), t). \tag{3.77}$$

3.4.7.3 Nonautonomous Hessenberg DAEs

Hessenberg DAEs of size two are defined in the nonautonomous context by a system of the form

$$y' = h(y, z, t) \tag{3.78a}$$

$$0 = g(y, t), \tag{3.78b}$$

with $h \in C^\infty(W_0, \mathbb{R}^r)$ and $g \in C^\infty(\hat{W}_0, \mathbb{R}^p)$; now W_0 is open in $\mathbb{R}^{r+p+1}$, whereas $\hat{W}_0$ is open in $\mathbb{R}^{r+1}$ and contains the (y, t)-projection of W_0.

Again, the set W_1 reads $\{(y,z,t) \in W_0 \,/\, g(y,t) = 0\}$. Proceeding as in 3.4.6.2, if g_y has maximal rank then we can derive a one-step reduction of (3.78) of the form

$$u' = h_1((u,\psi(u,t)),z,t) \qquad (3.79a)$$
$$\psi_u(u,t)u' = h_2((u,\psi(u,t)),z,t) - \psi_t(u,t), \qquad (3.79b)$$

where we have assumed w.l.o.g. that the last p components w of y yield a nonsingular derivative g_w; the variable u then stands for the first $r-p$ components of y and the implicit function theorem makes $g(y,t) = 0$ locally equivalent to $w = \psi(u,t)$.

Premultiplying (3.79) by

$$\begin{pmatrix} I_{r-p} & 0 \\ g_u((u,\psi(u,t)),t) & g_w((u,\psi(u,t)),t) \end{pmatrix}$$

we now get, via the implicit function theorem,

$$u' = h_1 \qquad (3.80a)$$
$$0 = g_u h_1 + g_w h_2 + g_t, \qquad (3.80b)$$

where we have removed the arguments of the different mappings for the sake of notational simplicity. Note that (3.80b) stands for the *hidden constraint* $g_y h + g_t = 0$, so that (3.80) reads, equivalently,

$$u' = h_1((u,\psi(u,t)),z,t) \qquad (3.81a)$$
$$0 = g_y((u,\psi(u,t)),t)h((u,\psi(u,t)),z,t) + g_t((u,\psi(u,t)),t). \qquad (3.81b)$$

System (3.81) represents a semiexplicit DAE. Mind that, in the constraint (3.81b), only h depends on the algebraic variable z; therefore, (3.81) is index one if and only if the derivative $g_y h_z$ is nonsingular. The point (y,z,t) will be then regular with geometric index two for the Hessenberg DAE (3.78) if and only if $g(y,t) = 0$, $g_y(y,t)h(y,z,t) + g_t(y,t) = 0$, and $g_y(y,t)h_z(y,z,t)$ is nonsingular. If the nonsingularity of $g_y h_z$ holds whenever the pair of conditions $g = 0$ and $g_y h + g_t = 0$ are met, then (3.78) will be said to be an index two DAE.

Finally, around an index two point the nonsingularity of the product $g_y h_z$ allows for a local description of (3.81b) of the form $z = \zeta(u,t)$. This leads to the local state space equation

$$u' = h_1((u,\psi(u,t)),\zeta(u,t),t). \qquad (3.82)$$

This type of reduction will be used in the derivation of state space models for index two circuit configurations in subsection 6.2.5.

3.4.7.4 Schur reduction and semiexplicit DAEs

In different semistate systems it is often the case that several variables are eliminated in a model reduction process which leads to somehow "intermediate" formulations. In many cases these eliminations are special (so-called *Schur*) reductions which, as detailed here, preserve the geometric index of the model. This will the case, in particular, in the semiexplicit circuit models considered in Chapter 6.

Let us then consider again the nonautonomous semiexplicit DAE (3.76). Split the variable z as (z_1, z_2), z_1 and z_2 being p_1- and p_2-dimensional, respectively, with $p_1 > 0$, $p_2 > 0$ and $p = p_1 + p_2$, and split accordingly the mapping g as (g_1, g_2) with $g_1 \in C^\infty(W_0, \mathbb{R}^{p_1})$ and $g_2 \in C^\infty(W_0, \mathbb{R}^{p_2})$. This makes it possible to rewrite (3.76) in the form

$$y' = h(y, z_1, z_2, t) \tag{3.83a}$$
$$0 = g_1(y, z_1, z_2, t) \tag{3.83b}$$
$$0 = g_2(y, z_1, z_2, t). \tag{3.83c}$$

Assume that the derivative g_{2z_2} is nonsingular whenever $g_2(y, z_1, z_2, t) = 0$. By the implicit function theorem, locally around any point satisfying (3.83c) this set can be described in terms of a local map α as $z_2 = \alpha(y, z_1, t)$. Inserting this into (3.83a)-(3.83b), we get the (locally defined) model

$$y' = h(y, z_1, \alpha(y, z_1, t), t) \equiv \tilde{h}(y, z_1, t) \tag{3.84a}$$
$$0 = g_1(y, z_1, \alpha(y, z_1, t), t) \equiv \tilde{g}(y, z_1, t). \tag{3.84b}$$

The importance of this kind of reduction (which will be called a *Schur reduction*) relies on the fact that, under the nonsingularity assumption on g_{2z_2}, the geometric index of (3.83) and (3.84) at regular points coincides, as stated in Proposition 3.7 below. More precisely, a point (y, z_1, z_2, t) satisfying (3.83c) and where g_{2z_2} is nonsingular will be regular with geometric index ν for (3.83) if and only if (y, z_1, t) is so for (3.84).

Lemma 3.3. *Let*

$$E = \begin{pmatrix} E_{11} & E_{12} \\ E_{21} & E_{22} \end{pmatrix} \tag{3.85}$$

be a matrix in $\mathbb{R}^{p \times p}$, with $E_{11} \in \mathbb{R}^{p_1 \times p_1}$, $E_{12} \in \mathbb{R}^{p_1 \times p_2}$, $E_{21} \in \mathbb{R}^{p_2 \times p_1}$, $E_{22} \in \mathbb{R}^{p_2 \times p_2}$, $p = p_1 + p_2$. Assume that E_{22} is nonsingular, and let $S_E(E_{22})$ stand for the Schur complement of E_{22} in E, that is,

$$S_E(E_{22}) = E_{11} - E_{12} E_{22}^{-1} E_{21}. \tag{3.86}$$

Then $\mathrm{rk}\, E = \mathrm{rk}\, S_E(E_{22}) + p_2$. In particular, the matrix E is nonsingular if and only if $S_E(E_{22})$ is nonsingular.

This result (see e.g. [137]) can be directly derived from the identity

$$\begin{pmatrix} I & -E_{12}E_{22}^{-1} \\ 0 & E_{22}^{-1} \end{pmatrix} \begin{pmatrix} E_{11} & E_{12} \\ E_{21} & E_{22} \end{pmatrix} = \begin{pmatrix} E_{11} - E_{12}E_{22}^{-1}E_{21} & 0 \\ E_{22}^{-1}E_{21} & I \end{pmatrix}, \quad (3.87)$$

since the first matrix is nonsingular and the rank of the last one equals $\operatorname{rk}(E_{11} - E_{12}E_{22}^{-1}E_{21}) + p_2$.

Remark 3.6. The property $\operatorname{rk} E = \operatorname{rk} S_E(E_{22}) + p_2$ holds also, with the same proof, if $E_{11} \in \mathbb{R}^{p_1 \times p_3}$, $E_{21} \in \mathbb{R}^{p_2 \times p_3}$, with $p_3 > p_1$.

Proposition 3.7. *Consider the semiexplicit DAE (3.83) and assume that the derivative $g_{2z_2}(y, z_1, z_2, t)$ is nonsingular whenever $g_2(y, z_1, z_2, t) = 0$. Then the geometric index of (3.83) at regular points equals that of the Schur reduction (3.84).*

Proof. We work locally around a given point (y, z_1, z_2, t) satisfying $g_2(y, z_1, z_2, t) = 0$ without further explicit mention. Let us consider the index one case separately. In the light of the results in 3.4.7.2, (3.83) and (3.84) have geometric index one if and only if g_z and $\tilde{g}_{z_1}$ are nonsingular, respectively. The equivalence of both conditions follows easily from Lemma 3.3 by setting $E = g_z$, $E_{11} = g_{1z_1}$, $E_{12} = g_{1z_2}$, $E_{21} = g_{2z_1}$ and $E_{22} = g_{2z_2}$, since the implicit function theorem yields

$$\tilde{g}_{z_1} = g_{1z_1} + g_{1z_2}\alpha_{z_1} = g_{1z_1} - g_{1z_2}g_{2z_2}^{-1}g_{2z_1} = S_{g_z}(g_{2z_2}). \quad (3.88)$$

Assume in the sequel that (y, z, t) is not index one for (3.83). We show below that the maximal rank condition

$$\operatorname{rk}(g_y \; g_z) = p \quad (3.89)$$

holds for (3.83) if and only if

$$\operatorname{rk}(\tilde{g}_y \; \tilde{g}_{z_1}) = p_1 \quad (3.90)$$

is met in the Schur reduction (3.84) and, furthermore, that a common one-step reduction may be defined for both DAEs in this situation.

The equivalence of (3.89) and (3.90) follows from the identity

$$(\tilde{g}_y \; \tilde{g}_{z_1}) = \left(g_{1y} - g_{1z_2}g_{2z_2}^{-1}g_{2y} \quad g_{1z_1} - g_{1z_2}g_{2z_2}^{-1}g_{2z_1}\right), \quad (3.91)$$

which owes to the implicit function theorem, and the fact that the matrix displayed in (3.91) is the Schur reduction of g_{2z_2} in

$$\begin{pmatrix} g_{1y} & g_{1z_1} & g_{1z_2} \\ g_{2y} & g_{2z_1} & g_{2z_2} \end{pmatrix} = (g_y \; g_z).$$

3.4. Quasilinear DAEs: The geometric index

We may then apply the result stated in Remark 3.6 to conclude that the conditions depicted in (3.89) and (3.90) are indeed equivalent.

In turn, the chance to derive the same one-step reductions for (3.83) and (3.84) is supported on the following. Provided that (3.89) is met, let $\bar{y}$, $\bar{z}$ be certain components of y, z_1 such that the matrix

$$\begin{pmatrix} g_{1\bar{y}} & g_{1\bar{z}} & g_{1z_2} \\ g_{2\bar{y}} & g_{2\bar{z}} & g_{2z_2} \end{pmatrix} \quad (3.92)$$

is nonsingular, $\hat{y}$ and $\hat{z}$ being the remaining components of y and z_1. This is always possible because of the nonsingularity of g_{2z_2}, and without loss of generality we may assume that $\hat{y}$ and $\hat{z}$ are the first components of y and z_1, respectively. We can hence describe the set $g_1 = 0$, $g_2 = 0$ locally as

$$\bar{y} = \psi_1(\hat{y}, \hat{z}, t) \quad (3.93a)$$
$$\bar{z} = \psi_2(\hat{y}, \hat{z}, t) \quad (3.93b)$$
$$z_2 = \psi_3(\hat{y}, \hat{z}, t), \quad (3.93c)$$

which yields the one-step reduction of (3.83)

$$\hat{y}' = h_1((\hat{y}, \psi_1(\hat{y}, \hat{z}, t)), (\hat{z}, \psi_2(\hat{y}, \hat{z}, t), \psi_3(\hat{y}, \hat{z}, t)), t) \quad (3.94a)$$
$$(\psi_1(\hat{y}, \hat{z}, t))' = h_2((\hat{y}, \psi_1(\hat{y}, \hat{z}, t)), (\hat{z}, \psi_2(\hat{y}, \hat{z}, t), \psi_3(\hat{y}, \hat{z}, t)), t), \quad (3.94b)$$

where $(\psi_1(\hat{y}, \hat{z}, t))'$ stands for $\psi_{1\hat{y}}\hat{y}' + \psi_{1\hat{z}}\hat{z}' + \psi_{1t}$.

Additionally, proceeding as above the nonsingularity of (3.92) is easily proved equivalent to that of

$$(\tilde{g}_{\bar{y}} \; \tilde{g}_{\bar{z}}),$$

which in turn leads to the local description

$$\bar{y} = \beta_1(\hat{y}, \hat{z}, t) \quad (3.95a)$$
$$\bar{z} = \beta_2(\hat{y}, \hat{z}, t) \quad (3.95b)$$

of the set $g_1(y, z_1, \alpha(y, z_1, t), t) = 0$. The corresponding one-step reduction of (3.84) reads

$$\hat{y}' = \tilde{h}_1((\hat{y}, \beta_1(\hat{y}, \hat{z}, t)), (\hat{z}, \beta_2(\hat{y}, \hat{z}, t)), t) \quad (3.96a)$$
$$(\beta_1(\hat{y}, \hat{z}, t))' = \tilde{h}_2((\hat{y}, \beta_1(\hat{y}, \hat{z}, t)), (\hat{z}, \beta_2(\hat{y}, \hat{z}, t)), t). \quad (3.96b)$$

Now, since $z_2 = \alpha(y, z_1, t))$ locally describes the manifold $g_2 = 0$, from (3.93) and (3.95) it must be $\psi_1(\hat{y}, \hat{z}, t) = \beta_1(\hat{y}, \hat{z}, t)$, $\psi_2(\hat{y}, \hat{z}, t) = \beta_2(\hat{y}, \hat{z}, t)$, and $\alpha((\hat{y}, \beta_1(\hat{y}, \hat{z}, t)), (\hat{z}, \beta_2(\hat{y}, \hat{z}, t)), t) = \psi_3(\hat{y}, \hat{z}, t)$. Together with the identities defining $\tilde{h}$ and $\tilde{g}$ in (3.84), this shows that the one-step reductions (3.94) and (3.96) are locally coincident. Therefore, the geometric index of (3.83) at a regular point (y, z_1, z_2, t) with $z_2 = \alpha(y, z_1, t)$ equals that of the Schur reduction (3.84) at (y, z_1, t). □

3.5 Dynamical aspects

The reduction procedure discussed in subsections 3.4.2 - 3.4.4 opens a way to characterize local dynamical aspects near regular points of autonomous quasilinear DAEs. The key aspect in this direction is that, around an index ν point with $r_\nu > 0$, the mapping $x = \varphi(u)$ defined in (3.42) is a local embedding which maps the phase portrait of the state space equation (3.41) onto that of the DAE (3.13); this means that the dynamical behavior is exactly the same for both systems.

Now, if a given local property is known to be preserved in the reduction process, it is useful to have tools allowing one to assess this property directly in the DAE setting, without the need to compute explicitly a state reduction. This idea will be illustrated by examining linear stability properties of regular equilibria in quasilinear DAEs, in terms of the matrix pencil associated with the linearized problem; this problem will be revisited in the circuit context in Chapter 6 (Section 6.3). Nevertheless, the scope of this approach seems to be broader; for instance, it might be used as well in local bifurcation analyses of parametrized DAEs, under a suitable extension of the above reduction framework to parametrized problems which is however beyond the scope of this book.

Qualitative properties of regular equilibria

We refer the reader to [3, 6, 7, 110, 127] for comprehensive discussions of qualitative aspects of explicit ODEs and of differential equations on manifolds. For an explicit ODE $u' = f(u)$, with $f \in C^1(\Omega, \mathbb{R}^r)$, Ω open in $\mathbb{R}^r$, linear stability properties of an equilibrium u^* (i.e. a point in Ω satisfying $f(u^*) = 0$) are those which can be assessed just in terms of the spectrum of the Jacobian matrix $f'(u^*)$. In particular, the equilibrium is said to be *hyperbolic* or *exponentially stable* if all the eigenvalues of $f'(u^*)$ verify $\text{Re}\,\lambda \neq 0$ or $\text{Re}\,\lambda < 0$, respectively. Owing to Lyapunov's theorem (see e.g. Th. 15.6 in [3]), an exponentially stable equilibrium point is asymptotically stable, that is, it is a stable attractor of nearby trajectories.

In the quasilinear DAE context of (3.13), a given x^* is said to be an equilibrium point if $x(t) \equiv x^*$ is a solution, namely, if $f(x^*) = 0$. It will be said to be a *regular equilibrium* if, additionally, it is a regular point in the sense specified after Definition 3.5 (p. 106), i.e., if it is regular with any index $\nu \geq 0$; we will also assume that $r_\nu > 0$, so that x^* is not isolated in the regular manifold W^{reg}.

3.5. Dynamical aspects

Our present goal is to characterize the linear stability properties of a given regular equilibrium x^* for the local flow of the DAE or, equivalently, the linear stability properties of u^* for any reduction of the form (3.41). Note, incidentally, that the invariance property stated in Theorem 3.4 (p. 116) is consistent with the idea that the local qualitative features of the DAE should be independent of the specific form of the state space description obtained in the reduction. The point is, of course, to perform the above-mentioned characterization in the original problem setting, i.e. in terms of (3.13) and without computing explicitly a state space reduction.

This will be possible via the matrix pencil notion discussed in subsection 2.1.1; as shown below, the (pencil) spectrum of the linearized problem will characterize the linear stability properties of equilibria in regular contexts. This result can be seen as an extension of the corresponding property for the linear DAE (2.4) under an assumption of regularity on the pencil; in that setting the results follows immediately from the identity stated in (2.7).

Theorem 3.5. *Let x^* be a regular equilibrium of the DAE (3.13) with $r_\nu > 0$. Then the matrix pencil $\lambda A(x^*) - f'(x^*)$ is regular, and the spectrum at $u^* = \varphi^{-1}(x^*)$ of the Jacobian matrix of the vector field $A_\nu^{-1}(u)f_\nu(u)$ in the state reduction (3.41) equals the spectrum of the pencil $\lambda A(x^*) - f'(x^*)$.*

This result has appeared in different forms in the literature, cf. [228, 231]. Related results are discussed in [183, 184, 186, 240, 291]. It is anyway of interest to look at its proof, which follows from the one-step property stated in Lemma 3.4 below. We will make use of the fact that, if x^* is 0-regular and $f(x^*) = 0$, the tangent space to W_1^{reg} at x^* can be described as

$$\{v \in \mathbb{R}^m \ / \ H(x^*)f'(x^*)v = 0\}, \tag{3.97}$$

with $H(x)$ defined in Lemma 3.1. This can be easily seen by differentiating the identity $H(x)f(x) = 0$ which by (3.23) locally describes W_1^{reg}, and noting that $f(x^*) = 0$ at the equilibrium.

Lemma 3.4. *Consider the quasilinear DAE $A(x)x' = f(x)$ and a one-step local reduction $A_1(\xi)\xi' = f_1(\xi)$ around a 0-regular equilibrium $x^* = \varphi_1(\xi^*)$ with $r_1 > 0$. Then, for any $\lambda \in \mathbb{C}$, $\det(\lambda A(x^*) - f'(x^*)) = 0$ if and only if $\det(\lambda A_1(\xi^*) - f_1'(\xi^*)) = 0$.*

Proof. Recall from (3.27) that the operators in the local reduction $A_1(\xi)\xi' = f_1(\xi)$ given in (3.26) read $A_1(\xi) = P_1(\varphi_1(\xi))A(\varphi_1(\xi))\varphi_1'(\xi)$ and $f_1(\xi) = P_1(\varphi_1(\xi))f(\varphi_1(\xi))$; since $f(\varphi_1(\xi^*)) = 0$, the latter leads to

$$f_1'(\xi^*) = P_1(\varphi_1(\xi^*))f'(\varphi_1(\xi^*))\varphi_1'(\xi^*).$$

Assume that $\det(\lambda A(x^*) - f'(x^*)) = 0$ and let $v \in \mathbb{C}^m - \{0\}$ be such that
$$(\lambda A(x^*) - f'(x^*))v = 0. \tag{3.98}$$
Premultiplication by $H(x^*)$ yields $H(x^*)f'(x^*)v = 0$, which shows that $v \in T_{x^*}W_1^{\text{reg}}$. Since φ_1 is a local parametrization of $W_1^{\text{reg}} \stackrel{\text{loc}}{=} W_1$, we can write $v = \varphi_1'(\xi^*)w$ for some $w \in \mathbb{C}^{r_1} - \{0\}$. Inserting $v = \varphi_1'(\xi^*)w$ into (3.98) and premultiplying this equation by $P_1(\varphi_1(\xi^*))$ we arrive at
$$(\lambda A_1(\xi^*) - f_1'(\xi^*))w = 0, \tag{3.99}$$
proving that $\det(\lambda A_1(\xi^*) - f_1'(\xi^*)) = 0$.

Conversely, the assumption that $\det(\lambda A_1(\xi^*) - f_1'(\xi^*)) = 0$ means that (3.99) holds for some $w \in \mathbb{C}^{r_1} - \{0\}$. Define $v = \varphi_1'(\xi^*)w$, and remark that v does not vanish and belongs to $T_{x^*}W_1^{\text{reg}}$ because φ_1 is a parametrization of W_1^{reg}. Rewrite (3.99) as
$$(\lambda P_1(x^*)A(x^*) - P_1(x^*)f'(x^*))v = 0. \tag{3.100}$$
By the characterization of $T_{x^*}W_1^{\text{reg}}$ depicted in (3.97) we have $H(x^*)f'(x^*)v = 0$, meaning that $f'(x^*)v \in \text{im}\, A(x^*)$. Since additionally $A(x^*)v \in \text{im}\, A(x^*)$, and $P_1(x^*)$ is an isomorphism $\text{im}\, A(x^*) \to \mathbb{R}^{r_1}$, from (3.100) we get that $(\lambda A(x^*) - f'(x^*))v = 0$, showing that (3.98) is met and therefore $\det(\lambda A(x^*) - f'(x^*)) = 0$. $\square$

In Lemma 3.4 we do not need to assume that the involved pencils are regular. However, this is necessarily the case at regular points, as stated in Theorem 3.5 and proved below.

Proof of Theorem 3.5. The spectrum preservation proved in Lemma 3.4 applies to all reduction steps. Since the Jacobian matrix of the state space reduction (3.41) at the equilibrium has only a finite number of eigenvalues, so it does the spectrum of the original matrix pencil which is thereby a regular one. The assertion then follows in a straightforward manner from an iterative application of Lemma 3.4. $\square$

It is worth emphasizing that a somehow converse result which might be conjectured at this stage does not hold; namely, a DAE may certainly display a regular pencil at points where the geometric index is not defined. This shows that the regularity of the pencil does not suffice to support the geometric index notion in nonlinear problems. A simple example illustrating this is defined by the DAE
$$\begin{aligned} xx' &= y \\ 0 &= x + y, \end{aligned} \tag{3.101}$$

which displays a regular index one pencil everywhere, whereas already the first constant rank requirement in the geometric index definition fails around the origin.

Finally, from the theorem of Lyapunov mentioned above and the fact that the map φ in (3.42) is a local embedding, we immediately get the following result on asymptotic stability of equilibria.

Corollary 3.1. *If x^* is a regular equilibrium of (3.13), and $\mathrm{Re}\,\lambda < 0$ for all $\lambda \in \sigma(\{A(x^*), -f'(x^*)\})$, then u^* is an asymptotically stable equilibrium for the reduced equation (3.41), and hence so it is x^* for the flow defined by the DAE on the regular manifold.*

The results in this Section illustrate that, in practice, it is not necessary to perform a state space reduction in order to characterize local dynamic properties of a quasilinear DAE. In particular, from Theorem 3.5 it follows that the linear stability properties of a regular equilibrium (y^*, z^*) of a semiexplicit index one DAE (3.1) or a Hessenberg index two system (3.6) are characterized by the spectra of the pencils $\{A, -J_1\}$ and $\{A, -J_2\}$, with

$$A = \begin{pmatrix} I_r & 0 \\ 0 & 0 \end{pmatrix}, \ J_1 = \begin{pmatrix} h_y & h_z \\ g_y & g_z \end{pmatrix}\bigg|_{(y^*, z^*)}$$

and

$$A = \begin{pmatrix} I_r & 0 \\ 0 & 0 \end{pmatrix}, \ J_2 = \begin{pmatrix} h_y & h_z \\ g_y & 0 \end{pmatrix}\bigg|_{(y^*, z^*)}$$

respectively. In general, this approach will make it possible to simplify cumbersome analyses in circuit theory, where the conditions arising in qualitative studies are often mixed-up with those needed to compute a state space description: this issue will be extensively addressed in Chapter 6.

3.6 Reduction methods for fully nonlinear DAEs

We briefly consider in the last two Sections of this Chapter the fully implicit or fully nonlinear DAE (1.2). The main reason for the shorter treatment of fully nonlinear problems is their unusual appearance in applications; for instance, all the differential-algebraic circuit models addressed in Chapters 5 and 6 are quasilinear. Furthermore, (1.2) can be naturally rewritten in the quasilinear (actually semiexplicit) form (1.7); this makes the results of Sections 3.4 and 3.5 applicable to nonlinear problems. Therefore, in this Section we just present some general ideas on geometric methods for fully implicit problems, referring the reader to the bibliography where more detailed discussions can be found. Section 3.7 will summarize the differentiation index framework, whereas a detailed discussion of projector-based methods for nonlinear DAEs can be found in [157].

The framework of Rabier and Rheinboldt summarized in subsection 3.4.1 can be extended, with some differences and a substantial technical effort, to fully nonlinear autonomous DAEs

$$F(x, x') = 0, \tag{3.102}$$

as detailed in [221]. We present below a very rough summary of this work, aimed at providing the reader with some hints on how the ideas of 3.4.1 can be applied to fully nonlinear problems.

Briefly, the framework of [221] assumes that F in (3.102) is a sufficiently differentiable mapping $\mathbb{R}^m \times \mathbb{R}^m \to \mathbb{R}^m$. Condition G1 in subsection 3.4.1 above (p. 95) can be recast for (3.102) by requiring $F_p(x, p)$ to have constant rank on an appropriate submanifold of the ambient space, whereas the submersion requirement for F stated in G2 applies also in this fully nonlinear context. Technically, $M_0 = F^{-1}(0)$ is assumed to be a π-submanifold (cf. Section 3 in [221]) of $T\mathbb{R}^m \simeq \mathbb{R}^m \times \mathbb{R}^m$. As in the quasilinear setting, W_1 stands for the projection of M_0 onto the first factor $\mathbb{R}^m$, and by the working assumptions can be guaranteed to have a manifold structure.

If $x(t)$ is a local solution of (3.102) with $(x(0), x'(0)) = (x_0, p_0) \in M_0$, then the pair $(x(t), x'(t))$ must belong to $M_0 \cap TW_1 = M_1$; this is called a reduction of M_0 and, if non-empty, is proved to be a π-submanifold of M_0. Note that this is essentially the same idea supporting the discussion in subsection 3.4.1. The procedure can be iterated in a way such that, if $M_\nu = M_{\nu+1}$, the M_k-manifold sequence becomes stationary: the minimum integer ν for which $M_\nu = M_{\nu+1}$ holds is then termed the index of the fully nonlinear problem (3.102). A vector field is locally defined on the projection of $\bigcap_{k \geq 0} M_k$ and solvability results follow.

Prior to this framework, it is worth mentioning the paper [220] by the same authors, where among other results the connection of these reduction techniques with matrix pencils and linear DAEs is detailed, as well as the paper of Reich [230].

3.7 The differentiation index and derivative arrays

Much research on DAEs is based upon the differentiation index concept, mainly developed by S. Campbell. A glimpse at the idea behind the differ-

3.7. The differentiation index and derivative arrays

entiation index was given in Section 1.2, and its application to semiexplicit problems has been briefly discussed in Sections 3.1 and 3.2. We sketch below how this notion, together with the closely related one of a *derivative array*, applies to general nonlinear DAEs. We follow [45, 46], where the reader is referred to for additional details. See also [30] and [10, 41–43, 54, 94].

Consider the fully nonlinear, nonautonomous DAE

$$F(t, x, x') = 0, \qquad (3.103)$$

where $F : \Omega \to \mathbb{R}^m$ is sufficiently differentiable, with Ω open in $\mathbb{R}^{2m+1}$. In order to understand the idea supporting the derivative array notion, think of a solution $x(t)$ to (3.103) and derive $F(t, x(t), x'(t))$ with respect to time to get

$$\frac{d}{dt} F(t, x, x') = F_t(t, x, x') + F_x(t, x, x')x' + F_{x'}(t, x, x')x'' = 0, \qquad (3.104)$$

which in turn can be derived with respect to t. Iteratively, the i-th derivative yields a formal expression on the variables $(t, x, x', x'', \ldots, x^{(i)}, x^{(i+1)})$. Altogether, the expressions up to the k-th derivative define the *derivative array equations*

$$F_{[k]}(t, x, x', x'', \ldots, x^{(k)}, x^{(k+1)}) = \begin{bmatrix} F(t, x, x') \\ \frac{d}{dt} F(t, x, x') \\ \vdots \\ \frac{d^k}{dt^k} F(t, x, x') \end{bmatrix} = 0 \qquad (3.105)$$

in the understanding that $\frac{d^i F}{dt^i}$ is written in terms of $(t, x, x', \ldots, x^{(i+1)})$ for $i = 1, \ldots, k$, as it is done in (3.104) for $i = 1$.

Let us now formally replace x' by $v \in \mathbb{R}^m$ and $(x'', \ldots, x^{(k)}, x^{(k+1)})$ by $w \in \mathbb{R}^{km}$ in the derivative array (3.105); think of (t, x, v, w) as a variable lying on an extended space $\Omega^e \in \mathbb{R}^{(k+2)m+1}$ whose projection onto the first $2m+1$ components yields Ω, e.g. $\Omega^e = \Omega \times \mathbb{R}^{km}$. The mapping $F_{[k]} : \Omega^e \to \mathbb{R}^{(k+1)m}$ which results from the replacement of $x', x'', \ldots, x^{(k+1)}$ by $v, w_1, \ldots, w_k$ (with $w_i \in \mathbb{R}^m$) in (3.103), (3.104) and eventual subsequent expressions has the form

$$F_{[k]}(t, x, v, w) = \begin{bmatrix} F(t, x, v) \\ F_t(t, x, v) + F_x(t, x, v)v + F_v(t, x, v)w_1 \\ F_{tt}(t, x, v) + F_{tx}(t, x, v)v + \ldots + F_v(t, x, v)w_2 \\ \vdots \\ F_{t^{(k)}t}(t, x, v) + \ldots + F_v(t, x, v)w_k \end{bmatrix}. \qquad (3.106)$$

In this setting, a given (t,x) is said to be *consistent* if there exists a $(t,x,v,w) \in \mathbf{\Omega}^e$ for which $F_{[k]}(t,x,v,w) = 0$. Note that for a fixed consistent (t,x) the vector (t,x,v,w) solving $F_{[k]} = 0$ need not be unique in $\mathbf{\Omega}^e$. But it may happen that, for a given consistent (t,x), these solutions are projected onto the same vector $(t,x,v) \in \mathbf{\Omega}$; this means that the derivative array equations $F_{[k]} = 0$ uniquely determine v from (t,x). If this happens for every consistent (t,x), the smallest nonnegative integer k with this property is called the *differentiation index* of (3.103).

As acknowledged in [45], this notion assumes the specification of the set $\mathbf{\Omega}^e$. If (3.103) is solvable (cf. [46]), and with $v = g(t,x)$ denoting the above-mentioned relation between (t,x) and v, the ODE $x' = g(t,x)$ is a differential equation on the manifold defined by the solutions of the DAE. A relation $x' = h(t,x)$ holding on an open neighborhood of this manifold is called a *completion*: see [43, 46] and references therein.

In these terms, by a straightforward application of the implicit function theorem one can see that (3.103) has differentiation index zero on $\mathbf{\Omega}$ if $F_v(t,x,v)$ is nonsingular for all $(t,x,v) \in \mathbf{\Omega}$ such that $F(t,x,v) = 0$. Similarly, it is not difficult to check that the conditions depicted in Propositions 3.5 and 3.6 (pp. 119 and 121) guarantee that the semiexplicit problem (3.1) and the Hessenberg DAE of size two (3.6) have differentiation index one and two, respectively; the set $\mathbf{\Omega}$ may be simply defined as $W_0 \times \mathbb{R}^m$ with $m = r+p$, whereas $\mathbf{\Omega}^e$ can be chosen as $W_0 \times \mathbb{R}^m \times \mathbb{R}^m$ and $W_0 \times \mathbb{R}^m \times \mathbb{R}^{2m}$ for (3.1) and (3.6), respectively. The underlying ODEs (3.2) and (3.8) define completions of both types of DAEs. Note also that, if the algebraic conditions in Propositions 3.5 and 3.6 hold at a given point (y,z), then the differentiation index notion applies on some neighborhood of this point and we may hence speak of the index on that neighborhood. Analogous statements hold for Hessenberg DAEs of higher sizes; in particular, the Hessenberg DAE (1.18) has differentiation index k if the matrix in (1.19) is nonsingular.

It is finally worth mentioning the related notion of the *uniform differentiation index* which, for sufficiently differentiable problems and in terms of certain algebraic conditions on the derivative array (3.106), supports solvability results and a rather general numerical approach for nonlinear, higher index DAEs. The reader is referred to [30, 45–47, 51, 52] for details in this regard.

Chapter 4

Singularities

The word 'singularity' is used pervasively in mathematics. Even within a given research field, this term often has different meanings. DAE theory is not an exception; the expression 'singular system' has been used to mean at least 'differential-algebraic system' (see [39, 40]), 'higher index differential-algebraic system' (e.g. in [118]) or, closer to the standpoint in this Chapter, 'differential-algebraic system without an index'. Roughly speaking, singular DAEs or, more precisely, singular points of DAEs will be locally defined as those for which the assumptions supporting an index notion fail.

In this direction, the goals of the present Chapter are to make this notion precise and to adapt the projector-based methods and reduction techniques of Chapters 2 and 3 to singular linear and quasilinear DAEs, allowing for the analysis of the dynamical phenomena associated with singular problems. In Section 4.1 we elaborate on the ideas supporting the singularity concept in DAE theory. Singular points of linear time-varying DAEs are then analyzed in Sections 4.2 and 4.3, whereas Section 4.4 addresses singularities of quasilinear systems.

4.1 What is a singular DAE?

Several approaches to the analysis of differential-algebraic equations are based on an iterative or recursive definition of an index, and describe the DAE solutions in terms of some kind of related ODE. As detailed in Chapters 2 and 3, this is the case of projector-based techniques and reduction methods, which unravel the DAE behavior in terms of an inherent or reduced ODE, respectively. The assumptions supporting the index definition usually imply that, at least locally, this ODE can be written in an explicit form, and existence and uniqueness of solutions for initial value problems can be then derived under sufficient smoothness requirements.

Sometimes, although the analysis procedure can be completed, it ends up with an ODE which cannot be rewritten in an explicit manner. A rough picture of this can be obtained from Theorem 3.2 on page 107; it may well happen that the reduction (3.40) is feasible around a given point, but yields a leading matrix $A_\nu(u)$ which is singular on a hypersurface of Ω_ν including u^*. This would be an instance of what shall be called a *quasilinear ODE*, which can be rewritten in explicit manner on a dense subset of the reduced state space. Some of these "last-step" singularities are analyzed by Rabier and Rheinboldt in [222, 223, 228]. Analogous phenomena can be depicted in the linear time-varying context.

A more pathological behavior occurs when the assumptions supporting the reduction of the DAE as discussed in Chapter 3, or those basing the tractability index definition in Chapter 2, fail before the last step of the procedure. Even a taxonomy of these situations is hard to find in the differential-algebraic literature.

In this Chapter, we introduce general singularity notions which accommodate these phenomena, for both linear time-varying and quasilinear DAEs. Our approach will label as *singular* those points where the algebraic assumptions supporting the DAE analysis fail, accommodating also the above-mentioned last-step singularities. As detailed in Section 4.4, this framework will also include singular points arising in the context of quasilinear ODEs and semiexplicit index one DAEs, which have attracted quite a lot of attention in the specialized literature. Furthermore, as it should be expected, singularities will be proved independent of the operators arising in the analysis, as well as invariant with respect to the appropriate local equivalence relations.

For these singular problems, the analysis methods as presented in previous Chapters do not apply. We need to figure out working hypotheses which, relaxing the requirements on the problem, still make it possible to characterize the solutions in terms of an inherent ODE or a reduction. These hypotheses will somehow capture the essential requirements for the methods to work; speaking again in general terms, these requirements will not rely on constant rank or transversality conditions but on the chance to continue through the singularity certain spaces associated with the DAE.

The above-mentioned working hypotheses will end up with an ODE which only on certain subsets can be written in explicit form. We need to distinguish at this point between linear time-varying and quasilinear DAEs. The projector-based analysis of linear time-varying DAEs developed

4.2. Singularities of properly stated linear time-varying DAEs

in Sections 4.2 and 4.3 will lead to a *scalarly implicit* inherent ODE of the form

$$\omega(t)u' + F(t)u = g(t). \qquad (4.1)$$

Here, the scalar function $\omega(t)$ does not have zeros in the regular set, which will be dense in the working interval, and typically vanishes at singular points. This way, the projector framework drives the analysis of singularities in linear DAEs to the singular ODE setting defined by (4.1). Singular problems of the form (4.1) have been extensively studied in the ODE literature (see e.g. [71, 127, 300]) and will not be further analyzed here.

The analysis of singular points x^* in quasilinear autonomous DAEs in Section 4.4 will proceed via a local reduction to

$$\tilde{A}_\nu(u)u' = \tilde{f}_\nu(u), \ u \in \Omega_\nu \subseteq \mathbb{R}^{\tilde{r}_\nu}, \qquad (4.2)$$

in which the matrix $\tilde{A}_\nu(u)$ is nonsingular on a dense subset of a neighborhood of $u^* = \varphi^{-1}(x^*)$. Problems of the form (4.2), which can be rewritten as an explicit ODE on such a dense set, will be called *quasilinear ODEs* and only recently have been analyzed in detail; for this reason, the dynamic phenomena which can be displayed by (4.2) will be surveyed in 4.4.1.

Summing up, singular points will be defined by the failure of algebraic assumptions supporting the DAE analysis procedure. We will not try to address related dynamic aspects such as impasse phenomena, singularity-crossing, multiplicity of solutions, etc., directly in the differential-algebraic setting; instead, we will classify the singularities and figure out broad working assumptions which drive the analysis to the somehow simpler setting of singular ODEs of the form (4.1) or (4.2).

Note finally that the analysis of singularities in fully nonlinear DAEs $F(x, x') = 0$ with $x \in \mathbb{R}^m$, $m \geq 2$, defines an open research direction which might extend the results of [7, 34, 72, 289], focused on scalar problems.

4.2 Singularities of properly stated linear time-varying DAEs

We drive in this Section the attention back to linear DAEs of the form

$$A(t)(D(t)x(t))' + B(t)x(t) = q(t), \ t \in \mathcal{J}, \qquad (4.3)$$

with $A(t) \in C(\mathcal{J}, \mathbb{R}^{m \times n})$, $D(t) \in C(\mathcal{J}, \mathbb{R}^{n \times m})$, $B(t) \in C(\mathcal{J}, \mathbb{R}^{m \times m})$, $q(t) \in C(\mathcal{J}, \mathbb{R}^m)$ and $\mathcal{J} \subseteq \mathbb{R}$ an open interval. Smoothness assumptions will be discussed later. Roughly speaking, singular points of (4.3) will be those where the matrix chain defining the tractability index in Chapter 2 cannot be constructed beyond a certain step; formally, a point $t^* \in \mathcal{J}$ will be called singular if no neighborhood of t^* admits an index or, equivalently, if there is no regularity interval including t^* (cf. Definition 2.7 on p. 74).

Previous approaches [139, 225, 228] do not detail the extent to which these singular points can be handled. As discussed in Section 4.1, the projector-based analysis procedure will be adapted here in order to accommodate singularities, typically ending up with a singular ODE, although some (informally called *harmless*) cases may result in a nonsingular linear ODE. We work with mild smoothness assumptions, driving the results beyond the analytic setting of [139, 225]. Our approach extends the discussion of [250] for index one DAEs in standard form, the results in this Section being directly based on [195–197].

Singular points will be classified in 4.2.1 according to a taxonomy that reflects the phenomenon from which the singularity stems. The different types of singular points will be proved independent of projectors and invariant with respect to rescaling and linear, time-dependent coordinate changes. Additionally, it will shown that for sufficiently smooth problems, all singular points fall in the types A and B defined in the taxonomy here introduced. In 4.2.2 we present a working scenario which accommodates singular points of (4.3). Solutions of the singular DAE will be unveiled through the scalarly implicit decoupling presented in Theorem 4.3. These ideas will be adapted to accommodate singularities of standard form linear DAEs in Section 4.3.

4.2.1 Classification of singular points

As indicated in subsection 2.2.7, the notion of a *regular point* introduced in Definition 2.7 naturally raises the problem of the behavior of (4.3) around points within $\mathcal{J} - \mathcal{J}_{\mathrm{reg}}$.

Definition 4.1. Points in $\mathcal{J} - \mathcal{J}_{\mathrm{reg}}$ will be called *singular* for (4.3).

Thereby, a point $t^* \in \mathcal{J}$ is singular for the DAE (4.3) if it is not regular, that is, if there is no regularity interval $\mathcal{I} \subseteq \mathcal{J}$ with $t^* \in \mathcal{I}$. Note that we replace the term 'critical' used in [196, 197] by 'singular' for later consistency with the quasilinear setting of Section 4.4 and, specially, to avoid any confusion with the so-called 'noncritical' singularities arising there.

Our aim is to study these singular points in terms of the conditions (a), (b) and (c) on page 40, supporting the chain construction based on P-projectors, or (equivalently) in terms of the corresponding conditions on page 45 for the Π-framework. Specifically, we will focus on singularities

4.2. Singularities of properly stated linear time-varying DAEs

arising from the failure of conditions (a) or (b). This means that the DAE coefficients will be assumed to be smooth enough as to avoid singular points arising from the failure of condition (c) (cf. Theorem 4.2 below).

Definition 4.2 is based on this viewpoint. The taxonomy of singular points will reflect not only the step at which the singularity shows up (via the integer k), but also the nature of the singularity, the labels A and B corresponding to the failure of assumptions (a) and (b), respectively.

Definition 4.2. Consider the DAE (4.3). A point $t^* \in \mathcal{J} - \mathcal{J}_{\text{reg}}$ is said to be a singularity of

(i) *type 0* if (4.3) is not properly stated on any open interval $\mathcal{I}$ with $t^* \in \mathcal{I}$;
(ii) *type k-A*, $k \geq 1$, if there exists an open interval $\mathcal{I}$, with $t^* \in \mathcal{I}$, where the DAE is nice up to level $k-1$, but G_k has a rank drop at t^*;
(iii) *type k-B*, $k \geq 1$, if there exists an open interval $\mathcal{I}$, with $t^* \in \mathcal{I}$, where the DAE is nice up to level $k-1$ and G_k has constant rank, but $(N_0(t^*) \oplus \ldots \oplus N_{k-1}(t^*)) \cap N_k(t^*) \neq \{0\}$.

Here, a continuous matrix mapping $G : \mathcal{I} \to \mathbb{R}^{m \times m}$, with $\mathcal{I} \subseteq \mathbb{R}$, is said to have a rank drop at $t^* \in \mathcal{I}$ if each neighborhood of t^* contains points t for which $\text{rk}\, G(t) \neq \text{rk}\, G(t^*)$. According to the notion of a properly stated DAE displayed in Definition 2.2, type 0 singular points may be due to rank drops at t^* either in $A(t)$ or in $D(t)$, and also to the lack of transversality of the spaces $\ker A(t^*)$ and $\text{im}\, D(t^*)$ (see (2.36)). Any of these situations yields a rank drop in G_0 at the point t^*.

For $k \geq 1$, the term 'k-singular point' or 'k-singularity' will be used to refer either to type k-A or to type k-B singularities. Similarly, a singular point will be said to be of type A or B if it has type k-A or k-B with arbitrary k; the terms 'A-singular' and 'B-singular' will also be used for singular points of types A and B, respectively.

It is worth clarifying that, following Theorem 2.2, the P- and Π-frameworks discussed in subsections 2.2.3 and 2.2.4, respectively, do not make any difference in the classification of singularities, although they will certainly make it in later analyses (see Remark 4.1 on page 148). Indeed, for $k \geq 1$ the G_i-matrices and N_i-spaces in Definition 4.2 can be understood to be constructed from an admissible projector sequence $\{Q_0, \ldots, Q_{k-1}\}$ or, equivalently, from an admissible sequence $\{\Pi_0, \ldots, \Pi_{k-1}\}$; cf. Definitions 2.3 and 2.6. Note also that the admissibility requirement implicit in the notion of a nice DAE implies that the products $DP_0 \cdots P_i D^-$ (or $D\Pi_i D^-$) are in C^1 for $0 \leq i \leq k-1$; the case $i = 0$ stands in particular for

$R = DD^-$ which is in C^1 because of the proper statement of the DAE on $\mathcal{I}$ for k-singularities with $k \geq 1$.

Theorem 4.1 below proves that these notions are actually independent of the (admissible) choice of projectors, and that they are invariant under linear time-varying coordinate changes $x(t) = F(t)y(t)$ and premultiplication by a nonsingular, continuous matrix mapping $C(t)$. Following [196], we prove this result within the P-framework. Based on Definition 4.2, we are allowed to work locally around a given singular point; this means that the projectors are assumed to be defined in the interval $\mathcal{I}$ arising in Definition 4.2. This will often be assumed without explicit mention in the sequel.

Theorem 4.1. *The notions of a singular point of type 0, k-A and k-B are independent of the (admissible) choice of projectors.*

Additionally, if t^* is a singular point of type 0, k-A or k-B for (4.3), then t^* is a singular point of the same type for the DAE

$$\tilde{A}(t)(\tilde{D}(t)y(t))' + \tilde{B}(t)y(t) = C(t)q(t), \quad t \in \mathcal{J}, \qquad (4.4)$$

where

$$\tilde{A}(t) = C(t)A(t), \quad \tilde{D}(t) = D(t)F(t), \quad \tilde{B}(t) = C(t)B(t)F(t) \qquad (4.5)$$

for nonsingular matrix-valued maps $C(t), F(t) \in C(\mathcal{J}, \mathbb{R}^{m \times m})$.

Proof. Note first that there is no projector involved in the notion of a type 0 singularity. For k-singular points with $k \geq 1$ we follow the proof of Theorem 3.3 in [196], based in turn on the results presented in [191]. Let $\{Q_0, \ldots, Q_{k-1}\}$ and $\{\hat{Q}_0, \ldots, \hat{Q}_{k-1}\}$, be admissible (up to level $k-1$) projector sequences on $\mathcal{I}$, and denote by $\{G_i\}$, $\{\hat{G}_i\}$ the corresponding matrices constructed from these projector sequences. We make use of the fact that $\operatorname{rk} G_k = \operatorname{rk} \hat{G}_k$, which is due to the identity $\hat{G}_k = G_k Z_k$, with a nonsingular factor defined recursively for $i \leq k$ as $Z_i = (I + Q_{i-1}\hat{Q}_{i-1}P_{i-1} + \sum_{j=0}^{i-2} Q_j Z_{ij} P_0 \cdots P_{i-2}) Z_{i-1}$, for certain Z_{ij}; see Theorem 2.3 in [191]. Therefore, rank changes occur exactly at the same points for both matrices G_k and $\hat{G}_k$, showing that the notion of a singular point of type k-A is independent of the choice of the admissible projector sequence.

From Theorem 2.3 in [191] it follows that $N_0 \oplus \ldots \oplus N_i = \hat{N}_0 \oplus \ldots \oplus \hat{N}_i$ holds for all $i \leq k-1$. Now, from the identity $N_k = Z_k \hat{N}_k$ and the expression defining Z_k, we derive

$$N_k \subseteq (N_0 \oplus \ldots \oplus N_{k-1}) + \hat{N}_k = (\hat{N}_0 \oplus \ldots \oplus \hat{N}_{k-1}) + \hat{N}_k$$

4.2. Singularities of properly stated linear time-varying DAEs

and, conversely,
$$\hat{N}_k \subseteq (\hat{N}_0 \oplus \ldots \oplus \hat{N}_{k-1}) + N_k = (N_0 \oplus \ldots \oplus N_{k-1}) + N_k.$$
We then get $(N_0 \oplus \ldots \oplus N_{k-1}) + N_k = (\hat{N}_0 \oplus \ldots \oplus \hat{N}_{k-1}) + \hat{N}_k$ and, therefore, $\dim(N_0 \oplus \ldots \oplus N_{k-1}) \cap N_k = \dim(\hat{N}_0 \oplus \ldots \oplus \hat{N}_{k-1}) \cap \hat{N}_k$. Type B singular points are defined by the case in which this dimension is no longer zero, and this occurs simultaneously for both projector sequences $\{Q_0, \ldots, Q_{k-1}\}$ and $\{\hat{Q}_0, \ldots, \hat{Q}_{k-1}\}$. Therefore, the type B singularity notion is also independent of the choice of admissible projectors.

Regarding the asserted invariance with respect to the transformed DAE (4.4), it is immediate to check from (4.5) that $\ker \tilde{A}(t) = \ker A(t)$ and $\operatorname{im} \tilde{D}(t) = \operatorname{im} D(t)$. Hence, the proper statement of (4.4) is obviously equivalent to that of (4.3) and therefore the type 0 singularity notion is invariant. In turn, for k-singular points with $k \geq 1$ we follow again [196]; the projectors $\tilde{Q}_i = F^{-1} Q_i F$ constructed in [189] for (4.4) yield the identities $\tilde{G}_i = CG_i F$. The rank of G_i is therefore transferred to $\tilde{G}_i$ and type A singularities are hence invariant. Additionally, $\tilde{N}_i = \ker \tilde{G}_i = F^{-1} N_i$, so that the loss of transversality in the N_i spaces defining type B singularities is also transferred to $\tilde{N}_i$ and the proof is completed. □

The importance of the types A and B introduced in Definition 4.2 relies on the fact that they provide a complete description of the singular set $\mathcal{J} - \mathcal{J}_{\text{reg}}$ in sufficiently smooth (that is, C^{m-1}) problems. As in [196], the proof of this result is based on the P-framework; note that the smoothness properties of the projectors P_i below, based in turn on those of Q_i, would be trivially transferred to the products $\Pi_i = P_0 \cdots P_i$ (cf. Theorem 2.2) within the Π-framework.

Theorem 4.2. *Assume that the coefficients $A(t)$, $D(t)$, $B(t)$ in the linear DAE (4.3) belong to C^{m-1}. Then every singular point is of type 0, k-A, or k-B, with $1 \leq k \leq m$.*

Proof. Let us first remark that, around any point in $\mathcal{J}$ which is not a 0-singularity, the maps $D^-(t)$, $R(t)$ and $P_0(t)$ can be chosen from C^{m-1} according to the requirements (2.37), (2.38) and (2.39). In this situation, the locally defined matrix mapping $G_1 = G_0 + BQ_0$ also belongs to C^{m-1}. Now, except at type 0, 1-A and 1-B singularities, we may locally choose a preadmissible Q_1 in C^{m-1}, so that Q_1 will actually be admissible. Then B_1 and so G_2 will be in the class C^{m-2}. At subsequent levels, the construction can be locally extended in an admissible manner up to the mappings G_{m-1}, Q_{m-1} which can be taken from C^1. This means that B_{m-1} and G_m will be continuous. Finally, at points which are neither regular with index $\leq m$ nor singular of type $k < m$ or m-A, the space $N_0 \oplus \ldots \oplus N_{m-1}$ must fill $\mathbb{R}^m$ and will therefore intersect N_m non-trivially. This necessarily yields a singular point of type m-B, thus completing the proof. □

Type A singularities

Some instances of A-singularities can be easily introduced in terms of the circuit model (2.124) considered in subsection 2.2.6. Note first that isolated zeros of $C_1(t)$, $C_2(t)$ or $L(t)$ lead to rank deficiencies in $D(t)$ and would hence define type 0 singularities of the DAE.

Assume now, as in 2.2.6.2, that the circuit parameters verify the conditions $C_1(t) \neq 0 \neq C_2(t)$, $L(t) \neq 0$ and $R_1(t) = 0$ on the working interval $\mathcal{J}$. The reader can check that, in this setting, isolated zeros of $R_2(t)$ yield rank deficiencies in the matrix G_1 within (2.127), therefore making such zeros singular points of type 1-A for (2.124).

In turn, if the condition $C_1(t) + C_2(t) \neq 0$ fails to hold at an isolated point where $R_2(t)$ does not vanish, then the matrix G_1 has locally constant rank and the spaces N_0 and N_1 are still transversal, but now the matrix G_2 in (2.131) undergoes a rank deficiency at this point, which can be therefore labeled as a singularity of type 2-A.

Further remarks on the singularities of this system can be found in [196].

Type B singularities

The Hessenberg DAE

$$x' + y + \alpha(t)z = q_1(t) \quad (4.6\text{a})$$
$$y' - x = q_2(t) \quad (4.6\text{b})$$
$$x = q_3(t) \quad (4.6\text{c})$$

models an electrical circuit considered in [196, 197]. This system can be written in the properly stated form (4.3) using

$$A = \begin{pmatrix} 1 & 0 \\ 0 & 1 \\ 0 & 0 \end{pmatrix}, \; D = \begin{pmatrix} 1 & 0 & 0 \\ 0 & 1 & 0 \end{pmatrix}, \; B = \begin{pmatrix} 0 & 1 & \alpha(t) \\ -1 & 0 & 0 \\ 1 & 0 & 0 \end{pmatrix}.$$

The product $G_0 = AD$ reads

$$G_0 = \begin{pmatrix} 1 & 0 & 0 \\ 0 & 1 & 0 \\ 0 & 0 & 0 \end{pmatrix},$$

4.2. Singularities of properly stated linear time-varying DAEs

and then $N_0 = \ker G_0 = \mathrm{span}\left[\begin{pmatrix} 0 \\ 0 \\ 1 \end{pmatrix}\right]$. Setting

$$D^- = \begin{pmatrix} 1 & 0 \\ 0 & 1 \\ 0 & 0 \end{pmatrix}, \quad Q_0 = M_0 = \begin{pmatrix} 0 & 0 & 0 \\ 0 & 0 & 0 \\ 0 & 0 & 1 \end{pmatrix},$$

we get

$$G_1 = \begin{pmatrix} 1 & 0 & \alpha(t) \\ 0 & 1 & 0 \\ 0 & 0 & 0 \end{pmatrix}, \quad N_1 = \ker G_1 = \mathrm{span}\left[\begin{pmatrix} -\alpha(t) \\ 0 \\ 1 \end{pmatrix}\right].$$

Mind that the matrix $G_1(t)$ has constant rank $r_1 = 2$, and that the intersection $N_0(t) \cap N_1(t)$ becomes non-trivial when $\alpha(t)$ vanishes. This means that the zeros of $\alpha(t)$ define singular points of type 1-B for (4.6).

Again, the reader is referred to [196, 197] for additional details concerning the singular points of this DAE.

4.2.2 Decoupling

The existence of singular points in $\mathcal{J}$ precludes the construction of a tractability chain for (4.3) defined on the whole working interval. This means that the analysis of linear DAEs via projector methods, in the terms discussed in Chapter 2, is not feasible for systems with singularities. These problems require the introduction of suitable working hypotheses allowing us to adapt the framework of Chapter 2 to singular settings. For later detailed reasons (see Remark 4.1), the Π-framework of subsection 2.2.4 will display some important advantages over the construction based on P-projectors and will therefore be preferred.

4.2.2.1 Working hypotheses

Broadly speaking, the framework of Chapter 2 can be adapted to singular problems with a dense subset of regular points by relaxing the proper statement of the DAE and replacing the working conditions (a) and (b) on page 45 by the existence of a continuation of the characteristic spaces $N_0 \oplus \ldots \oplus N_i$, preserving the smoothness property on $D\Pi_i D^-$ displayed in (c). Hereafter, if there is no further specification, 'continuation' will mean 'C^0-continuation'. Under mild conditions on $D(t)$ and $A(t)$, the existence of a such a continuation meeting the smoothness requirements can be expressed in terms of the existence of a C^1-continuation of the spaces $N_0 \oplus \ldots \oplus N_i$; see Proposition 4.1 below. This fact will provide in particular a nice framework for the standard form problems considered in Section 4.3, and provides additional support for the idea that the characteristic spaces $N_0 \oplus \ldots \oplus N_i$ somehow define the "core" of the projector approach. Although the working hypotheses will be stated globally on $\mathcal{J}$, it should be clear how to reformulate them locally around a singular point.

These working assumptions will make it possible to unravel the behavior of the DAE through a singular inherent ODE, which will actually display a *scalarly implicit* form, as detailed in Theorem 4.3.

The setting for our analysis of singular problems assumes that the regular set is dense in the working interval, as stated below. This includes (but is not restricted to) DAEs with isolated singularities. Assumption P1 will play a key role in the decoupling of singular problems discussed in 4.2.2.2.

Assumption P1. *The set $\mathcal{J}_{\text{reg}}$ of regular points is dense in $\mathcal{J}$.*

Assumption P2 below captures the requirements on the matrix mappings $A(t)$, $D(t)$ within the DAE formulation (4.3). We will restrict the attention to problems with a constant rank matrix $D(t)$, via the C^1 structure on $\operatorname{im} D(t)$ stated in Assumption P2. This is reasonable in many applications since D is intended to capture the components of the semistate vector x which actually need to be differentiated, although some problems would need a broader setting allowing for rank changes in D. The type 0 singularities accommodated in the present framework are hence due to rank drops in $A(t)$ for which the space $X(t)$ arising below is well-defined.

Assumption P2. *There exists a time-varying space $X(t) \subseteq \ker A(t)$ such that $X(t) \oplus \operatorname{im} D(t) = \mathbb{R}^n$ for all $t \in \mathcal{J}$, and both spaces $X(t)$ and $\operatorname{im} D(t)$ are spanned by C^1 bases on $\mathcal{J}$.*

At regular points, where the DAE is (locally) properly stated, the space $X(t)$ must equal $\ker A(t)$; this space can be thus seen as a C^1 continuation (assumed to exist) to the whole of $\mathcal{J}$ of the kernel of $A(t)$ at regular points. The C^1-space (cf. Remark 2.1 on page 36) $X(t)$ is also said to be a C^1-restriction of $\ker A(t)$.

In virtue of Assumption P2, there exists a C^1 projector $R(t)$ along $X(t)$ onto $\operatorname{im} D(t)$ on the whole of $\mathcal{J}$. Because of the relation $X(t) \subseteq \ker A(t)$, the identity $A(t)R(t) = A(t)$ holds on $\mathcal{J}$; this can be checked from the fact that $I - R(t)$ projects onto a subspace of $\ker A(t)$ for all $t \in \mathcal{J}$, yielding

4.2. Singularities of properly stated linear time-varying DAEs

$A(t)(I - R(t)) = 0$ for $t \in \mathcal{J}$. Furthermore, under the density hypothesis on the regular set stated in Assumption P1, the space $X(t)$ and the projector $R(t)$ are uniquely defined by the requirements depicted in Assumption P2.

The constant rank assumption on $D(t)$ implies that $\ker D(t)$ has a continuous basis on $\mathcal{J}$. Set $K_0(t) = \ker D(t)$. Note that $\ker D(t) \subseteq \ker A(t)D(t) = \ker G_0(t)$ on $\mathcal{J}$, so that $K_0(t)$ is a continuous restriction of $N_0(t) = \ker G_0(t)$. Moreover, at regular points we have $\ker A(t)D(t) = \ker D(t)$, and therefore $K_0(t)$ actually equals $N_0(t) = \ker G_0(t)$ on $\mathcal{J}_{\text{reg}}$; $K_0(t)$ can be therefore seen as a continuation to the whole of $\mathcal{J}$ of the space $N_0(t) = \ker G_0(t)$ at regular points. The existence of a continuous projector $\Pi_0(t)$ along $K_0(t)$ follows from the C^0-structure of $K_0(t)$; in particular, the orthogonal projector along $K_0(t)$ is continuous on $\mathcal{J}$.

Following these remarks, Assumption P2 also yields the existence of a continuous (on $\mathcal{J}$) mapping $D^-(t)$ satisfying the relations (2.37), (2.38) and (2.46), i.e.

$$DD^-D = D, \quad D^-DD^- = D^-, \quad DD^- = R, \quad D^-D = \Pi_0, \qquad (4.7)$$

where we have removed the dependence on t for notational simplicity.

The projectors $\Pi_0(t)$ and $R(t)$ defined above extend to the whole of $\mathcal{J}$ the corresponding "regular" projectors $\Pi_0(t)$ and $R(t)$ defined on $\mathcal{J}_{\text{reg}}$; this means that $\Pi_0(t)$ projects along $N_0(t) = \ker G_0(t)$ on $\mathcal{J}_{\text{reg}}$, and $R(t)$ projects along $\ker A(t)$ onto $\operatorname{im} D(t)$ on $\mathcal{J}_{\text{reg}}$. This continuation is C^1 in the case of $R(t)$. The assumption that analogous continuations exist in subsequent levels is the key requirement for the projector framework to be applicable to singular problems, as depicted below.

Assumption P3. *There exist projector mappings $\Pi_1(t), \ldots, \Pi_{m-1}(t)$ continuous on $\mathcal{J}$, with $D\Pi_i D^-$ continuously differentiable on $\mathcal{J}$, which satisfy $\ker \Pi_i = N_0 \oplus \ldots \oplus N_i$ and $\operatorname{im} \Pi_i \subseteq \operatorname{im} \Pi_{i-1}$ on $\mathcal{J}_{\text{reg}}$.*

Here i ranges from 1 to $m-1$, with Π_0 (arising in the relation $\operatorname{im} \Pi_1 \subseteq \operatorname{im} \Pi_0$ for $i = 1$) defined above. If a nonsingular matrix G_i is met at some point for $i = \nu$ (cf. Proposition 4.2), then it must be understood that $N_0 \oplus \ldots \oplus N_i = N_0 \oplus \ldots \oplus N_{\nu-1}$ and $\Pi_i = \Pi_{\nu-1}$ for $\nu \leq i \leq m-1$.

Proposition 4.1. *Provided that there exists a space $X(t)$ satisfying the requirements depicted in Assumption P2, then Assumption P3 holds if*

(i) *$D(t)$ is in $C^1(\mathcal{J}, \mathbb{R}^{n \times m})$ with constant rank; and*
(ii) *the space $N_0(t) \oplus \ldots \oplus N_i(t)$ admits a C^1-continuation $K_i(t)$ on $\mathcal{J}$, for $i = 1, \ldots, m-1$.*

Proof. Because of the C^1 assumption on $D(t)$, the space $K_0(t) = \ker D(t)$ is in C^1 and so it is the orthogonal projector $\Pi_0(t)$ along $K_0(t)$. From the relations defining $D^-(t)$ in (4.7) it follows that this matrix-valued map is in C^1. Now, for $i = 1, \ldots, m-1$, let $\Pi_i(t)$ be the orthogonal projector along $K_i(t)$; this orthogonal projector is again in C^1 due to the C^1 nature of $K_i(t)$. The products $D(t)\Pi_i(t)D^-(t)$ are therefore in C^1 for $i = 1, \ldots, m-1$. Finally, since $K_i = N_0 \oplus \ldots \oplus N_i$ at regular points, the relations $\ker \Pi_i = N_0 \oplus \ldots \oplus N_i$ and $\operatorname{im} \Pi_i \subseteq \operatorname{im} \Pi_{i-1}$ on $\mathcal{J}_{\text{reg}}$ are also satisfied. □

Note that we use the notation $K_i(t)$ for the continuation on $\mathcal{J}$ of the space $N_0(t) \oplus \ldots \oplus N_i(t)$; the latter is well-defined only on the regular set $\mathcal{J}_{\text{reg}}$.

Remark 4.1. Assumptions P1-P3 accommodate not only type A but also type B singular points, in contrast to the framework presented in [196]. The working assumptions there can be roughly described as requiring the individual spaces $N_i(t)$ to admit a continuation through the singular points, retaining the transversality and smoothness properties (b)-(c) on page 40. This means that the projectors $P_i(t)$ (or equivalently $Q_i(t) = I - P_i(t)$) can be continued through the singularity in an admissible manner, making it possible to work with the products $P_0(t) \cdots P_i(t)$ which satisfy Assumption P3 above. As proved in Proposition 4.2 of [196], the assumptions there rule out the existence of B-singular points. The key idea is that the admissibility condition reads $Q_i Q_j = 0$ for $i > j$, and relies upon the transversality condition $(N_0 \oplus \ldots \oplus N_{i-1}) \cap N_i = \{0\}$; trying to force $Q_i Q_j = 0$ through a type B singular point yields an unbounded projector and therefore rules out the P-based matrix chain construction. The setting here considered, based on Π-projectors, does not display this limitation: the Π-projectors may admit a continuation through the singular points in situations in which the P-projectors do not.

The importance of Assumption P3 relies on the fact that, together with Assumptions P1 and P2, allows one to build the chain $\{G_i\}$ in this singular context according to the rules specified in (2.50), that is, by setting $M_0 = I - \Pi_0$, $G_1 = G_0 + B_0 M_0$, $M_i = \Pi_{i-1} - \Pi_i$, and

$$B_i = (B_{i-1} - G_i D^-(D\Pi_i D^-)'D)\Pi_{i-1}, \quad G_{i+1} = G_i + B_i M_i. \quad (4.8)$$

Moreover, this construction makes it possible to adapt the decoupling discussed in Chapter 2 to singular problems, as detailed in Theorem 4.3 below. Note that this construction particularizes to a tractability chain at regular points.

4.2. Singularities of properly stated linear time-varying DAEs

This framework excludes index changes, as proved below.

Proposition 4.2. *Under Assumptions P1-P3, all regular points of the DAE (4.3) have the same tractability index ν.*

Proof. The result follows from the fact that the projectors $\Pi_0, \ldots, \Pi_{m-1}$ define a tractability chain on $\mathcal{J}_{\text{reg}}$. At regular points, the projectors P_i along $N_i = \ker G_i$ are well-defined by Theorem 2.2, and satisfy $\operatorname{im} P_i = \ker \Pi_{i-1} \oplus \operatorname{im} \Pi_i$ (cf. (2.51)). The assumed continuity on Π_i comprises a constant dimension hypothesis on its kernel and image; in turn, this implies a constant rank condition on P_i in $\mathcal{J}_{\text{reg}}$ and hence on G_i, since P_i projects along $\ker G_i$. This means that the nonsingular mapping G_i defining the index will be reached at the same step on the whole of $\mathcal{J}_{\text{reg}}$. □

4.2.2.2 Decoupling

Theorem 2.3 characterizes the solutions of a regular linear DAE in terms of an explicit inherent ODE. We show below that Assumptions P1-P3 make it possible to describe the behavior of singular linear DAEs by means of an *implicit* inherent ODE (cf. (4.10)). The leading coefficient of this scalarly implicit ODE, which somehow captures the singular behavior of the DAE, does not vanish at regular points and therefore the inherent ODE amounts locally (i.e. on regular subintervals) to an explicit problem. For the analysis of singular ODEs such as (4.10) the reader is referred to [71, 127, 300].

Within Theorem 4.3, the matrix chain construction follows (4.8). According to Proposition 4.2, this chain results in a matrix mapping $G_\nu(t)$ which is nonsingular on $\mathcal{J}_{\text{reg}}$, with $0 \leq \nu \leq m$. We denote $\omega_\nu(t) = \det G_\nu(t)$, and let $\operatorname{Adj} G_\nu$ be the adjoint of G_ν, namely the transpose of the matrix of cofactors of G_ν (also called the *adjugate* matrix, cf. [137]). In the sequel we remove explicit dependences on t, and assume that the excitation q is smooth enough as to permit all the involved differentiations (cf. Remark 2.9 on page 53).

Theorem 4.3. *Under Assumptions P1-P3, a given mapping*

$$x \in C_D^1(\mathcal{I}, \mathbb{R}^m) = \{x \in C(\mathcal{I}, \mathbb{R}^m) \ / \ Dx \in C^1(\mathcal{I}, \mathbb{R}^n)\}$$

solves (4.3) in a subinterval $\mathcal{I} \subseteq \mathcal{J}$ if and only if it can be written as

$$x = D^- u + v_{\nu-1} + \ldots + v_1 + v_0, \quad (4.9)$$

where $u \in C^1(\mathcal{I}, \mathbb{R}^n)$ is a solution of the scalarly implicit inherent ODE

$$\omega_\nu u' - \omega_\nu (D\Pi_{\nu-1} D^-)' u + D\Pi_{\nu-1} \operatorname{Adj} G_\nu B D^- u = D\Pi_{\nu-1} \operatorname{Adj} G_\nu q \quad (4.10)$$

lying on the locally invariant space im $D\Pi_{\nu-1}D^-$, whereas the solution components v_k verify

$$\omega_\nu^{\nu-k} v_k = -\tilde{\mathcal{K}}_k D^- u + \sum_{j=k+1}^{\nu-1} \tilde{\mathcal{N}}_{kj}(Dv_j)' + \sum_{j=k+2}^{\nu-1} \tilde{\mathcal{M}}_{kj} v_j + \tilde{\mathcal{L}}_k q \quad (4.11)$$

for $k = \nu - 1, \ldots, 0$, with $v_k \in C_D^1(\mathcal{I}, \mathbb{R}^m)$ for $k > 0$ and $v_0 \in C(\mathcal{I}, \mathbb{R}^m)$.

Setting $\tilde{Q}_i = \operatorname{Adj} G_\nu B_i M_i$ and $\tilde{P}_i = \omega_\nu I - \tilde{Q}_i$ for $0 \leq i \leq \nu - 1$, the coefficients of (4.11) are continuous on $\mathcal{J}$ and read

$$\tilde{\mathcal{K}}_k = M_k \tilde{P}_{k+1} \cdots \tilde{P}_{\nu-1} \operatorname{Adj} G_\nu B$$
$$+ M_k(\omega_\nu \tilde{P}_{k+1} \cdots \tilde{P}_{\nu-1} - \omega_\nu^{\nu-k} I) D^- (D\Pi_{\nu-1} D^-)' D$$

$$\tilde{\mathcal{N}}_{kj} = \omega_\nu^{\nu-j} M_k \tilde{P}_{k+1} \cdots \tilde{P}_{j-1} \tilde{Q}_j D^-$$

$$\tilde{\mathcal{M}}_{kj} = M_k \left(\sum_{i=k+1}^{j-1} \omega_\nu^{\nu-i} \tilde{P}_{k+1} \cdots \tilde{P}_{i-1} (\tilde{Q}_i - \tilde{P}_i) \right) D^- (DM_j D^-)' D$$

$$\tilde{\mathcal{L}}_k = M_k \tilde{P}_{k+1} \cdots \tilde{P}_{\nu-1} \operatorname{Adj} G_\nu,$$

where the products $\tilde{P}_{k+1} \cdots \tilde{P}_l$ amount to the identity I whenever $l < k+1$.

Once the working hypotheses depicted in Assumptions P1-P3 have been figured out, the proof of this result follows in a natural way. The key idea is that the decoupling presented in Theorem 2.3 is valid within $\mathcal{I} \cap \mathcal{J}_{\text{reg}}$ and can be rewritten in scalarly implicit form via the identity $\det G_\nu \cdot I = \operatorname{Adj} G_\nu \cdot G_\nu$. We will make repeated use of the fact that any continuous mapping vanishing on a dense subset of a given interval actually vanishes on the whole interval.

Proof of Theorem 4.3. Let $x \in C_D^1(\mathcal{I}, \mathbb{R}^m)$ be a solution of (4.3) in $\mathcal{I} \subseteq \mathcal{J}$. Since $x(t)$ solves (4.3) on $\mathcal{I}$, the components $u = Dx$ and $v_k = M_k x$, $k = \nu - 1, \ldots, 0$, satisfy on $\mathcal{I} \cap \mathcal{J}_{\text{reg}}$ the relations depicted in (2.68) and (2.69). Multiplying (2.68) by ω_ν we immediately get that (4.10) holds on $\mathcal{I} \cap \mathcal{J}_{\text{reg}}$; since all involved operators are continuous, and $\mathcal{I} \cap \mathcal{J}_{\text{reg}}$ is dense in $\mathcal{I}$, this relation remains valid on $\mathcal{I}$. The same reasoning makes it possible to derive (4.11) via the multiplication of (2.69) by $\omega_\nu^{\nu-k}$, for $k = \nu - 1, \ldots, 0$. Remark that $\tilde{Q}_i = \operatorname{Adj} G_\nu B_i M_i$ and $\tilde{P}_i = \omega_\nu I - \tilde{Q}_i$ are continuous on the whole of $\mathcal{J}$ and, at regular points, the projectors Q_i, P_i are well-defined and verify $\omega_\nu Q_i = \tilde{Q}_i$, $\omega_\nu P_i = \tilde{P}_i$. Using these identities together with $\omega_\nu G_\nu^{-1} = \operatorname{Adj} G_\nu$, the expressions for $\tilde{\mathcal{K}}_k$, $\tilde{\mathcal{N}}_{kj}$, $\tilde{\mathcal{M}}_{kj}$ and $\tilde{\mathcal{L}}_k$ are obtained

4.2. Singularities of properly stated linear time-varying DAEs

in a straightforward manner from the ones depicted for $\mathcal{K}_k$, $\mathcal{N}_{kj}$, $\mathcal{M}_{kj}$ and $\mathcal{L}_k$ in Theorem 2.3.

Note that $u = Dx$, as well as $Dv_k = DM_k x = DM_k D^- Dx$ for $k = \nu - 1, \ldots, 1$, are in $C^1(\mathcal{I}, \mathbb{R}^n)$; again, we have implicitly used the fact that the identity $M_k D^- D = M_k \Pi_0 = M_k$ holds on $\mathcal{I} \cap \mathcal{J}_{\mathrm{reg}}$ and therefore on $\mathcal{I}$.

The space $\operatorname{im} D\Pi_{\nu-1} D^- = \operatorname{im} D\Pi_{\nu-1}$ is invariant for the scalarly implicit ODE (4.10) since $y = (I - D\Pi_{\nu-1} D^-)u$ satisfies the homogeneous equation

$$\omega_\nu [y' + (D\Pi_{\nu-1} D^-)' y] = 0$$

on $\mathcal{I}$. Taking into account that $\omega_\nu \neq 0$ on a dense set, we derive that $y' + (D\Pi_{\nu-1} D^-)' y = 0$ on $\mathcal{I}$, and therefore a vanishing initial condition for y, which owes to $u = D\Pi_{\nu-1} D^- u$, has as unique solution the trivial one on the whole $\mathcal{I}$.

On the other hand, assuming that $u \in C^1(\mathcal{I}, \mathbb{R}^n)$ together with $v_{\nu-1}, \ldots, v_1 \in C_D^1(\mathcal{I}, \mathbb{R}^m)$ and $v_0 \in C(\mathcal{I}, \mathbb{R}^m)$ satisfy the relations depicted in (4.10)-(4.11), we can parallelize the proof of Theorem 2.3 to show that $x = D^- u + v_{\nu-1} + \ldots + v_0 \in C_D^1(\mathcal{I}, \mathbb{R}^m)$. Additionally, the identity $A(Dx)' + Bx - q = 0$ holds on the dense (in $\mathcal{I}$) set $\mathcal{I} \cap \mathcal{J}_{\mathrm{reg}}$; this is due to the fact that the decoupling (4.10)-(4.11) amounts to the one in Theorem 2.3 at regular points. Now, since the map $A(Dx)' + Bx - q$ is continuous, $A(Dx)' + Bx - q = 0$ remains true on $\mathcal{I}$, and therefore $x = D^- u + v_{\nu-1} + \ldots + v_0$ is a solution of the properly stated DAE (4.3) in $C_D^1(\mathcal{I}, \mathbb{R}^m)$, thus completing the proof. □

Remark 4.2. The stronger requirements within the A-framework of [196], referred to in Remark 4.1, make it possible to derive a decoupling with a singular coefficient ω_ν (that is, without increasing exponents) in front of the algebraic components v_k; cf. (4.11). Find details in Theorem 4.3 of [196]. In general, the chance to reduce further the exponents of ω_ν within (4.10)-(4.11) is an open question.

4.2.2.3 Campbell's example and harmless singularities

As detailed above, Assumptions P1-P3 allow for the construction of a matrix chain for singular linear DAEs and for the characterization of their behavior in terms of the decoupling discussed in Theorem 4.3. The leading coefficient $\omega_\nu(t) = \det G_\nu(t)$ within (4.10)-(4.11) will vanish if and only if G_ν is singular; remember that ν is given by Proposition 4.2. It might be conjectured that a singular point t^* should always result in a singular matrix $G_\nu(t^*)$ but this is not true, as shown by example (4.12) below. We will say that t^* is a *harmless* singular point for (4.3) if Assumptions P1-P3 hold and the resulting matrix mapping $G_\nu(t)$ is nonsingular at t^*.

The following DAE is considered, with $q(t) = 0$, in [40] (p. 120):

$$\begin{pmatrix} 1 & 0 & 0 \\ 0 & 0 & t \\ 0 & 0 & 0 \end{pmatrix} x' + x = q(t). \tag{4.12}$$

This equation can be written as

$$\begin{pmatrix} 1 & 0 & 0 \\ 0 & 0 & t \\ 0 & 0 & 0 \end{pmatrix} \left[\begin{pmatrix} 1 & 0 & 0 \\ 0 & 0 & 0 \\ 0 & 0 & 1 \end{pmatrix} x \right]' + x = q(t), \tag{4.13}$$

which has the form depicted in (4.3) with

$$A = \begin{pmatrix} 1 & 0 & 0 \\ 0 & 0 & t \\ 0 & 0 & 0 \end{pmatrix}, \quad D = \begin{pmatrix} 1 & 0 & 0 \\ 0 & 0 & 0 \\ 0 & 0 & 1 \end{pmatrix}, \quad B_0 = B = I. \tag{4.14}$$

The point $t^* = 0$ is a type 0 singularity since $A(t)$ has a rank drop there. The space $\ker A(t)$ admits however the smooth continuation

$$X(t) = \mathrm{span}\left[\begin{pmatrix} 0 \\ 1 \\ 0 \end{pmatrix} \right]$$

and Assumption P2 is easily checked to hold. The space $K_0(t)$ equals $X(t)$. Set $G_0 = AD = A$ and $D^- = R = \Pi_0 = D$, the latter given in (4.14). This yields

$$M_0 = I - \Pi_0 = \begin{pmatrix} 0 & 0 & 0 \\ 0 & 1 & 0 \\ 0 & 0 & 0 \end{pmatrix}, \quad G_1 = G_0 + B_0 M_0 = \begin{pmatrix} 1 & 0 & 0 \\ 0 & 1 & t \\ 0 & 0 & 0 \end{pmatrix}.$$

The matrix G_1 has constant rank, with

$$N_1(t) = \mathrm{span}\left[\begin{pmatrix} 0 \\ -t \\ 1 \end{pmatrix} \right],$$

yielding the following continuation of the space $N_0(t) \oplus N_1(t)$:

$$K_1(t) = \mathrm{span}\left[\begin{pmatrix} 0 \\ 1 \\ 0 \end{pmatrix}, \begin{pmatrix} 0 \\ 0 \\ 1 \end{pmatrix} \right].$$

4.2. Singularities of properly stated linear time-varying DAEs

Straightforward computations lead to

$$\Pi_1 = \begin{pmatrix} 1 & 0 & 0 \\ 0 & 0 & 0 \\ 0 & 0 & 0 \end{pmatrix}, \quad G_2 = \begin{pmatrix} 1 & 0 & 0 \\ 0 & 1 & t \\ 0 & 0 & 1 \end{pmatrix}.$$

The set of regular points is $\mathbb{R} - \{0\}$, meeting Assumption P1. The DAE has tractability index two at regular points. Assumption P3 can also be checked to hold with Π_0, Π_1 defined above and $\Pi_2 = \Pi_1$. The fact that Assumptions P1-P3 hold on $\mathbb{R}$ supports the construction of the matrix chain on the whole working interval and, in particular, at the origin.

Since the matrix G_2 is nonsingular at the origin, this point is a harmless singularity. We show below, for illustrative purposes, how the decoupling in Theorem 2.3 (which applies here because of the harmless nature of the singularity) describes the solutions of the DAE; note, however, that in this case the solutions can be directly computed from (4.12). Solutions can be easily checked to be given by

$$x = D^- u + v_1 + v_0 = \begin{pmatrix} u_1(t) \\ 0 \\ 0 \end{pmatrix} + \begin{pmatrix} 0 \\ 0 \\ q_3(t) \end{pmatrix} + \begin{pmatrix} 0 \\ q_2(t) - tq_3'(t) \\ 0 \end{pmatrix}$$

provided that $u_1(t)$ comes from the inherent ODE which, on the invariant space

$$\operatorname{im} D\Pi_1 D^- = \operatorname{span} \left[\begin{pmatrix} 1 \\ 0 \\ 0 \end{pmatrix} \right],$$

reads $u_1' + u_1 = q_1(t)$.

The full characterization of harmless singularities is an open problem. Certainly, type ν-A singular points can never be harmless, ν being the index of the DAE in $\mathcal{J}_{\text{reg}}$ under Assumptions P1-P3 (cf. Proposition 4.2). As shown in Proposition 4.4 of [196], type $(\nu-1)$-A singularities cannot be harmless, either. This implies in particular that the harmless phenomenon is specific to higher (≥ 2) index contexts. Remark, incidentally, that the index of (4.12) is two on $\mathbb{R} - \{0\}$.

Finally, the reader is referred to [196, 197] for the discussion of other examples involving singularities of linear DAEs with properly stated leading term.

4.3 Singularities of standard form linear time-varying DAEs

The analysis of singular points of properly stated DAEs performed above can be applied to the standard form problem

$$A(t)x'(t) + E(t)x(t) = q(t), \quad t \in \mathcal{J}, \tag{4.15}$$

as detailed in this Section. In (4.15), $A(t)$, $E(t) \in C(\mathcal{J}, \mathbb{R}^{m \times m})$ and $q(t) \in C(\mathcal{J}, \mathbb{R}^m)$, with $\mathcal{J} \subseteq \mathbb{R}$ an open interval. Smoothness assumptions on $A(t)$ and $E(t)$ will be discussed later and, as in Section 4.2, the excitation $q(t)$ is assumed to be smooth enough as to admit all the involved differentiations (cf. Remark 2.9 on page 53).

Roughly speaking, the idea will be to reformulate (4.15) in the properly stated form (4.16) below, and then apply the taxonomy and results of Section 4.2. Note that the set $\mathcal{J}_{\text{reg}}$ of regular points of (4.15) is well-defined and independent of choice of the projector P in (4.16); this is a direct consequence of Proposition 2.7 (p. 76).

4.3.1 *Classification*

Definition 4.2 is not applicable to the standard form DAE (4.15) at level 0 since the notion of a type 0 singular point there is crucially based on the properly stated form of the problem. Incidentally, we define below type 0 singularities for (4.15) as those for which a local proper statement is not feasible.

Definition 4.3. A point $t^* \in \mathcal{J} - \mathcal{J}_{\text{reg}}$ is said to be a singular point of *type 0* for (4.15) if $\ker A(t)$ does not admit a C^1 basis on any open subinterval $\mathcal{I} \subseteq \mathcal{J}$ with $t^* \in \mathcal{I}$.

Remark 4.3. If $A(t)$ is in C^1, then t^* is a singular point of type 0 if and only if $A(t)$ has a rank drop at t^*; cf. Remark 2.1 on page 36.

Singular points which are not type 0 belong to some open interval where $\ker A(t)$ is in C^1 and hence admit locally a proper statement of the form (2.142), i.e.

$$A(t)(P(t)x(t))' + B(t)x(t) = q(t), \tag{4.16}$$

where $P(t)$ is a C^1 projector along $\ker A(t)$, and $B(t) = E(t) - A(t)P'(t)$.

Type k singular points, with $k \geq 1$, can be now defined in terms of (4.16) following Definition 4.2.

4.3. Singularities of standard form linear time-varying DAEs

Definition 4.4. A point $t^* \in \mathcal{J} - \mathcal{J}_{\text{reg}}$ which is not type 0 is said to be a singular point of *type k-A* or *k-B*, $k \geq 1$, for (4.15), if it has type k-A or k-B, respectively, for (4.16).

Around a k-singular point, there exists a subinterval where the DAE is nice up to level $k-1$ (cf. Definition 4.2). This notion comprises a smoothness requirement on $D\Pi_i D^-$ for $i \leq k-1$, so that smoothness assumptions are implicit in Definition 4.4. As in Theorem 2.4, the standard form of the DAE makes this smoothness requirement amount to a C^1 condition on the characteristic spaces $N_0 \oplus \ldots \oplus N_i$.

Proposition 4.3. *The notions introduced in Definition 4.4 do not depend on the specific choice of $P(t)$ in (4.16).*

Proof. This result parallelizes, in the singular setting, Proposition 2.7; the proof of the latter shows in turn that Theorem 4.6 in [189] applies. In the proof of item (i) within that Theorem, it is shown that the matrices G_i and $\tilde{G}_i$ arising from different choices P, $\tilde{P}$ of the C^1 projector along ker $A(t)$ verify a relation of the form $\tilde{G}_i = G_i \mathcal{F}_i$ for a given nonsingular factor $\mathcal{F}_i$. This shows that the notion of a k-A singular point does not depend on the choice of P. Additionally, due to the form of $\mathcal{F}_i$ one can check, proceeding as in the proof of Theorem 4.1, that an eventual nontrivial intersection $(N_0 \oplus \ldots \oplus N_{k-1}) \cap N_k$ occurs simultaneously for P and $\tilde{P}$, which proves that k-B singularities are also independent of the actual choice of this projector. □

The concepts introduced in Definition 4.4 are also independent of later choices of the projectors and invariant with respect to time-varying coordinate changes and nonsingular rescaling, according to Theorem 4.1. Additionally, the taxonomy depicted above accounts for all the singularities in sufficiently smooth standard form problems, as stated below.

Theorem 4.4. *Let $A(t) \in C^m(\mathcal{J}, \mathbb{R}^{m \times m})$, $E(t) \in C^{m-1}(\mathcal{J}, \mathbb{R}^{m \times m})$ in (4.15). Then every singular point is of type 0, k-A, or k-B, with $1 \leq k \leq m$.*

Proof. If a given singular point is not type 0, then $P(t) = D(t)$ can be taken from C^m, so that $B = E - AP'$ is in C^{m-1}. The hypotheses of Theorem 4.2 then hold for (4.16) and the result follows in a straightforward manner. □

4.3.2 Decoupling

In a neighborhood of a k-singular point with $k \geq 1$, a matrix chain can be constructed for (4.15) up to the k-th step as in Section 2.3, that is, via the relations $G_0 = A$, $B_0 = B = E - AP'$, $G_1 = G_0 + B_0 M_0$, and

$$B_i = (B_{i-1} - G_i \Pi_0 \Pi_i') \Pi_{i-1} \quad (4.17\text{a})$$

$$G_{i+1} = G_i + B_i M_i \quad (4.17\text{b})$$

for $1 \leq i \leq k-1$. Here, Π_0 equals P, whereas Π_i is a C^1 projector along the space $N_0 \oplus \ldots \oplus N_i$ verifying $\operatorname{im} \Pi_i \subseteq \operatorname{im} \Pi_{i-1}$, with $M_i = \Pi_{i-1} - \Pi_i$. In particular, Π_i can be chosen as the *orthogonal* projector along $N_0 \oplus \ldots \oplus N_i$, provided that $P = \Pi_0$ is orthogonal. The characteristic spaces $N_0 \oplus \ldots \oplus N_i$ are in C^1 and allow for such a C^1 choice of Π_i.

Such a k-singularity will be defined either by a rank drop in G_k or by the loss of transversality of $N_k = \ker G_k$ with respect to $N_0 \oplus \ldots \oplus N_{k-1}$. This precludes the construction of the matrix chain beyond this step in the terms assumed in Chapter 2. However, as a particularization of the ideas introduced in Section 4.2, if the space $N_0 \oplus \ldots \oplus N_k$ admits a C^1-continuation through a k-singular point, the construction of the chain and a singular decoupling will still be feasible. Notably, this single requirement comprises (at step k) all the conditions stated in Assumption P3 (p. 147).

Theorem 4.5. *The reformulation (4.16) of the standard form DAE (4.15) is feasible and meets Assumptions P1-P3 on pages 146-147 if*

(i) *the set $\mathcal{J}_{\mathrm{reg}}$ of regular points is dense in $\mathcal{J}$;*
(ii) *there exists a C^1-space $X(t)$ such that $X(t) = \ker A(t)$ on $\mathcal{J}_{\mathrm{reg}}$; and*
(iii) *the space $N_0(t) \oplus \ldots \oplus N_i(t)$ admits a C^1-continuation $K_i(t)$ on $\mathcal{J}$, for $i = 1, \ldots, m-1$.*

If these requirements hold, then

$$x \in C_P^1(\mathcal{I}, \mathbb{R}^m) = \{ x \in C(\mathcal{I}, \mathbb{R}^m) \,/\, Px \in C^1(\mathcal{I}, \mathbb{R}^m) \}$$

solves (4.16) in a subinterval $\mathcal{I} \subseteq \mathcal{J}$ if and only if it can be written as

$$x = u + v_{\nu-1} + \ldots + v_1 + v_0, \quad (4.18)$$

where $u \in C^1(\mathcal{I}, \mathbb{R}^m)$ is a solution of the scalarly implicit inherent ODE

$$\omega_\nu u' - \omega_\nu \Pi_{\nu-1}' u + \Pi_{\nu-1} \operatorname{Adj} G_\nu B u = \Pi_{\nu-1} \operatorname{Adj} G_\nu q \quad (4.19)$$

lying on the locally invariant space $\operatorname{im} \Pi_{\nu-1}$, and the components v_k verify

$$\omega_\nu^{\nu-k} v_k = -\tilde{\mathcal{K}}_k D^- u + \sum_{j=k+1}^{\nu-1} \tilde{\mathcal{N}}_{kj} v_j' + \sum_{j=k+2}^{\nu-1} \tilde{\mathcal{M}}_{kj} v_j + \tilde{\mathcal{L}}_k q \quad (4.20)$$

4.3. Singularities of standard form linear time-varying DAEs

for $k = \nu - 1, \ldots, 0$, with $v_k \in C^1(\mathcal{I}, \mathbb{R}^m)$ for $k > 0$ and $v_0 \in C(\mathcal{I}, \mathbb{R}^m)$. With $\tilde{Q}_k = \operatorname{Adj} G_\nu B_k M_k$, $\tilde{P}_k = \omega_\nu I - \tilde{Q}_k$, the coefficients of (4.20) read

$$\tilde{\mathcal{K}}_k = M_k \tilde{P}_{k+1} \cdots \tilde{P}_{\nu-1} \operatorname{Adj} G_\nu B + M_k(\omega_\nu \tilde{P}_{k+1} \cdots \tilde{P}_{\nu-1} - \omega_\nu^{\nu-k} I)\Pi'_{\nu-1}$$

$$\tilde{\mathcal{N}}_{kj} = \omega_\nu^{\nu-j} M_k \tilde{P}_{k+1} \cdots \tilde{P}_{j-1} \tilde{Q}_j$$

$$\tilde{\mathcal{M}}_{kj} = M_k \left(\sum_{i=k+1}^{j-1} \omega_\nu^{\nu-i} \tilde{P}_{k+1} \cdots \tilde{P}_{i-1}(\tilde{Q}_i - \tilde{P}_i) \right) \Pi_0 M'_j$$

$$\tilde{\mathcal{L}}_k = M_k \tilde{P}_{k+1} \cdots \tilde{P}_{\nu-1} \operatorname{Adj} G_\nu,$$

where the products $\tilde{P}_{k+1} \cdots \tilde{P}_l$ amount to the identity I whenever $l < k+1$.

Proof. Assumption P1 is explicitly stated in (i). Let P be the orthogonal projector along the space $X(t)$, which is in C^1 because so it is $X(t)$. Since the identity $AP = A$ holds on the dense subset $\mathcal{J}_{\text{reg}}$, it remains true on the whole of $\mathcal{J}$. This means that $X(t) \subseteq \ker A(t)$ for all $t \in \mathcal{J}_{\text{reg}}$ and also that the reformulation (4.16) is feasible in this context; the identity $D(t) = P(t)$ shows that Assumption P2 holds for (4.16). Assumption P3 is trivially met via Proposition 4.1.

On the other hand, the characterization of solutions via (4.18), (4.19) and (4.20) follows from Theorem 4.3 exactly as Theorem 2.5 is derived from Theorem 2.3. □

Remark 4.4. Once the existence of the C^1-space $X(t)$ (supporting the introduction of $P(t)$) is assumed, the space $N_0(t) = \ker G_0(t) = \ker A(t)P(t)$ trivially equals $\ker A(t)$ in standard form problems; this means that, *a posteriori*, the assumption on $X(t)$ in (ii) can be reformulated as the existence of a C^1-continuation $K_0(t)$ (namely, $X(t)$) of $N_0(t)$, akin to the statement in (iii) for $i \geq 1$.

As in Section 4.2 the meaning of ν in Theorem 4.5 emanates from Proposition 4.2, and the identities $N_0 \oplus \ldots \oplus N_i = N_0 \oplus \ldots \oplus N_{\nu-1}$, $\Pi_i = \Pi_{\nu-1}$ must be understood for $\nu \leq i \leq m - 1$.

4.3.3 Analytic problems

Analytic, standard form problems virtually fill the scope of other approaches to singular linear time-varying DAEs: see [139, 225] for related discussions in the context of the strangeness and geometric indices, respectively. The above-presented projector framework does not display this restriction. We show below that it applies in particular to standard form, analytic linear DAEs, so that it covers in particular the set of problems considered in these references. For analogous results involving analytic, properly stated DAEs the reader is referred to [197].

In the proof of Theorem 4.6 below we will make repeated use of the following property: an analytic matrix-valued map defined on an open interval $\mathcal{J}$ has constant rank except on a discrete set (i.e. a set composed of isolated points) $\mathcal{S}$ which does not have accumulation points in $\mathcal{J}$. Additionally, the orthogonal projectors along its kernel and along its image are analytic on $\mathcal{J} - \mathcal{S}$ and can be extended as analytic maps on the whole interval $\mathcal{J}$ (see Lemma 2.2 in [225]). This relies on the well-known property that zeros of non-trivial analytic maps are isolated (see e.g. [78]).

We will also use the property $\ker A \cap \ker B = \ker(A^\mathsf{T} A + B^\mathsf{T} B)$ for any two matrices A, B having the same order, and use implicitly the terminological abuse $N_0(t) = \ker A(t)$ supported on Remark 4.4.

Theorem 4.6. *If $A(t)$, $E(t)$ in the standard form DAE (4.15) are analytic on $\mathcal{J}$ and the regular set $\mathcal{J}_{\mathrm{reg}}$ is non-empty, then $\mathcal{J}_{\mathrm{reg}}$ is actually dense in $\mathcal{J}$, and the spaces*

$$N_0(t) \oplus \ldots \oplus N_i(t)$$

admit an analytic continuation $K_i(t)$ on the whole of $\mathcal{J}$ for $i = 0, \ldots, m-1$.

In this situation, the solution characterization described in Theorem 4.5 applies, all the operators within (4.19)-(4.20) being analytic.

Proof. Let P be the analytic continuation to $\mathcal{J}$ of the orthogonal projector along $\ker A$ at maximal rank points. Denoting by $\mathcal{J}_0$ the set of type 0 singular points, we then have that the set of nice at level 0 points $\mathcal{J} - \mathcal{J}_0$ is dense in $\mathcal{J}$. As in Theorem 4.5, the reformulation (4.16) is supported on the fact that $A = AP$ (or, equivalently, $A(I - P) = 0$) holds on the whole of $\mathcal{J}$. Set then $D = R = P$, $G_0 = AP = A$, $B_0 = B = E - AP'$, $\Pi_0 = P$ and $M_0 = I - \Pi_0$. The time-varying space $K_0(t) = \ker P(t) = \ker \Pi_0(t)$ is an analytic continuation of $N_0(t) = \ker A(t) = \ker G_0(t)$.

The analytic matrix mapping $G_1 = G_0 + B_0 M_0$ meets rank deficiencies on a set of isolated points; its intersection with $\mathcal{J} - \mathcal{J}_0$ defines the set $\mathcal{J}_{1A}$ of type 1-A singular points. Write $N_1 = \ker G_1$. From the absence of accumulation points in $\mathcal{J}$ for both $\mathcal{J}_0$ and $\mathcal{J}_{1A}$ it follows that $\mathcal{J} - (\mathcal{J}_0 \cup \mathcal{J}_{1A})$ is dense in $\mathcal{J}$.

Now, $\ker(\Pi_0^\mathsf{T} \Pi_0 + G_1^\mathsf{T} G_1) = \ker(\Pi_0 + G_1^\mathsf{T} G_1)$ equals the intersection $N_0 \cap N_1$ except maybe at type 0 and type 1-A singular points. We have

4.3. Singularities of standard form linear time-varying DAEs

used the fact that orthogonal projectors are symmetric and then $\Pi_0^T \Pi_0 = \Pi_0 \Pi_0 = \Pi_0$. Therefore, excluding types 0 and 1-A, the intersection $N_0 \cap N_1$ is trivial if and only if the analytic matrix map $\Pi_0 + G_1^T G_1$ has maximal rank: since the regular set is non-empty, this maximal rank is met at some point and hence on the whole interval except again on a discrete set without accumulation points in $\mathcal{J}$. Its intersection with $\mathcal{J} - (\mathcal{J}_0 \cup \mathcal{J}_{1A})$ defines the set of singular points of type 1-B and, again, the set of nice at level 1 points $\mathcal{J} - (\mathcal{J}_0 \cup \mathcal{J}_{1A} \cup \mathcal{J}_{1B})$ is dense in $\mathcal{J}$.

Additionally, note that the orthogonal projector Q_1 onto N_1 at nice at level 1 points can be extended as an analytic function on the whole interval. Since $I - \Pi_0$ is, at nice at level 0 points, the orthogonal projector onto N_0, we can express at nice at level 1 points:

$$N_0 \oplus N_1 = (N_0^\perp \cap N_1^\perp)^\perp = (\ker((I - \Pi_0)(I - \Pi_0) + Q_1 Q_1))^\perp$$
$$= (\ker(I - \Pi_0 + Q_1))^\perp = \operatorname{im}(I - \Pi_0 + Q_1),$$

where again we have used the fact that orthogonal projectors are symmetric. This means that, at nice at level 1 points, the direct sum $N_0 \oplus N_1$ can be expressed as the image space of the analytic matrix-valued mapping $I - \Pi_0 + Q_1$; therefore, there exists an analytic matrix map Π_1 which defines the orthogonal projector along $N_0 \oplus N_1$ at nice at level 1 points. The time-varying set $K_1(t) = \ker \Pi_1(t)$ then provides an analytic continuation of $N_0(t) \oplus N_1(t)$ on $\mathcal{J}$.

Set then $M_1 = \Pi_0 - \Pi_1$, $B_1 = (B_0 - G_1 \Pi_0 \Pi_1')\Pi_0$, $G_2 = G_1 + B_1 M_1$ and proceed analogously in order to show that singular points defined by rank deficiencies on G_2 and by the condition $\ker(\Pi_1 + G_2^T G_2) \neq \{0\}$ define a discrete set without accumulation points in $\mathcal{J}$, and that there exists an analytic continuation of the orthogonal projector Π_2 along $(N_0 \oplus N_1) \oplus N_2 = \operatorname{im}(I - \Pi_1 + Q_2)$ and thereby an analytic continuation $K_2(t)$ of the space $N_0(t) \oplus N_1(t) \oplus N_2(t)$.

The proof is completed by repeating the procedure up to the step in which a nonsingular G_ν is met at some point. This must happen on a dense subset of $\mathcal{J}$ and the density of $\mathcal{J}_{\text{reg}}$ follows. Finally, the fact that Theorem 4.5 applies, as well as the analyticity of the operators within (4.19)-(4.20), are straightforward. □

Note that the assumption that the regular set is non-empty cannot be removed from Theorem 4.6 since, for instance, in a constant coefficient DAE defined by a singular matrix pencil all points in $\mathcal{J}$ would be singular.

4.4 Singularities of autonomous quasilinear DAEs

Let us drive our attention back to the quasilinear DAE (3.13), that is,

$$A(x)x' = f(x), \qquad (4.21)$$

where $A(x) \in C^\infty(W_0, \mathbb{R}^{m \times m})$ and $f(x) \in C^\infty(W_0, \mathbb{R}^m)$; the set W_0 is open in $\mathbb{R}^m$. As in Chapter 3, the smoothness requirement can be easily relaxed to work in C^k settings with finite k, and we will occasionally do so.

A seminal reference in this context is the paper [219] by P. Rabier. Broadly speaking, this paper characterizes the behavior of quasilinear ODEs near generic singularities, specifically those which would be later identified with *impasse points*. Here and in the sequel we are using the expression 'quasilinear ODE' to mean problems of the form (4.21) in which $A(x)$ is nonsingular on a dense subset of W_0, the term singularity denoting in this case the set of points with singular $A(x)$. Later developments in this setting can be found in [165, 166, 202, 203, 236, 240, 254, 276, 277, 309]; cf. subsection 4.4.1.

Singular problems have also been addressed in the context of semiexplicit index one DAEs [22, 23, 48, 61, 62, 173, 174, 233, 236, 249, 288]; regarding parametrized systems and, in particular, singularity-induced bifurcations see [19–21, 241, 243, 246, 275, 296–298, 308]. Singular bifurcations in higher index Hessenberg DAEs are studied in [242]. Note that the term 'index' must be used with care in singular cases; cf. Remark 4.11 (p. 178). We will not tackle singular semiexplicit index one DAEs in as much detail as quasilinear ODEs for the reasons discussed below, although some additional remarks on singular semiexplicit index one problems can be found in subsection 4.4.5.

Notably, Chua and Deng [61, 62] studied impasse points in semiexplicit index one equations almost simultaneously to the above-mentioned characterization of the same phenomenon in quasilinear ODEs by Rabier. Only later, within the joint work of Rabier and Rheinboldt [222, 223], the relation between both settings would become clear; the local behavior of a broad family of "singular" quasilinear DAEs, including in particular semiexplicit index one systems, can be characterized in terms of a reduced quasilinear ODE, in the same way as regular DAEs can be locally reduced to an explicit ODE. This idea will be the *leitmotiv* of the present Section, where we show that the notion of a singular quasilinear DAE can be extended in a way which, allowing for a quasilinear ODE reduction, spreads the scope of [222, 223]. The working scenario which allows for this is detailed in subsection 4.4.3 and compiled in Theorem 4.11 (p. 178).

4.4. Singularities of autonomous quasilinear DAEs

Beyond quasilinear ODEs and semiexplicit index one DAEs, a general singularity notion for autonomous quasilinear DAEs is hard to find in the literature. The analysis of singular phenomena in quasilinear DAEs with arbitrary index virtually amounts to the above-mentioned papers of Rabier and Rheinboldt [222, 223]; see also Chapter 7 of [228]. The singular framework of [222, 223, 228] assumes however that the algebraic conditions supporting the reduction process hold at every reduction step, as detailed in 3.4.1. The procedure of Rabier and Rheinboldt ends up with the quasilinear ODE (3.19), in which the leading matrix $A_\nu(u)$ may be singular at some points. This approach accounts for impasse phenomena which, roughly speaking, arise only in the last reduction step; in the authors' terms, their approach does not accommodate features that affect the validity of the reduction procedure. Hence no general singularity notion, apart from the one underlying those last-step singular points, is presented either in [222, 223, 228]. The *k-singularity* notion introduced in 4.4.2 will accommodate, in an invariant manner, singular points arising at any step.

Another type of singular phenomena not included in the framework of Rabier and Rheinboldt stems from the global constant rank condition G1 (p. 95) imposed on the sets W_k. There may well exist points satisfying the condition $f(x) \in \operatorname{im} A(x)|_{T_x W_{k-1}}$ with a rank deficiency in $A(x)|_{T_x W_{k-1}}$. As acknowledged in Remark 26.2 of [228], the impasse framework of Rabier and Rheinboldt suffice to ascertain that $A_\nu(u(t))$ is invertible in (3.19) if $x(t) = \varphi(u(t))$ solves the original DAE (4.21); the trajectories (4.36) and (4.37) for the DAE (4.35), or those depicted in (4.62) for the example (4.61) in 4.4.6.2, will show that this singularity-crossing behavior is therefore out of the scope of their framework. The *inner singularity* concept here introduced for points in $W_k - W_k^{\text{reg}}$ will account for this type of phenomena. The taxonomy of singularities presented in this work will also define as *boundary* singularities the points in $\overline{W_k} - W_k$, generalizing an idea already used in [228] (Section 42).

The key aspect of singular DAEs is the fact that the failure of the regularity conditions before the last reduction step will typically preclude a smooth structure on the sets W_i, since around a k-singularity they may, for $i > k$, no longer coincide with W_i^{reg}. The framework discussed in 4.4.3 will introduce milder assumptions which make it possible to replace, locally around a k-singularity, the sequence (3.44) by another one of the form

$$W_0 \supset W_1^{\text{reg}} \supset \ldots \supset W_k^{\text{reg}} \supset \tilde{W}_{k+1} \supset \ldots \supset \tilde{W}_\nu \overset{\text{loc}}{=} \tilde{W}_{\nu+1} \qquad (4.22)$$

(cf. (4.48)), the new sets $\tilde{W}_i$ being manifolds and allowing for a reduction of the problem to a quasilinear ODE. As detailed in Theorem 4.11, the local dimensions of the manifolds in (4.22) will be now defined by the sequence

$$m = r_0 > r_1 > \ldots > r_k > \tilde{r}_{k+1} > \ldots > \tilde{r}_\nu = \tilde{r}_{\nu+1}. \qquad (4.23)$$

Finally, singular DAEs raise the question of the extent to which these problems actually define *new* dynamic phenomena, not with respect to explicit ODEs but regarding the quasilinear ODE context. The broad scope of the working hypotheses in subsection 4.4.3 somehow answers this in the negative, in the sense that quasilinear ODEs capture the main dynamic phenomena expected for singular DAEs. In particular, in 4.4.4 we show how to lift the taxonomy of noncritical singularities discussed in subsection 4.4.1 to singular quasilinear DAEs, regardless of the index.

This way, we show that the already-mentioned analogy between impasse points in quasilinear ODEs and DAEs [61, 62, 219, 222, 223] extends to other phenomena associated with K, I and IK singularities [277]. The main idea underlying these results is that different problems concerning singularity-crossing phenomena, multiplicity of solutions, bifurcations in parameterized systems, etc., need not be addressed in the DAE context but can be driven to the quasilinear ODE setting. Additionally, this may help to bridge the gap between several theories which, using different terminologies, actually describe the same dynamic phenomena [21–23, 166, 236, 241, 243, 277, 296, 298].

4.4.1 Quasilinear ODEs and impasse phenomena

As indicated above, the term 'quasilinear ODE' will be used to mean problems of the form (4.21) in which the index zero set $W^{\text{ind}0}$ is dense in W_0. We will also use this idea in a local way. In the quasilinear ODE context, a point x^* is often called *regular* for (4.21) if it is index zero, that is, if $A(x^*)$ is nonsingular; this is consistent with the terminology introduced in Chapter 3, since the condition that $W^{\text{ind}0}$ is dense in W_0 implies that $W^{\text{reg}} = W^{\text{ind}0}$.

Within the quasilinear ODE literature, the point x^* is usually said to be a *singularity* if $A(x^*)$ is singular, this notion being implicitly supported on the fact that x^* is in the closure of the regular set by the density hypothesis mentioned above. To avoid later ambiguities we will refer below to these points as *0-singularities*, anticipating a notion which will be defined in general in subsection 4.4.2. Nevertheless, '0-singularity' can be thought of as a synonym for 'singularity' throughout subsection 4.4.1.

4.4. Singularities of autonomous quasilinear DAEs

4.4.1.1 Noncritical singularities

Definition 4.5. A point $x^* \in W_0$ is said to be a *noncritical 0-singularity* of (4.21) if

$$\det A(x^*) = 0, \ (\det A)'(x^*) \neq 0. \tag{4.24}$$

The noncritical condition at x^* implies that the set of rank-deficient points is a hypersurface locally around x^*, and also that $\dim \ker A(x^*) = 1$, i.e., $\operatorname{rk} A(x^*) = m - 1$ (cf. for instance Proposition 2.1 in [219]). Noncritical singularities are termed *regular impasse points* in [277]. We prefer to retain the original name introduced in [219] and, for reasons detailed later, to reserve the term 'impasse point' in this setting for singularities satisfying conditions (4.25) and (4.26) below.

4.4.1.2 Impasse points

Definition 4.6. A noncritical 0-singularity x^* is said to be an *impasse point* for (4.21) if

$$f(x^*) \notin \operatorname{im} A(x^*) \tag{4.25}$$

and

$$(\det A)'(x^*)v \neq 0 \text{ for } v \in \ker A(x^*) - \{0\}. \tag{4.26}$$

The condition depicted in (4.26), comprises in particular the noncritical one at a singular point x^*. It means that $\ker A(x^*)$ is transversal to the singular set $\{x \in W_0 \ / \ \det A(x) = 0\}$ at x^*. If x^* is a noncritical singularity, or under the milder requirement $\operatorname{rk} A(x^*) = m-1$, the transversality condition (4.26) can be equivalently formulated (see e.g. Lemma 39.2 in [228]) as

$$(A'(x^*)v)v \notin \operatorname{im} A(x^*) \text{ for } v \in \ker A(x^*) - \{0\}. \tag{4.27}$$

In this situation a pair of trajectories are known to collapse at x^* in (either backward or forward) finite time with infinite speed, hence displaying a true impasse behavior in the sense already used in circuit theory [61, 62]: see Theorem 4.7 below. It is worth emphasizing at this stage that impasse points can also be displayed in index one and higher index contexts; Definition 4.13 on p. 180 will account for these cases.

This trajectory behavior at impasse points can be explained in terms of the system obtained after multiplying (4.21) by the adjoint matrix $\operatorname{Adj} A(x)$, i.e. the transposed matrix of cofactors of $A(x)$. This yields (cf. [219])

$$\det A(x) x' = \operatorname{Adj} A(x) f(x), \tag{4.28}$$

where we have used the property $\operatorname{Adj} A \cdot A = \det A \cdot I$. Note that (4.28) becomes useless for DAEs in which $A(x)$ is everywhere singular.

With the notation

$$g(x) = \operatorname{Adj} A(x) f(x), \tag{4.29a}$$
$$\omega(x) = \det A(x), \tag{4.29b}$$

and making the dependence on t explicit for later clarity, (4.28) reads

$$\omega(x(t)) x'(t) = g(x(t)). \tag{4.30}$$

System (4.30) depicts exactly the same trajectory behavior as (4.21) on $W^{\mathrm{ind}0}$. In turn, a time reparametrization converts trajectories of

$$\tilde{x}'(s) = g(\tilde{x}(s)) \tag{4.31}$$

on $W^{\mathrm{ind}0}$ into those of (4.30). Indeed, let $\tilde{x}(s)$ be a solution of (4.31) satisfying $\tilde{x}(s) \in W^{\mathrm{ind}0}$ for $s \in (s_0, s_1)$, with $s_0 < 0 < s_1$. Define

$$t = \gamma(s) = \int_0^s \omega(\tilde{x}(\tau)) d\tau. \tag{4.32}$$

The condition $\tilde{x}(s) \in W^{\mathrm{ind}0}$ for all s implies that $\omega(\tilde{x}(s)) = \det A(\tilde{x}(s))$ does not change sign in (s_0, s_1) and then γ is a diffeomorphism of (s_0, s_1) onto some interval (t_0, t_1). If $\omega > 0$, $\gamma(s)$ is increasing and this transformation preserves orientation, whereas if $\omega < 0$ then $\gamma(s)$ is decreasing and the orientation is reversed. In any case, $x(t) = \tilde{x}(\gamma^{-1}(t))$ solves (4.30), since

$$x'(t) = \frac{d\tilde{x}}{ds}(\gamma^{-1}(t)) \frac{d\gamma^{-1}}{dt}(t) = \frac{g(\tilde{x}(\gamma^{-1}(t)))}{\omega(\tilde{x}(\gamma^{-1}(t)))} = \frac{g(x(t))}{\omega(x(t))},$$

the orientation of trajectories being reversed if $\omega < 0$.

As shown below, this idea can be adapted in order to characterize the local behavior of (4.21) at impasse points, thereby simplifying the original proof of Theorem 4.1 in [219]. We make use of the fact that the condition $f(x^*) \notin \operatorname{im} A(x)$ in (4.25) is equivalent to $g(x^*) = \operatorname{Adj} A(x^*) f(x^*) \neq 0$, because of the identities $\operatorname{Adj} A \cdot A = 0$ and $\operatorname{rk} \operatorname{Adj} A = 1$ holding under $\operatorname{rk} A = m - 1$. Note also that $A \cdot \operatorname{Adj} A = 0$ supports reformulating (4.26) as

$$(\det A)'(x^*) g(x^*) \neq 0, \tag{4.33}$$

since $g(x^*) = \operatorname{Adj} A(x^*) f(x^*) \in \ker A(x^*) - \{0\}$. The fact that (4.33) implies in particular that $g(x^*) \neq 0$ makes the condition in (4.33) actually equivalent to (4.25)-(4.26) at a singular point.

4.4. Singularities of autonomous quasilinear DAEs

Theorem 4.7. *Let A, f in (4.21) be in C^1. Assume that a given noncritical 0-singularity x^* is an impasse point, and write $g(x) = \text{Adj } A(x) f(x)$. Then there exists a $\delta > 0$ such that (4.21) has two distinct solutions*

(a) in $C^1((0,\delta), W_0) \cap C^0([0,\delta), W_0)$ if $(\det A)'(x^*) g(x^*) > 0$, or

(b) in $C^1((-\delta, 0), W_0) \cap C^0((-\delta, 0], W_0)$ if $(\det A)'(x^*) g(x^*) < 0$,

with $x(0) = x^*$, whose derivatives blow up at $t = 0$.

Proof. In both cases, there exists a sufficiently small $\alpha > 0$ such that (4.31) has a solution $\tilde{x}(s)$ in $C^1((-\alpha, \alpha), W_0)$ with $\tilde{x}(0) = x^*$; α is assumed to be diminished further if required below.

Under item (a), consider the restrictions γ_1 and γ_2 of the function defined in (4.32) to the subintervals $(-\alpha, 0)$ and $(0, \alpha)$, respectively. The condition $(\det A)'(x^*) g(x^*) > 0$ makes it possible to assume that α is small enough as to guarantee that $\omega = \det A$ verifies $\omega(\tilde{x}(s)) > 0$ if $s \in (0, \alpha)$ and $\omega(\tilde{x}(s)) < 0$ if $s \in (-\alpha, 0)$; therefore, γ_1 maps $(-\alpha, 0)$ onto a subinterval $(0, \delta_1)$ for some $\delta_1 > 0$, and γ_2 maps $(0, \alpha)$ onto $(0, \delta_2)$ with $\delta_2 > 0$ as well. Set $\delta = \min\{\delta_1, \delta_2\}$ and, for $t \in (0, \delta)$,

$$x_1(t) = \tilde{x}(\gamma_1^{-1}(t)),$$
$$x_2(t) = \tilde{x}(\gamma_2^{-1}(t)).$$

Both mappings are easily checked to be solutions of (4.21) in $C^1((0, \delta), W_0)$. Furthermore, for $i = 1, 2$ we have

$$\lim_{t \to 0^+} x_i(t) = \tilde{x}(0) = x^*,$$

so that $x_i(0) = x^*$ extends both solutions to $C^0([0, \delta), W_0)$. Finally,

$$\lim_{t \to 0^+} \|x_i'(t)\| = \lim_{t \to 0^+} \frac{\|g(x(t))\|}{|\omega(x(t))|} = \infty$$

since $g(x^*) \neq 0$, $\omega(x^*) = 0$.

The proof for item (b) proceeds in an entirely analogous manner; it is only worth detailing that now γ in (4.32) maps both subintervals $(-\alpha, 0)$ and $(0, \alpha)$ onto certain subintervals $(\delta_1', 0)$ and $(\delta_2', 0)$ with $\delta_1' < 0$, $\delta_2' < 0$; the corresponding trajectories of (4.21) are defined for negative t and meet the singularity in forward time. □

Both solutions in Theorem 4.7 actually belong to $C^k((0, \delta), W_0)$, with $k \in \{2, 3, \ldots, \infty\}$, if A and f are in C^k.

A simple instance of this behavior is given by the ODEs

$$2xx' = \pm 1, \qquad (4.34)$$

displaying the pair of solutions

$$x = \pm\sqrt{t} \in C^\infty((0,\infty), \mathbb{R}) \cap C^0([0,\infty), \mathbb{R})$$

for the "+" case, and

$$x = \pm\sqrt{-t} \in C^\infty((-\infty,0), \mathbb{R}) \cap C^0((-\infty,0], \mathbb{R})$$

for the "−" sign in (4.34).

Definition 4.7. A noncritical 0-singularity is called a *backward* or a *forward* impasse point if $(\det A)'(x^*)g(x^*) > 0$ or < 0, respectively.

The reason for this terminology should be clear from Theorem 4.7: two trajectories emanate from (resp. terminate at) a backward (resp. forward) impasse point with infinite speed; cf. (4.34). By the condition $(\det A)'(x^*)g(x^*) \neq 0$, when $m \geq 2$ both solutions are transversal to the singular set. Backward and forward impasse points are called *inaccessible* and *accessible*, respectively, in [223, 228]. Local normal forms for quasilinear ODEs near impasse points are discussed in [202, 236, 277, 309].

4.4.1.3 *Image singularities and singularity crossing phenomena*

Although the local behavior at impasse points is the generic one at singularities of quasilinear ODEs, these systems may experience other singular phenomena.

Definition 4.8. A noncritical 0-singularity x^* is called an *I singularity* for (4.21) if $f(x^*) \in \operatorname{im} A(x^*)$ and $(\det A)'(x^*)v \neq 0$ for $v \in \ker A(x^*) - \{0\}$.

The term '*I* singularity' is taken from [277], *I* coming from 'image'. These singular points are characterized by the failure of (4.25), and correspond to the so-called *pseudoequilibrium* points in the semiexplicit index one context of [296–298]. This correspondence will be further examined in subsections 4.4.4 and 4.4.5. These points may accommodate a variety of dynamic phenomena, which includes the existence of smooth trajectories crossing the singular set, non-uniqueness of solutions, the presence of equilibria on the singular manifold, or different bifurcations in parameterized problems; see [19–23, 166, 173, 241–243, 246, 249, 276, 277, 296–298, 308–310] and references therein.

4.4. Singularities of autonomous quasilinear DAEs

The above-mentioned singularity-crossing phenomenon, as well as the non-uniqueness of solutions at certain singular points, can be illustrated by the simple example defined on $\mathbb{R}^2$ by

$$(x+y)x' = x \tag{4.35a}$$

$$y' = -1. \tag{4.35b}$$

The singular set is defined by the straight line $x + y = 0$, which is entirely composed of noncritical 0-singularities. Singularities with $y = -x > 0$ are forward impasse points, whereas those verifying $y = -x < 0$ are backward impasse points. An I singularity is located at the origin, and the reader can easily check that actually two solutions cross smoothly the singular set at this point, namely

$$x(t) = 0, \ y(t) = -t, \tag{4.36}$$

and

$$x(t) = 2t, \ y(t) = -t. \tag{4.37}$$

The behavior of (4.35) around the origin actually reflects a more general property which can also be explained in terms of the system (4.31). For (4.35) this system reads

$$x' = x \tag{4.38a}$$

$$y' = -x - y. \tag{4.38b}$$

The straight line $x = 0$ comprises a one-dimensional *stable manifold* (see for instance Section IX.6 of [127] or pp. 264-269 of [3]) for (4.38). This invariant manifold accommodates the two trajectories $x = 0$, $y = e^{-t}$ and $x = 0$, $y = -e^{-t}$ which converge to the origin. By a suitable adaptation of the time reparametrization depicted in (4.32), it is possible to show that both trajectories can be smoothly joined at the origin to yield the solution (4.36) of the original DAE (4.35). This solution crosses the singularity through the origin in finite time; note that the orientation is reversed for the trajectory on $x = 0$, $y < 0$ since $\omega(x,y) = x + y < 0$ there. This property does not rely on the linear structure of (4.38): see details in [173]. An analogous reasoning applies to the *unstable manifold* $x + 2y = 0$ of (4.38), yielding the second smooth solution (4.37) which crosses as well the singularity through the origin in finite time.

Normal forms for I singularities are discussed in [203, 277]. A more detailed analysis of the flow of singular DAEs near I singularities can be found in [23]; note that the results are stated there for semiexplicit index one problems and the terminology is different. Different results concerning *singular equilibria*, which are particular instances of I singularities, can be found in [22, 240, 245, 249, 254] and references therein.

4.4.1.4 *K and IK singularities*

Finally, the lack of transversality of $\ker A(x^*)$ and the singular set defines K and IK singularities.

Definition 4.9. Let x^* be a noncritical 0-singularity of (4.21) verifying $(\det A)'(x^*)v = 0$ for all $v \in \ker A(x^*)$. Then x^* is called a K *singularity* if $f(x^*) \notin \operatorname{im} A(x^*)$ or an IK *singularity* if, on the contrary, $f(x^*) \in \operatorname{im} A(x^*)$.

This terminology is also taken from [277], K coming from 'kernel'. These singular points, characterized by the failure of (4.26), are essentially the analog in the quasilinear ODE context of the so-called *semi-singular* points discussed for semiexplicit DAEs in [296]; cf. subsection 4.4.5. Related normal forms can be found in [236, 277].

4.4.1.5 *Invariance*

The framework introduced in subsections 4.4.2 and 4.4.3 will allow us to drive systematically the dynamical study of singularities in DAEs to the quasilinear ODE setting. The key step in this regard will be to lift the notions defined above to singular quasilinear DAEs with arbitrary index, as detailed in 4.4.4. In particular, the invariance of the corresponding notions for DAEs (cf. subsection 4.4.4) will be based upon the following invariance result in the index zero context.

Theorem 4.8. *The notions introduced for quasilinear ODEs in Definitions 4.5 - 4.9 are invariant with respect to the local equivalence relation of Definition 3.6.*

This result follows in a straightforward manner from the local equivalence notion introduced in Definition 3.6 (p. 112). Details are left to the reader.

4.4.2 Singular points of quasilinear DAEs

We turn now the attention to the quasilinear DAE (4.21), driving the analysis beyond the index zero scope of subsection 4.4.1.

4.4.2.1 *0-singular points*

The noncritical 0-singularities of Definition 4.5 are particular instances of the below-introduced 0-singular points. Recall that W_1 was defined in (3.14) as $\{x \in W_0 \;/\; f(x) \in \operatorname{im} A(x)\}$, whereas $F : W_0 \times \mathbb{R}^m \to \mathbb{R}^m$

4.4. Singularities of autonomous quasilinear DAEs

reads $F(x,p) = A(x)p - f(x)$ in (3.15). By $\overline{W_1}$ we mean the closure of W_1 in W_0; later on, for $k \geq 1$ $\overline{W_{k+1}}$ will stand for the closure of W_{k+1} in W_k.

Definition 4.10. A point $x^* \in W_0$ is called an *inner 0-singularity* for (4.21) if $x^* \in W_1$ and

(a) either $A(x)$ does not have constant rank in any neighborhood of x^*, or
(b) F is not a submersion at (x^*, p^*), for some p^* verifying $A(x^*)p^* = f(x^*)$.

The point $x^* \in W_0$ is called a *boundary 0-singularity* if $x^* \in \overline{W_1} - W_1$.

Finally, x^* is said to be a *0-singularity* if it is either an inner or a boundary 0-singularity.

Inner 0-singularities as defined above capture the failure of assumptions R1 or R2 in Definition 3.2 (p. 98). The noncritical I and IK singularities arising in Definitions 4.8 and 4.9 are particular cases of inner 0-singularities, whereas impasse points and K singularities in Definitions 4.6 and 4.9 are instances of boundary 0-singularities.

Nevertheless, the reader should at this stage avoid identifying the "0" in the 0-singularity notion introduced above with the index zero setting of 4.4.1; in this regard see also Remark 4.8 on p. 172. As detailed below, the integer prefix $k \geq 0$ in the k-singularity notion refers to the number of feasible reduction steps before a singularity shows up, and 0-singularities may well be displayed in an index one context (see e.g. (4.39) below or the cases tackled in 4.4.6.4 and 4.4.6.5) as well as in higher index problems. In these settings, inner 0-singularities correspond to points of W_1 where a one-step local reduction (3.26) is not feasible, at least in the terms stated in Theorem 3.1. This reflects the idea discussed in Section 4.1, according to which singular points of DAEs are those where the algebraic conditions supporting the reduction fail. Actually, around inner 0-singularities the set W_1 will typically lack a local manifold structure, precluding the application of the results discussed in [228, 236].

Note also that classifying a singularity as an impasse point or as an I, K, or IK singular point will be independent of the reduction step in which the singularity arises: cf. in this direction subsection 4.4.4.

Remark 4.5. Our analysis of inner singular points will be restricted to cases in which it is the constant rank condition in (a) within Definition 4.10 the one that fails. But certainly the failure of the submersion condition may be as well responsible for an inner singularity, as stated in (b). Just as a sample, in the DAE

$$x' = z \quad (4.39a)$$
$$y' = -y \quad (4.39b)$$
$$0 = x^2 + 2y^2 + 2xz, \quad (4.39c)$$

the double cone (4.39c) yields an inner singularity at the origin, where the submersion condition fails. Indeed, F is a submersion if and only if so it is $g(x, y, z) = x^2 + 2y^2 + 2xz$; the failure of this condition is then due to the vanishing of g' at the origin. The local dynamical behavior of (4.39) around the origin is analyzed in [249].

It is worth mentioning at this point that, under the constant rank assumption in (a), according to Lemma 3.1 the submersion condition in (b) does not depend on the choice of p^*. This holds for semiexplicit systems such as (4.39) but also for the general quasilinear DAE (4.21)

Analytic desingularization methods [56], which are beyond the scope of this book, might provide a valuable tool for the systematic analysis of singularities arising from the failure of the submersion condition in item (b) of Definition 4.10. See also [290] for related results in the polynomial setting.

4.4.2.2 *k-singular points*

If $x^* \in W_1$ is a 0-regular point of (4.21) with $m > r_1 > 0$, it may well happen that $\xi^* = \varphi_1^{-1}(x^*)$ be a 0-singular point for the local reduction (3.26). Iteratively, this idea supports the following definition of a k-singularity.

Definition 4.11. A point $x^* \in W_0$ is said to be an *inner* or a *boundary* *k-singularity* for the DAE (4.21), with $k \geq 1$, if it is $(k - 1)$-regular with $r_k > 0$ and $\zeta^* = (\varphi_1 \circ \cdots \circ \varphi_k)^{-1}(x^*)$ is an inner or a boundary 0-singularity, respectively, for the k-th reduction (3.35).

The point x^* is said to be a *k-singularity* if it is either an inner or a boundary k-singularity.

The term *k-singular point* is a synonym for '*k*-singularity', whereas a *singular point* or a *singularity* is any k-singularity with $k \geq 0$. Singular points of quasilinear DAEs are here defined in a way which accommodates in particular singularities of quasilinear ODEs, singular points of semiexplicit index one DAEs, and also the "last-step" singularities of Rabier and Rheinboldt (cf. Sections 41 and 42 in [228]); indeed, the impasse points within $W_\nu - W_{\nu+1}$ considered by Rabier and Rheinboldt can be shown to

4.4. Singularities of autonomous quasilinear DAEs

be boundary ν-singularities, whereas singular points in $\overline{W_\nu} - W_\nu$ would yield boundary $(\nu - 1)$-singularities.

Theorem 4.9. *The notions of an inner and a boundary k-singularity do not depend on the choice of the reduction pairs $(P_1, \varphi_1), \ldots, (P_k, \varphi_k)$ leading to the k-th reduction (3.35). Moreover, these notions are invariant with respect to the local equivalence relation of Definition 3.6.*

Proof. We already know from subsection 3.4.5 that any two k-th local reductions around $(k-1)$-regular points of two locally equivalent DAEs (or, in particular, of a given DAE) are locally equivalent. The claims for inner singularities then follow in a straightforward manner from items (i), (ii) and (iii) of Lemma 3.2 (p. 113). The statements for boundary singularities are due to the relation $\overline{W_1^a} \cap U_a = \phi(\overline{W_1^b} \cap U_b)$ which follows from item (i) of Lemma 3.2 and the fact that ϕ in that Lemma is a diffeomorphism. □

Remark 4.6. The reader can check that x^* is an inner or boundary k-singularity for the original DAE (4.21) if and only if it is an inner or boundary $(k - i)$-singularity, respectively, for an i-th local reduction of (4.21), with $1 \leq i \leq k - 1$.

It remains to figure out conditions under which the local behavior around singular points of quasilinear DAEs with arbitrary index can be characterized; the mere notion of an index in singular settings is still imprecise. These issues are addressed below.

4.4.3 A reduction framework for singular problems

We introduce here rather general working assumptions under which the reduction procedure of Chapter 3 can be adapted in order to accommodate singular problems. Although among singularities only inner ones may admit smooth solutions, these working assumptions hold also at boundary singularities, which may display for instance impasse phenomena.

4.4.3.1 Working hypotheses

Assumption S1 below is aimed to cover cases in which the constant rank assumption R1 in Definition 3.2 (page 98) fails after the k-th reduction step, that is, on $A_k(\zeta)$ (cf. (3.35)), $k = 0$ standing for rank deficiencies in the matrix $A(x)$. This can be the case for both inner and boundary k-singularities. Assumption S1 describes situation very often found in practice in which, despite the rank deficiency, im $A_k(\zeta)$ admits a smooth continuation (sometimes called an 'extension') $L_k(\zeta)$ on a neighborhood of the singularity, satisfying the identity $L_k(\zeta) = \text{im } A_k(\zeta)$ on some dense subset of that neighborhood.

By an r-dimensional C^∞-space $L(x)$ defined on an open set $U \subseteq \mathbb{R}^m$ (with $m \geq 2$; see Remark 2.1 on p. 36 for $m = 1$) we mean below an x-dependent linear space which, for every $x^* \in U$, is spanned on some neighborhood of x^* by r basis mappings depending smoothly on x; this defines an r-dimensional vector bundle structure on $\bigcup_{x \in U} \{x\} \times L(x)$ (cf. e.g. [1]). Note that, if $A(x)$ has constant rank on U, then im $A(x)$ (as well as ker $A(x)$) is a C^∞-space on U; see, in this regard, Proposition 3.4.18(ii) in [1].

Assumption S1. *Let x^* be a k-singularity for (4.21), with $k \geq 0$, and consider the k-th local reduction $A_k(\zeta)\zeta' = f_k(\zeta)$ displayed in (3.35). Write $x^* = \varphi_1 \circ \cdots \circ \varphi_k(\zeta^*)$. There exists an open neighborhood $\tilde{U}_k \subseteq \Omega_k \subseteq \mathbb{R}^{r_k}$ of ζ^* and, for some $\tilde{r}_{k+1} \leq r_k$, an $\tilde{r}_{k+1}$-dimensional C^∞-space $L_k(\zeta)$ defined on $\tilde{U}_k$ such that $\text{im } A_k(\zeta) = L_k(\zeta)$ on some dense subset of $\tilde{U}_k$.*

It is not difficult to check that the imposed density condition implies that the continuation L_k is unique. In Remark 4.9 below we show that the verification of this Assumption does not depend on the choice of the reduction sequence leading to (A_k, f_k). Note also that, for later consistency, we allow $\tilde{r}_{k+1}$ to vanish.

Remark 4.7. It may happen in particular that $\tilde{r}_{k+1} = r_k$: in this case Assumption S1 expresses that A_k is nonsingular on a dense subset of $\tilde{U}_k$, since $L_k(\zeta) = \mathbb{R}^{r_k}$ meets the requirements. We may speak in this situation of a "last-step" singularity. This is essentially the context considered by Rabier and Rheinboldt in [222, 223, 228]. There is no need for further reduction of the DAE, and Theorem 4.11 will apply. This is a particular instance of a singular index k problem, cf. Remark 4.11 on p. 178.

Remark 4.8. If $k = 0$ we assume $\zeta \equiv x$, $\Omega_0 = W_0$, $r_0 = m$, $A_0 = A$ and $f_0 = f$. If additionally Assumption S1 is met with $\tilde{r}_1 = m$ we are led to a singular index zero problem. This is the case for the noncritical singularities of quasilinear ODEs discussed in subsection 4.4.1, since (4.24) implies that $A(x)$ is singular only on a local hypersurface around x^* and then Assumption S1 holds with $L_0(x) = \mathbb{R}^m$.

4.4. Singularities of autonomous quasilinear DAEs

If $\tilde{r}_{k+1} < r_k$, from the structure of $L_k(\zeta)$ there must exist an open neighborhood $\hat{U}_k \subseteq \tilde{U}_k$ of $\zeta^* \in \Omega_k \subseteq \mathbb{R}^{r_k}$ and a smooth, maximal rank matrix-valued map $H_k(\zeta) \in \mathbb{R}^{(r_k - \tilde{r}_{k+1}) \times r_k}$ with $\ker H_k(\zeta) = L_k(\zeta)$ on $\hat{U}_k$, so that $v \in L_k(\zeta)$ if and only if $H_k(\zeta)v = 0$ for $\zeta \in \hat{U}_k$. Note that, if x^* is an inner k-singular point, the set $V_{k+1} = \{\zeta \in \Omega_k \ / \ f_k(\zeta) \in \operatorname{im} A_k(\zeta)\}$ defined in (3.37) cannot be guaranteed to admit a local parametrization near ζ^*, in the terms holding at k-regular points. Near (inner or boundary) k-singularities we will work instead with the set

$$\tilde{V}_{k+1} = \{\zeta \in \hat{U}_k \ / \ f_k(\zeta) \in L_k(\zeta)\} \\ = \{\zeta \in \hat{U}_k \ / \ H_k(\zeta)f_k(\zeta) = 0\} \subseteq \Omega_k, \tag{4.40}$$

which not even locally can be identified with V_{k+1}. But in the setting defined by Assumption S1, we have

$$V_{k+1} \cap \hat{U}_k \subseteq \overline{V_{k+1}} \cap \hat{U}_k \subseteq \tilde{V}_{k+1}. \tag{4.41}$$

Indeed, by the density hypothesis stated in Assumption S1 we have $\operatorname{im} A_k(\zeta) \subseteq L_k(\zeta) = \ker H_k(\zeta)$ for all $\zeta \in \hat{U}_k$, and then $V_{k+1} \cap \hat{U}_k \subseteq \tilde{V}_{k+1}$. The relations (4.41) then follow from the fact that $\tilde{V}_{k+1}$ is closed in $\hat{U}_k$.

Define, for later use,

$$\tilde{W}_{k+1} = \varphi_1 \circ \cdots \circ \varphi_k(\tilde{V}_{k+1}),$$

the obvious analog of (4.41) holding for W_{k+1}, $\overline{W_{k+1}}$ and $\tilde{W}_{k+1}$. The set $\tilde{W}_{k+1}$ can be checked to be independent of the actual choice of the parametrizations $\varphi_1, \ldots, \varphi_k$; cf. Remark 4.10 on page 178. Moreover, defining $L_k(\zeta) = \operatorname{im} A_k(\zeta)$ locally around k-regular points, the set $\tilde{W}_{k+1}$ constructed this way is locally coincident with W_{k+1}. Similar remarks hold for the later defined sets $\tilde{W}_i$, with $i \geq k+2$.

The relation depicted in (4.41) suggests that $\tilde{V}_{k+1}$ may also accommodate a reduction around boundary k-singularities. Under Assumption S2 below, $\tilde{V}_{k+1}$ will admit a local $\tilde{r}_{k+1}$-dimensional parametrization; note that for inner k-singularities the maximal rank condition in Assumption S2 will hold under the analog of the submersion condition R2 (p. 98) arising in the regular context, as stated in Proposition 4.4.

Assumption S2. *Let $x^* = \varphi_1 \circ \cdots \circ \varphi_k(\zeta^*)$ be a k-singularity for (4.21), with $k \geq 0$. If Assumption S1 holds with $\tilde{r}_{k+1} < r_k$, let $\hat{U}_k \subseteq \tilde{U}_k \subseteq \Omega_k$ be an open neighborhood of ζ^* such that $H_k \in C^\infty(\hat{U}_k, \mathbb{R}^{(r_k - \tilde{r}_{k+1}) \times r_k})$ verifies $\ker H_k(\zeta) = L_k(\zeta) \ \forall \zeta \in \hat{U}_k$. Then $H_k(\zeta)f_k(\zeta)$ is a submersion at ζ^*.*

Assumption S2 becomes unnecessary if $\tilde{r}_{k+1} = r_k$ for the reasons indicated in Remark 4.7. It is also worth mentioning that the maximal rank condition stated in Assumption S2 can be proved independent of the specific choice of H_k as long as it satisfies $\ker H_k = L_k$.

As stated above, Assumption S2 applies to both inner and boundary k-singular points. Inner ones verify $\zeta^* \in V_{k+1}$, and hence they admit solutions p^* to $A_k(\zeta^*)p^* - f_k(\zeta^*) = 0$: Assumption S2 then holds if the submersion condition in item R2 of Definition 3.2 is met in the current context, as acknowledged below. Denote by F_k the mapping $\Omega_k \times \mathbb{R}^{r_k} \to \mathbb{R}^{r_k}$ given by $F_k(\zeta, p) = A_k(\zeta)p - f_k(\zeta)$.

Proposition 4.4. *Let x^* be an inner k-singularity for (4.21). If Assumption S1 is met with $\tilde{r}_{k+1} < r_k$, and F_k is a submersion at (ζ^*, p^*) for some p^* satisfying $A_k(\zeta^*)p^* = f_k(\zeta^*)$, then Assumption S2 holds.*

The proof of this result parallelizes exactly the one showing that (ii) implies (i) in Lemma 3.1, since $H_k(\zeta)A_k(\zeta) = 0$ still holds in the present setting because of the relation $\mathrm{im}\, A_k(\zeta) \subseteq L_k(\zeta) = \ker H_k(\zeta)$ following from Assumption S1. Details are therefore omitted.

Remark 4.9. The verification of Assumptions S1 and S2 for a k-singular point with $k \geq 1$ does not depend on the choice of the reduction sequence $(P_1, \varphi_1), \ldots, (P_k, \varphi_k)$ leading to the k-th local reduction (3.35). This is a consequence of Proposition 3.3 together with the fact that the conditions arising in Assumptions S1 and S2 are invariant with respect to local equivalence. Indeed, if $A_k(\zeta)\zeta' = f_k(\zeta)$ and $B_k(\varsigma)\varsigma' = g_k(\varsigma)$ are locally equivalent via E_k, ϕ_k, and $L_{k,a}(\zeta)$ is a continuation of $\mathrm{im}\, A_k(\zeta)$ in the terms of Assumption S1, then $L_{k,b}(\varsigma) = E_k(\varsigma)(L_{k,a}(\phi_k(\varsigma)))$ is a continuation of $\mathrm{im}\, B_k(\varsigma) = \mathrm{im}\, E_k(\varsigma)A_k(\phi_k(\varsigma))\phi_k'(\varsigma)$: this means that if Assumption S1 holds for (A_k, f_k), then it holds for the locally equivalent DAE (B_k, g_k). Also, if Assumption S2 is met for $H_{k,a}(\zeta)$, then it is routine to check that it holds for $H_{k,b}(\varsigma) = H_{k,a}(\phi_k(\varsigma))E_k^{-1}(\varsigma)$.

4.4.3.2 Local reduction of singular quasilinear DAEs

Theorem 4.10 below, which generalizes the one-step local reduction of Theorem 3.1 to singular points as long as they meet Assumptions S1 and S2, is the key result for the extension of the reduction framework to singular problems.

4.4. Singularities of autonomous quasilinear DAEs

Theorem 4.10. *Let $x^* = \varphi_1 \circ \cdots \circ \varphi_k(\zeta^*)$ be a k-singularity for (4.21) satisfying Assumptions S1 and S2 with $0 < \tilde{r}_{k+1} < r_k$. Then there exists an open neighborhood $U_k \subseteq \hat{U}_k \subseteq \Omega_k \subseteq \mathbb{R}^{r_k}$ of ζ^* such that*

(i) *$\tilde{V}_{k+1} \cap U_k$ admits an $\tilde{r}_{k+1}$-dimensional parametrization $\zeta = \tilde{\varphi}_{k+1}(\eta)$ with surjective $\tilde{\varphi}_{k+1} : \Omega_{k+1} \to \tilde{V}_{k+1} \cap U_k$;*
(ii) *there exists a C^∞ matrix-valued map $\tilde{P}_{k+1} : U_k \to \mathbb{R}^{\tilde{r}_{k+1} \times r_k}$ verifying that $\tilde{P}_{k+1}(\zeta)|_{L_k(\zeta)}$ yields an isomorphism $L_k(\zeta) \to \mathbb{R}^{\tilde{r}_{k+1}}$ for all $\zeta \in U_k$.*

For any such $\tilde{\varphi}_{k+1}$, $\tilde{P}_{k+1}$, $\zeta(t)$ is a solution of the k-th reduction

$$A_k(\zeta)\zeta' = f_k(\zeta), \quad \zeta \in \Omega_k \subseteq \mathbb{R}^{r_k} \tag{4.42}$$

within U_k if and only if $\zeta(t) \in \tilde{V}_{k+1}$ for all t and $\eta(t) = \tilde{\varphi}_{k+1}^{-1}(\zeta(t))$ is a solution of

$$\tilde{A}_{k+1}(\eta)\eta' = \tilde{f}_{k+1}(\eta), \quad \eta \in \Omega_{k+1} \subseteq \mathbb{R}^{\tilde{r}_{k+1}} \tag{4.43}$$

with

$$\tilde{A}_{k+1}(\eta) = \tilde{P}_{k+1}(\tilde{\varphi}_{k+1}(\eta))A_k(\tilde{\varphi}_{k+1}(\eta))\tilde{\varphi}'_{k+1}(\eta), \tag{4.44a}$$
$$\tilde{f}_{k+1}(\eta) = \tilde{P}_{k+1}(\tilde{\varphi}_{k+1}(\eta))f_k(\tilde{\varphi}_{k+1}(\eta)). \tag{4.44b}$$

Proof. The existence of the smooth parametrization $\tilde{\varphi}_{k+1}$ follows from (4.40) together with Assumption S2, whereas that of $\tilde{P}_{k+1}$ is due to the smooth structure of $L_k(\zeta)$ in Assumption S1.

Assume that $\zeta(t)$ solves (4.42). Then $f_k(\zeta(t)) \in \operatorname{im} A_k(\zeta(t))$, that is, $\zeta(t) \in V_{k+1}$ and thus $\zeta(t) \in \tilde{V}_{k+1}$ for all t by (4.41). This means that $\eta(t)$ is well-defined by $\zeta(t) = \tilde{\varphi}_{k+1}(\eta(t))$: premultiplying (4.42) by $\tilde{P}_{k+1}(\tilde{\varphi}_{k+1}(\eta(t)))$ and inserting $\zeta(t) = \tilde{\varphi}_{k+1}(\eta(t))$, $\zeta'(t) = \tilde{\varphi}'_{k+1}(\eta(t))\eta'(t)$ in the resulting equation, we obtain (4.43).

Conversely, the assumption that (4.43) holds can be written as

$$\tilde{P}_{k+1}(\tilde{\varphi}_{k+1}(\eta))A_k(\tilde{\varphi}_{k+1}(\eta))\tilde{\varphi}'_{k+1}(\eta)\eta' = \tilde{P}_{k+1}(\tilde{\varphi}_{k+1}(\eta))f_k(\tilde{\varphi}_{k+1}(\eta))$$

or, in terms of $\zeta = \tilde{\varphi}_{k+1}(\eta)$,

$$\tilde{P}_{k+1}(\zeta)A_k(\zeta)\zeta' = \tilde{P}_{k+1}(\zeta)f_k(\zeta). \tag{4.45}$$

If we show that $A_k(\zeta)\zeta' \in L_k(\zeta)$, $f_k(\zeta) \in L_k(\zeta)$, the identity (4.45) would yield (4.42) due to the isomorphism $\tilde{P}_{k+1}(\zeta)|_{L_k(\zeta)} : L_k(\zeta) \to \mathbb{R}^{\tilde{r}_{k+1}}$. Indeed, the relation $A_k(\zeta)\zeta' \in L_k(\zeta)$ holds trivially due to $\operatorname{im} A_k(\zeta) \subseteq L_k(\zeta)$, whereas $\zeta = \tilde{\varphi}_{k+1}(\eta) \in \tilde{V}_{k+1}$ means that $f_k(\zeta) \in L_k(\zeta)$ by (4.40). $\square$

As it should be expected, local equivalence is preserved by this reduction, as detailed below. By a singular $(k+1)$-reduction we mean the reduction (4.43) derived from (4.21) via (4.42) in Theorem 4.10. Note however that the key reduction step here is the one from (4.42) to (4.43).

Proposition 4.5. *Any two singular $(k+1)$-reductions of locally equivalent quasilinear DAEs around k-singularities satisfying Assumptions S1 and S2 are locally equivalent.*

In particular, any two singular $(k+1)$-reductions of a quasilinear DAE around a k-singularity satisfying Assumptions S1 and S2 are locally equivalent.

Proof. Note that k-singularities are $(k-1)$-regular points. From the results of Chapter 3 we know that any two k-steps regular reductions, say $A_k(\zeta)\zeta' = f_k(\zeta)$ and $B_k(\varsigma)\varsigma' = g_k(\varsigma)$, of locally equivalent DAEs are locally equivalent. See also Remark 4.9.

Now, let $\tilde{A}_{k+1}(\eta)\eta' = \tilde{f}_{k+1}(\eta)$ and $\tilde{B}_{k+1}(v)v' = \tilde{g}_{k+1}(v)$ be derived from $A_k(\zeta)\zeta' = f_k(\zeta)$ and $B_k(\varsigma)\varsigma' = g_k(\varsigma)$ via certain reduction operators $\tilde{P}(\zeta)$, $\zeta = \tilde{\varphi}(\eta)$ and $\tilde{Q}(\varsigma)$, $\varsigma = \tilde{\psi}(v)$, respectively, as in Theorem 4.10. For the sake of notational simplicity we use $\tilde{P}$, $\tilde{\varphi}$, $\tilde{Q}$ and $\tilde{\psi}$ instead of $\tilde{P}_{k+1}$, $\tilde{\varphi}_{k+1}$, $\tilde{Q}_{k+1}$ and $\tilde{\psi}_{k+1}$. By $\tilde{V}^a_{k+1}$ and $\tilde{V}^b_{k+1}$ we denote the $\tilde{V}_{k+1}$-sets defined for (A_k, f_k) and (B_k, g_k) according to (4.40).

As in Proposition 3.3 (page 114), the diffeomorphism $\eta = \phi_{k+1}(v)$ is given by $\phi_{k+1} = \tilde{\varphi}^{-1} \circ \phi_k \circ \tilde{\psi}$: for it to be well-defined we only need to check that $\tilde{V}^b_{k+1}$ is mapped onto $\tilde{V}^a_{k+1}$ via ϕ_k. But the condition $g_k(\varsigma) \in L_{k,b}(\varsigma)$ defining $\tilde{V}^b_{k+1}$ reads $E_k(\varsigma)f_k(\phi_k(\varsigma)) \in \text{im}\, E_k(\varsigma)|_{L_{k,a}(\phi_k(\varsigma))}$, which amounts to $f_k(\zeta) \in L_{k,a}(\zeta)$ with $\zeta = \phi_k(\varsigma)$, that is, $\zeta = \phi_k(\varsigma) \in \tilde{V}^a_{k+1}$.

Parallelizing again Proposition 3.3, let $\hat{P}(\zeta)$ be a C^∞ matrix-valued map such that $\hat{P}(\zeta)\tilde{P}(\zeta)|_{L_{k,a}(\zeta)} = \text{id}|_{L_{k,a}(\zeta)}$. Together with the property $\text{im}\, A_k(\zeta) \subseteq L_{k,a}(\zeta)$, this yields

$$\hat{P}(\phi_k(\tilde{\psi}(v)))\tilde{P}(\tilde{\varphi}(\phi_{k+1}(v)))A_k(\tilde{\varphi}(\phi_{k+1}(v))) = A_k(\phi_k(\tilde{\psi}(v))).$$

Similarly $\hat{P}(\phi_k(\tilde{\psi}(v)))\tilde{P}(\tilde{\varphi}(\phi_{k+1}(v)))f_k(\tilde{\varphi}(\phi_{k+1}(v))) = f_k(\phi_k(\tilde{\psi}(v)))$ since $\phi_k(\tilde{\psi}(v)) \in \tilde{V}^a_{k+1}$ and then $f_k(\phi_k(\tilde{\psi}(v))) \in L_{k,a}(\phi_k(\tilde{\psi}(v)))$. Letting

$$E_{k+1}(v) = \tilde{Q}(\tilde{\psi}(v))E_k(\tilde{\psi}(v))\hat{P}(\phi_k(\tilde{\psi}(v))),$$

we finally derive $\tilde{B}_{k+1}(v) = E_{k+1}(v)\tilde{A}_{k+1}(\phi_{k+1}(v))\phi'_{k+1}(v)$ and $\tilde{g}_{k+1}(v) = E_{k+1}(v)\tilde{f}_{k+1}(\phi_{k+1}(v))$, as required. □

4.4. Singularities of autonomous quasilinear DAEs

In the setting defined by Theorem 4.10, a one-step singular reduction is again suitable for assessment for the reduction (4.43). Define

$$V_{k+2}^s = \{\eta \in \Omega_{k+1} \, / \, \tilde{f}_{k+1}(\eta) \in \operatorname{im} \tilde{A}_{k+1}(\eta)\} \subseteq \Omega_{k+1} \subseteq \mathbb{R}^{\tilde{r}_{k+1}} \quad (4.46)$$

$$W_{k+2}^s = \varphi_1 \circ \cdots \circ \varphi_k \circ \tilde{\varphi}_{k+1}(V_{k+2}^s) \subseteq \tilde{W}_{k+1}. \quad (4.47)$$

The superscript s keeps track of the fact that a singularity has shown up at some step, allowing us to distinguish the sets V_i^s from the V_i ones constructed around regular points, although (3.37) and (4.46) are formally identical.

Let $\eta^* \in \Omega_{k+1}$ be defined by the relation $x^* = \varphi_1 \circ \cdots \circ \varphi_k \circ \tilde{\varphi}_{k+1}(\eta^*)$. If $\eta^* \in \overline{V_{k+2}^s}$ we may check again if Assumptions S1 and S2 hold (replacing k by $k+1$, A_k by $\tilde{A}_{k+1}$, etc.) for the reduced problem (4.43), that is, if there exists an $\tilde{r}_{k+2}$-dimensional continuation $L_{k+1}(\eta)$ of $\operatorname{im} \tilde{A}_{k+1}(\eta)$ and, when $\tilde{r}_{k+2} < \tilde{r}_{k+1}$, if a matrix-valued map $H_{k+1}(\eta)$ verifying $\ker H_{k+1}(\eta) = L_{k+1}(\eta)$ for $\eta \in \hat{U}_{k+1}$ makes $H_{k+1}\tilde{f}_{k+1}$ a submersion at η^*. These properties are again independent of the reduction operators, i.e., if Assumptions S1 and S2 hold at step $k+2$ for a given choice of $\tilde{\varphi}_{k+1}$ and $\tilde{P}_{k+1}$ in Theorem 4.10, then they hold for any other choice of these reduction operators, and the same is true in subsequent steps; this is a consequence of Remark 4.9 and Proposition 4.5.

If these assumptions are met, then

$$\begin{aligned}\tilde{V}_{k+2} &= \{\eta \in \hat{U}_{k+1} \, / \, \tilde{f}_{k+1}(\eta) \in L_{k+1}(\eta)\} \\ &= \{\eta \in \hat{U}_{k+1} \, / \, H_{k+1}(\eta)\tilde{f}_{k+1}(\eta) = 0\} \subseteq \Omega_{k+1}\end{aligned}$$

as well as

$$\tilde{W}_{k+2} = \varphi_1 \circ \cdots \circ \varphi_k \circ \tilde{\varphi}_{k+1}(\tilde{V}_{k+2}) \subseteq \tilde{W}_{k+1}$$

will admit local $\tilde{r}_{k+2}$-dimensional C^∞ structures, a local reduction in terms of $\tilde{A}_{k+2}$, $\tilde{f}_{k+2}$ will be possible following Theorem 4.10, and the procedure can be performed one-step further. Note that in particular $\tilde{A}_{k+1}$ may well have constant rank, as in the example considered in 4.4.6.3. In this case, the $(k+2)$-th reduction step would essentially be a regular one, and the singularity may not have an effect beyond that step.

This way, instead of the sequence of manifolds (3.44) constructed in the regular setting, we build up a sequence of the form

$$W_0 \supset W_1^{\text{reg}} \supset \ldots \supset W_k^{\text{reg}} \supset \tilde{W}_{k+1} \supset \ldots \supset \tilde{W}_\nu \stackrel{\text{loc}}{=} \tilde{W}_{\nu+1}, \quad (4.48)$$

the local stabilization after the ν-th step holding in the setting of Theorem 4.11 below. The importance of this construction stems from the fact that

W_{k+1} and later on W_{k+2}^s and subsequent sets may fail to have a C^∞ structure near an inner k-singularity, whereas $\tilde{W}_{k+1}$, $\tilde{W}_{k+2}$, etc., display a local C^∞ structure, allowing for a local reduction of the DAE. These manifolds comprise in addition the closures $\overline{W_{k+1}}$, $\overline{W_{k+2}^s}$, etc., and therefore may also accommodate boundary singularities.

Remark 4.10. The sets $\tilde{W}_{k+1}, \ldots, \tilde{W}_\nu$, $W_{k+2}^s \ldots, W_\nu^s$ do not depend on the choice of the sequence of reduction pairs (P_1, φ_1), ..., (P_k, φ_k), $(\tilde{P}_{k+1}, \tilde{\varphi}_{k+1})$, ..., $(\tilde{P}_\nu, \tilde{\varphi}_\nu)$. Indeed, the independence of the $\tilde{W}_i$ manifolds follows from the local property $\tilde{V}_{k+1}^a = \phi_k(\tilde{V}_{k+1}^b)$ arising in the proof of Proposition 4.5, whereas that of the W_i^s sets only requires item (i) of Lemma 3.2.

The repeated application of the one-step singular reduction in Theorem 4.10 yields the following analog of Theorem 3.2; the meaning of U parallelizes the one explained there. In the particular case $k = \nu$, the symbols $\tilde{r}_\nu$, $\tilde{A}_\nu$, $\tilde{f}_\nu$ and $\tilde{W}_\nu$ below must be replaced by r_ν, A_ν, f_ν and W_ν. Since no singular reduction is required for these last-step singular points (cf. Remark 4.7 on p. 172), in this situation Theorem 4.11 virtually amounts to the statement about (3.40) within Theorem 3.2, consistently with the fact that the setting of Rabier and Rheinboldt discussed in [228] accommodates last-step singularities.

Theorem 4.11. *Let $x^* \in W_0$ be a k-singularity for (4.21), $k \geq 0$. Suppose that Assumptions S1 and S2 hold in steps $k+1, k+2, \ldots, \nu$ of the singular reduction process described above with*

$$m = r_0 > r_1 > \ldots > r_k > \tilde{r}_{k+1} > \tilde{r}_{k+2} > \ldots > \tilde{r}_\nu > 0, \quad (4.49)$$

and that Assumption S1 is met in step $\nu + 1$ with $\tilde{r}_\nu = \tilde{r}_{\nu+1}$. Let

$$\tilde{A}_\nu(u) u' = \tilde{f}_\nu(u), \quad u \in \Omega_\nu \subseteq \mathbb{R}^{\tilde{r}_\nu} \quad (4.50)$$

be a ν-th step reduction of (4.21) given by a sequence of reduction pairs (P_1, φ_1), ..., (P_k, φ_k), $(\tilde{P}_{k+1}, \tilde{\varphi}_{k+1})$, ..., $(\tilde{P}_\nu, \tilde{\varphi}_\nu)$ on a neighborhood Ω_ν of $u^* = (\varphi_1 \circ \cdots \circ \varphi_k \circ \tilde{\varphi}_{k+1} \circ \cdots \circ \tilde{\varphi}_\nu)^{-1}(x^*)$.

Then $x(t)$ is a solution of (4.21) within U if and only if $x(t) \in \tilde{W}_\nu$ for all t and $u(t) = (\varphi_1 \circ \cdots \circ \varphi_k \circ \tilde{\varphi}_{k+1} \circ \cdots \circ \tilde{\varphi}_\nu)^{-1}(x(t))$ solves (4.50).

Remark 4.11. The requirement that Assumption S1 holds in the last step with $\tilde{r}_\nu = \tilde{r}_{\nu+1} > 0$ amounts to saying that $\tilde{A}_\nu$ (or A_k if $\nu = k$) is nonsingular on some dense subset of $\tilde{U}_\nu \subseteq \Omega_\nu$. This means that points in this dense subset are regular with index ν. We speak of a k-singularity x^* as

4.4. Singularities of autonomous quasilinear DAEs

a *singular index* ν point when the hypotheses of this Theorem hold, but also when Assumptions S1 and S2 are met in steps $k+1, k+2, \ldots, \nu$ with $m = r_0 > \ldots > r_k > \tilde{r}_{k+1} > \ldots > \tilde{r}_\nu = 0$. In these situations the DAE (4.21) can be locally thought of as a singular index ν problem.

The difference between Theorem 4.11 and the regular index ν statement within Theorem 3.2 is that now $\tilde{A}_\nu(u^*)$ will be typically (although not always, as it happens at the below-defined harmless singularities) a singular matrix. This may be due to a rank deficiency arising at any reduction step, not necessarily at the last one. Theorem 4.11 hence drives the local analysis of a broad family of singular quasilinear DAEs not to the context of explicit ODEs, but to the quasilinear ODE setting discussed in subsection 4.4.1, as we aimed to. Note finally that any two ν-th step singular reductions of the form (4.50) of a given DAE are locally equivalent, by an inductive application of Proposition 4.5.

4.4.4 Dynamical aspects

Theorem 3.2 in Chapter 3 accounts for the fact that the local dynamical behavior of quasilinear DAEs near regular points can be described in terms of an explicit ODE. Theorem 4.11 can be seen as its counterpart in the singular setting, showing that the dynamics near singular points verifying Assumptions S1 and S2 is described by a quasilinear ODE, namely the reduction (4.50). Thereby Theorem 4.11 opens a way to lift the taxonomy of singularities of quasilinear ODEs described in Definitions 4.5-4.9 to quasilinear DAEs with arbitrary index. As indicated in subsection 4.4.1, different types of singularities in quasilinear ODEs account for different dynamic phenomena, which can be now systematically addressed in the DAE setting.

Note first that the rank deficiencies defining singular points of DAEs do not necessarily yield a singularity in the quasilinear ODE reduction. As for linear time-varying problems, the notion of a *harmless* singularity accommodates this behavior.

Definition 4.12. A k-singularity x^* of (4.21), $k \geq 0$, is said to be *harmless* either if the hypotheses of Theorem 4.11 hold and $A_\nu(u^*)$ is nonsingular, with $u^* = (\varphi_1 \circ \cdots \circ \varphi_k \circ \tilde{\varphi}_{k+1} \circ \cdots \circ \tilde{\varphi}_\nu)^{-1}(x^*)$, or if Assumptions S1 and S2 hold in steps $k+1, \ldots, \nu$ with $m = r_0 > \ldots > r_k > \tilde{r}_{k+1} > \ldots > \tilde{r}_\nu = 0$.

From Theorem 4.11 it follows that the local behavior around a harmless singularity is entirely analogous to the one near a regular point with index ν. Instances of harmless singular points can be found in 4.4.6.3 and 4.4.6.6 below.

Definition 4.13. Let $x^* \in W_0$ be a k-singularity of the quasilinear DAE (4.21), with $k \geq 0$, for which the hypotheses of Theorem 4.11 hold. Write $u^* = (\varphi_1 \circ \cdots \circ \varphi_k \circ \tilde{\varphi}_{k+1} \circ \cdots \circ \tilde{\varphi}_\nu)^{-1}(x^*)$.

If $A_\nu(u^*)$ is singular, x^* is said to be a *noncritical, backward* or *forward impasse, I, K* or *IK singularity* if so is u^* for the singular reduction (4.50).

It is worth indicating that, in noncritical cases, if the hypotheses of Theorem 4.11 hold up to step ν, then Assumption S1 is automatically met in step $\nu + 1$ with $\tilde{r}_\nu = \tilde{r}_{\nu+1}$, since $\det \tilde{A}_\nu$ will only vanish in a hypersurface around the singular point.

Theorem 4.12. *The notions of a harmless, noncritical, backward or forward impasse, I, K, and IK singularity of a quasilinear DAE do not depend on the specific choice of the reduction pairs* (P_1, φ_1), ..., (P_k, φ_k), $(\tilde{P}_{k+1}, \tilde{\varphi}_{k+1})$, ..., $(\tilde{P}_\nu, \tilde{\varphi}_\nu)$.

Moreover, these concepts are invariant with respect to the local equivalence relation of Definition 3.6.

The proof of this result can be directly derived from the corresponding property for quasilinear ODEs stated in Theorem 4.8, using the fact that singular reductions of locally equivalent DAEs are locally equivalent according to Proposition 4.5.

Definition 4.13, via Theorem 4.11, extends several dynamic notions from the theory of quasilinear ODEs to singular DAEs. Thereby, different dynamical phenomena involving impasse points, singularity-crossing phenomena, multiplicity of solutions, etc., need not be studied in the differential-algebraic setting but can instead be systematically driven to the somehow simpler context of quasilinear ODEs. Note that rich and seemingly different theories, directed to different structural forms but with the same underlying dynamic phenomena, have been developed in parallel in the last decades; besides [61, 62] and [219], compare e.g. [21–23, 241, 243, 296–298] with [236] or [166, 277, 309].

The chance to describe the behavior of autonomous singular DAEs in terms of quasilinear ODEs in the working scenario of subsection 4.4.3 suggests that new dynamical phenomena for singular problems should be

4.4. Singularities of autonomous quasilinear DAEs

sought in two contexts. The first one is defined by systems undergoing rank changes and for which Assumption S1 does not hold. Note that the density hypothesis within it excludes DAEs displaying index changes from the present framework. The second context is defined by the failing of Assumption S2 and, in particular, by inner singularities for which the submersion condition in Proposition 4.4 is not met; cf. Remark 4.5 on page 169.

4.4.5 Singular semiexplicit index one DAEs

When dealing with singular DAEs, depending on the structural form of the problem one certainly needs to compute the specific algebraic conditions which characterize the notions described in Definition 4.13. We undertake here this task for semiexplicit DAEs

$$y' = h(y, z) \tag{4.51a}$$
$$0 = g(y, z), \tag{4.51b}$$

with $h \in C^\infty(W_0, \mathbb{R}^r)$, $g \in C^\infty(W_0, \mathbb{R}^p)$, and W_0 open in $\mathbb{R}^{r+p}$.

As indicated in 3.4.6.1, $(y, z) \in W_0$ is 0-regular if $g(y, z) = 0$ and g is a submersion at (y, z), that is,

$$\operatorname{rk} g'(y, z) = p. \tag{4.52}$$

The failure of this submersion condition defines (inner) 0-singularities of (4.51), as it happens e.g. at the origin for the DAE (4.39) (cf. Remark 4.5, p. 169).

Let (y^*, z^*) be a 0-regular point. We know from Proposition 3.5 (p. 119) that (y^*, z^*) is regular with index one if and only if $g_z(y^*, z^*)$ is nonsingular. Assume in the sequel that $g_z(y^*, z^*)$ is singular in order to characterize, in terms of h and g in (4.51), the notions introduced in Definition 4.13.

In this regard, since g is a submersion at (y^*, z^*) we may split the (y, z) coordinates as

$$y = (\underbrace{y_1, \ldots, y_{\hat{r}}}_{\hat{y}}, \underbrace{y_{\hat{r}+1}, \ldots, y_r}_{\bar{y}}), \ z = (\underbrace{z_1, \ldots, z_{\bar{r}}}_{\hat{z}}, \underbrace{z_{\bar{r}+1}, \ldots, z_p}_{\bar{z}}),$$

in a way such that the derivative $(g_{\bar{y}} \ g_{\bar{z}})$ is nonsingular at (y^*, z^*), having assumed w.l.o.g. that the $\bar{y}$ and $\bar{z}$ variables are the last ones within y and z, respectively. Applying the implicit function theorem we may then describe the set $g = 0$ locally around (y^*, z^*) by means of certain relations of the form

$$\bar{y} = \psi_1(\hat{y}, \hat{z}) \tag{4.53a}$$
$$\bar{z} = \psi_2(\hat{y}, \hat{z}). \tag{4.53b}$$

This yields the local r-dimensional parametrization
$$u = (u_1, u_2) \to \varphi(u_1, u_2) = ((u_1, \psi_1(u_1, u_2)), (u_2, \psi_2(u_1, u_2)))$$
of the manifold $W_1 = \{(y, z) \in W_0 \;/\; g(y, z) = 0\}$; here u_1 and u_2 are $\hat{r}$- and $\bar{r}$-dimensional, respectively, with $\hat{r} + \bar{r} = r$. Set $u^* = \varphi^{-1}(y^*, z^*)$.

Via $P_1 = (I_r \; 0)$, the corresponding reduction $A_1(u)u' = f_1(u)$ of (4.51) reads
$$\begin{pmatrix} I_{\hat{r}} & 0 \\ \psi_{1u_1}(u_1, u_2) & \psi_{1u_2}(u_1, u_2) \end{pmatrix} \begin{pmatrix} u_1' \\ u_2' \end{pmatrix} = h(\varphi(u_1, u_2)). \tag{4.54}$$

We know from Proposition 3.5 that $A_1(u)$ (and thereby $\psi_{1u_2}(u)$) is nonsingular if and only if so it is $g_z(\varphi(u))$; mind that $\psi_{1u_2}(u)$ and $\varphi(u)$ stand, with notational abuse, for $\psi_{1u_2}(u_1, u_2)$ and $\varphi(u_1, u_2)$, respectively. For later use, it is anyway of interest to assess this property directly in terms of (4.54). Proceeding as in 3.4.7.2, via the implicit function theorem the coefficient matrix
$$A_1(u) = \begin{pmatrix} I_{\hat{r}} & 0 \\ \psi_{1u_1}(u) & \psi_{1u_2}(u) \end{pmatrix}$$
can be checked to be the Schur complement (cf. Lemma 3.3 on page 127) of the submatrix $(g_{\bar{y}} \; g_{\bar{z}})$ in
$$G(u) = \begin{pmatrix} I_{\hat{r}} & 0 & 0 & 0 \\ 0 & 0 & I_{\bar{r}} & 0 \\ g_{\hat{y}} & g_{\hat{z}} & g_{\bar{y}} & g_{\bar{z}} \end{pmatrix},$$
the derivatives being evaluated at $\varphi(u)$. This implies that $\psi_{1u_2}(u)$ is nonsingular if and only if so it is $(g_{\hat{z}}(\varphi(u)) \; g_{\bar{z}}(\varphi(u)))$, i.e., $g_z(\varphi(u))$.

Furthermore, from (3.87) it follows that $\det G = \det(g_{\bar{y}} \; g_{\bar{z}}) \cdot \det A_1$, and therefore
$$\det A_1(u) = \alpha(u) \det g_z(\varphi(u)) \tag{4.55}$$
around u^*, $\alpha(u)$ being a non-vanishing scalar factor.

Now, if g_z is singular at $(y^*, z^*) = \varphi(u^*)$, then the noncritical condition $(\det A_1)'(u^*) \neq 0$ yields, in terms of the original coordinates, the maximal rank requirement
$$\text{rk} \left. \begin{pmatrix} (\det g_z)' \\ g' \end{pmatrix} \right|_{(y^*, z^*)} = p + 1. \tag{4.56}$$

Indeed, from (4.55) and the fact that $\det A_1(u^*) = \det g_z(\varphi(u^*)) = 0$ we have $(\det A_1)'(u^*) = \alpha(u^*)(\det g_z)'(\varphi(u^*))\varphi'(u^*)$. The non-vanishing of

4.4. Singularities of autonomous quasilinear DAEs

$(\det A_1)'(u^*)$ is then equivalent to that of $(\det g_z)'(\varphi(u^*))\varphi'(u^*)$, which in turn yields (4.56) since $\operatorname{im}\varphi'(u^*)$ spans $T_{\varphi(u^*)}W_1 = \ker g'(\varphi(u^*))$.

Therefore, (y^*, z^*) is a noncritical 1-singular point of the semiexplicit DAE (4.51) if and only if $g(y^*, z^*) = 0$, $\det g_z(y^*, z^*) = 0$ and (4.56) holds. Note that (4.52) follows from (4.56). Also, using just $(\det g_z)'(y^*, z^*) \neq 0$ from (4.56) we conclude that it must be $\operatorname{rk} g_z(y^*, z^*) = p - 1$.

Analogously, the condition $(\det A_1)'(u^*)v \neq 0$ for $v \in \ker A_1(u^*) - \{0\}$ reads

$$(\det g_z)_z(y^*, z^*)v \neq 0 \text{ for } v \in \ker g_z(y^*, z^*) - \{0\}. \tag{4.57}$$

This is due to the fact that the condition depicted in (4.57) is equivalent to $\ker g_z(y^*, z^*) \cap \ker(\det g_z)_z(y^*, z^*) = \{0\}$; now, the existence of a non-vanishing vector v within $\ker(\det A_1)'(u^*) \cap \ker A_1(u^*)$ would be equivalent to that of a non-vanishing $\overline{v} = \varphi'(u^*)v \in T_{\varphi(u^*)}W_1$ belonging to $\ker(\det g_z)'(\varphi(u^*)) \cap \ker A$ with $A = \text{block-diag}(I_r, 0_p)$, that is

$$\overline{v} \in \ker \begin{pmatrix} g_y(y^*, z^*) & g_z(y^*, z^*) \\ (\det g_z)_y(y^*, z^*) & (\det g_z)_z(y^*, z^*) \\ I_r & 0 \end{pmatrix} - \{0\},$$

which would yield $\ker g_z(y^*, z^*) \cap \ker(\det g_z)_z(y^*, z^*) \neq \{0\}$.

As we did in (4.27), the requirement stated in (4.57) can be rewritten as

$$(g_{zz}(y^*, z^*)v)v \notin \operatorname{im} g_z(y^*, z^*) \text{ for } v \in \ker g_z(y^*, z^*) - \{0\}.$$

In turn, the condition $f_1(u^*) \notin \operatorname{im} A_1(u^*)$ is proved equivalent to

$$g_y(y^*, z^*)h(y^*, z^*) \notin \operatorname{im} g_z(y^*, z^*). \tag{4.58}$$

This is a consequence of the fact that $g(\varphi(u^*)) = 0$ yields $f(\varphi(u^*)) \in \operatorname{im} A$, and then $f_1(u^*) \notin \operatorname{im} A_1(u^*)$ can be equivalently rewritten as $f(\varphi(u^*)) \notin \operatorname{im} A\varphi'(u^*) = \operatorname{im} A|_{T_{\varphi(u^*)}W_1}$. The identity $T_{\varphi(u^*)}W_1 = \ker g'(\varphi(u^*))$ readily yields (4.58).

From these relations it follows that a noncritical 1-singularity (y^*, z^*) is an impasse point of (4.51) if both (4.57) and (4.58) hold. The reader can check that, moreover, it is a backward or a forward impasse point if

$$-(\det g_z)_z(y^*, z^*)\operatorname{Adj} g_z(y^*, z^*)g_y(y^*, z^*)h(y^*, z^*) > 0 \text{ or } < 0, \tag{4.59}$$

respectively.

The conditions characterizing (y^*, z^*) as a K, I or IK 1-singular point can also be trivially derived from (4.57) and (4.58): a noncritical 1-singular point (y^*, z^*) is a K singularity if (4.58) is met but (4.57) does not hold;

these singularities are *semi-singular* points in [296]. Analogously, it will be an I singularity if (4.57) is met but (4.58) fails to hold, corresponding to a *pseudoequilibrium point* in [296]. Finally, IK singularities are defined by the simultaneous failing of (4.57) and (4.58). Note also that a noncritical 1-singularity will be a boundary (resp. inner) singularity if and only if the image condition (4.58) holds (resp. does not hold).

Akin to the analysis of singularities in semiexplicit index one DAEs here performed, it may be of interest to compute the conditions characterizing different types of noncritical singularities in other structural forms, for instance in Hessenberg systems. This task, which should be guided by the specific forms of the differential-algebraic models arising in applications, is left to the reader.

4.4.6 *Examples*

4.4.6.1 *Impasse points: A simple instance*

A simple example of an impasse point in an index one context is given by
$$y' = \pm 1 \quad (4.60a)$$
$$0 = y - z^2, \quad (4.60b)$$
with $(y,z) \in W_0 = \mathbb{R}^2$. Since the derivative g_z of $g(y,z) = y - z^2$ reads $-2z$, the origin accommodates the unique singularity of the problem within the parabola defined by (4.60b). We have $\det g_z = -2z$ and
$$\mathrm{rk}\left.\begin{pmatrix} (\det g_z)' \\ g' \end{pmatrix}\right|_{(0,0)} = \mathrm{rk}\left.\begin{pmatrix} 0 & -2 \\ 1 & -2z \end{pmatrix}\right|_{(0,0)} = 2,$$
so that the origin is a noncritical 1-singularity, according to (4.56). Using the identities $(\det g_z)_z = -2 \neq 0$ and $g_y h = \pm 1 \neq 0$, it follows from (4.57) and (4.58) that $(0,0)$ is an impasse point. Additionally $\mathrm{Adj}\, g_z(0,0) = (\,1\,)$ and then (4.59) reads
$$-(\det g_z)_z(0,0) \mathrm{Adj}\, g_z(0,0) g_y(0,0) h(0,0) = \pm 2.$$
Therefore, the "+" (resp. "−") sign in (4.60a) makes the origin a backward (resp. forward) impasse point. This explains that the pair of trajectories
$$y = t,\ z = \pm\sqrt{t},\ t \geq 0$$
emanate from the origin with infinite speed (in $z'(t)$) for the "+" case in (4.60a), whereas the two solutions
$$y = -t,\ z = \pm\sqrt{-t},\ t \leq 0$$
terminate at the origin for the "−" case, with infinite speed in $z'(t)$ as well. Note, incidentally, that a singular reduction of (4.60) is defined by (4.34).

4.4.6.2 Singularity crossing in a singular index one problem

Let us further probe some of the ideas discussed in subsection 4.4.5 by means of the example

$$x' = y \qquad (4.61a)$$
$$y' = 1 \qquad (4.61b)$$
$$0 = x - y^2 + z^2, \qquad (4.61c)$$

where (x, y, z) denote the Euclidean coordinates in $W_0 = \mathbb{R}^3$. The set W_1 is defined by the hyperbolic paraboloid (4.61c). The leading matrix has constant rank $r_1 = 2$, and $g(x, y, z) = x - y^2 + z^2$ is a submersion on the whole of W_0. Therefore, all points in W_1 are 0-regular, a global parametrization of the paraboloid being given by $(x, y, z) = \varphi_1(y, z) = (y^2 - z^2, y, z)$. Letting

$$P_1 = \begin{pmatrix} 0 & 1 & 0 \\ -1 & 2y & 0 \end{pmatrix},$$

the corresponding reduction can be easily checked to read

$$\begin{pmatrix} 1 & 0 \\ 0 & 2z \end{pmatrix} \begin{pmatrix} y' \\ z' \end{pmatrix} = \begin{pmatrix} 1 \\ y \end{pmatrix}.$$

In these coordinates, we have $V_2 = \{(y, z) \in \mathbb{R}^2 \ / \ z \neq 0\} \cup \{(0, 0)\}$ and hence $W_2 = \{(x, y, z) \in \mathbb{R}^3 \ / \ x = y^2 - z^2, \ z \neq 0\} \cup \{(0, 0, 0)\}$. The matrix $A_1(y, z)$ has rank 2 on $V_2 - \{(0, 0)\}$ but rank 1 at $(0, 0)$; this implies that points satisfying $x = y^2 - z^2$, $z \neq 0$ are regular with index one, consistently with the condition $\det g_z = 2z \neq 0$ if $z \neq 0$.

In turn, the origin $(0, 0, 0) \in \mathbb{R}^3$ is an inner 1-singular point for the original DAE (4.61) which, furthermore, is an I singularity. The punctured parabola $x = y^2$, $y \neq 0$, $z = 0$ defines the set $\overline{W_2} - W_2$ of boundary 1-singularities, which can be easily checked to be backward or forward impasse points if $y > 0$ or $y < 0$, respectively, by the conditions discussed in subsection 4.4.5. Note that Assumption S1 holds globally with $L_1(y, z) = \mathbb{R}^2$ and $\tilde{r}_2 = 2 = r_1$, so that all points in the parabola $x = y^2$, $z = 0$ are singular with index one.

This example is also of interest for comparative purposes with the framework of Rabier and Rheinboldt. The failing of the constant rank assumption at the I singularity located at the origin avoids (4.61) from being index one in the sense of [228]. Actually, the exclusion of inner singularities of the impasse framework of [228] precludes the description of singularity crossing phenomena. In this example, it can be easily checked that a pair of solutions cross smoothly the singular set through the origin, namely

$$x = t^2/2, \ y = t, \ z = \pm t/\sqrt{2}. \tag{4.62}$$

Along these solutions the matrix $A_1(y,z)$ undergoes a rank deficiency at the origin, in contrast to the results holding in the framework of [228] and referred to at the beginning of Section 4.4 (cf. page 161).

4.4.6.3 Singular reduction

The example discussed below illustrates how the manifolds $\tilde{W}_i$ make it possible to overcome the absence of a smooth structure on W_i. This may be due to rank deficiencies in the leading matrix of the problem, either of (4.21) or of subsequent reductions. In particular, as already pointed out, inner singularities arising before the last reduction step may have a dramatic effect in the framework of Rabier and Rheinboldt due to the loss of a smooth manifold structure on the sets W_i.

With this aim, consider the DAE

$$x' = \alpha(x,y,z) \tag{4.63a}$$
$$xy' = z \tag{4.63b}$$
$$0 = y \tag{4.63c}$$

on $W_0 = \mathbb{R}^3$, α being a smooth function $W_0 \to \mathbb{R}$. Now the set W_1 reads $\{(x,y,z) \in W_0 \ / \ x \neq 0, \ y = 0\} \cup \{(0,0,0)\}$, and the leading matrix undergoes a rank deficiency at $x = 0$. In particular, the origin is an inner 0-singularity, whereas points of the form $(0,0,z)$ with $z \neq 0$ are boundary 0-singularities. Note that there is no way to apply the framework of [228] neither globally nor locally around the origin since W_1 does not have a manifold structure.

Assumptions S1 and S2 (as well as Proposition 4.4 at the origin) can be checked to hold at the fist step with $L_0 = \mathbb{R}^2 \times \{0\}$ and $\tilde{r}_1 = 2$, $\tilde{W}_1$ being given by $y = 0$. Setting $P_1 = (I \ \ 0)$, the corresponding singular reduction in (x,z) coordinates is defined by

$$\tilde{A}_1(x,z) = \begin{pmatrix} 1 & 0 & 0 \\ 0 & 1 & 0 \end{pmatrix} \begin{pmatrix} 1 & 0 & 0 \\ 0 & x & 0 \\ 0 & 0 & 0 \end{pmatrix} \begin{pmatrix} 1 & 0 \\ 0 & 0 \\ 0 & 1 \end{pmatrix} = \begin{pmatrix} 1 & 0 \\ 0 & 0 \end{pmatrix}$$

$$\tilde{f}_1(x,z) = \begin{pmatrix} \alpha(x,0,z) \\ z \end{pmatrix},$$

that is, $x' = \alpha(x,0,z)$, $0 = z$. Now $V_2^s = \tilde{V}_2 = \{(x,z) \in \mathbb{R}^2 \ / \ z = 0\}$, the matrix $\tilde{A}_1$ having constant rank $\tilde{r}_2 = 1$. The second-step reduction yields finally $x' = \alpha(x,0,0)$ with $\tilde{r}_3 = 1 = \tilde{r}_2$. Points $(x,0,0)$ with $x \neq 0$ are regular with index two, whereas the origin is a (harmless) singular point with index two in the sense specified in Remark 4.11. The rank deficiency at the origin does not have any effect beyond the first reduction step.

In the original setting, $W_2^s = \tilde{W}_2 = \{(x,y,z) \in W_0 \,/\, y = 0,\ z = 0\}$ comprises the solutions of the DAE (4.63), which behaves as an index two problem with a regular flow on $\tilde{W}_2$.

4.4.6.4 Semi-implicit DAEs

Quasilinear DAEs of the form

$$B(x)x' = h(x) \tag{4.64a}$$
$$0 = g(x), \tag{4.64b}$$

with $B: W_0 \to \mathbb{R}^{r \times (r+p)}$, $h: W_0 \to \mathbb{R}^r$, $g: W_0 \to \mathbb{R}^p$ sufficiently smooth and W_0 open in $\mathbb{R}^{r+p}$, are sometimes called *semi-implicit* and have been considered by several authors, cf. [236, 238, 304]. In particular, the family of DAEs

$$C(y,z)y' = h(y,z) \tag{4.65a}$$
$$0 = g(y,z), \tag{4.65b}$$

will be used in the sequel to illustrate some phenomena not displayed in the semiexplicit context of 4.4.5, 4.4.6.1 and 4.4.6.2. Mind that (4.63) in 4.4.6.3 actually has the form depicted in (4.65). For the DAE (4.65) we assume that $C \in C^\infty(W_0, \mathbb{R}^{r \times r})$, $h \in C^\infty(W_0, \mathbb{R}^r)$ and $g \in C^\infty(W_0, \mathbb{R}^p)$.

We will restrict the attention to cases in which $\det C(y,z)$ is a submersion on the set of points Σ where $C(y,z)$ is singular; this implies that Σ is a smooth codimension one submanifold of W_0. Furthermore, g will be assumed to be a submersion on $\tilde{W}_1 = \{(y,z) \in W_0 \,/\, g(y,z) = 0\}$, thereby making this set a smooth r-dimensional manifold. The notation $\tilde{W}_1$ for the above-introduced set will be justified later. We will additionally suppose that

$$\mathrm{rk}\left(\begin{array}{c} (\det C)' \\ g' \end{array} \right)\bigg|_{(y,z)} = p + 1$$

whenever $(y,z) \in \Sigma_1 = \Sigma \cap \tilde{W}_1$. This means that Σ and $\tilde{W}_1$ intersect transversally, and then

$$\Sigma_1 = \{(y,z) \in W_0 \,/\, g(y,z) = 0,\ \det C(y,z) = 0\}$$

is a codimension one submanifold of $\tilde{W}_1$. In particular this makes

$$\tilde{W}_1 - \Sigma_1 = \{(y,z) \in W_0 \ / \ g(y,z) = 0, \ \det C(y,z) \neq 0\} \qquad (4.66)$$

dense in $\tilde{W}_1$.

In this setting we have

$$W_1 = \{(y,z) \in W_0 \ / \ g(y,z) = 0, \ h(y,z) \in \operatorname{im} C(y,z)\} \qquad (4.67)$$

whereas, by the working assumptions, the set W_1^{reg} of 0-regular points can be easily checked to coincide with $\tilde{W}_1 - \Sigma_1$ (cf. (4.66)). Inner 0-singular points are given by $g(y,z) = 0$, $h(y,z) \in \operatorname{im} C(y,z)$ and $\det C(y,z) = 0$. Additionally, because of the fact that $\tilde{W}_1 - \Sigma_1 = W_1^{\text{reg}} \subseteq W_1$ is dense in $\tilde{W}_1$, which is closed in W_0, we have

$$\overline{W_1} = \tilde{W}_1, \qquad (4.68)$$

meaning that the set $\overline{W_1} - W_1$ of boundary 0-singular points is defined by $g(y,z) = 0$, $h(y,z) \notin \operatorname{im} C(y,z)$.

Assumption S1 holds globally with $L_0(y,z) = \mathbb{R}^r \times \{0\}$. This fact supports the above-introduced notation $\tilde{W}_1$ for the set $g = 0$. Moreover, setting $H = \begin{pmatrix} 0 & I_p \end{pmatrix}$ the product $H(y,z)f(y,z)$ arising in Assumption S2 amounts to $g(y,z)$, and therefore this Assumption holds by the above-stated submersion hypothesis on g.

Hence, a singular reduction can be locally performed around any 0-singularity in $\tilde{W}_1$. Let $g(y^*, z^*) = 0$ and, allowed by the fact that g is a submersion, assume as in subsection 4.4.5 that the matrix of partial derivatives $(g_{\bar{y}} \ g_{\bar{z}})$ is nonsingular at (y^*, z^*) for certain variables $\bar{y}$, $\bar{z}$, say $\bar{y} = (y_{\hat{r}+1}, \ldots, y_r)$, $\bar{z} = (z_{\bar{r}+1}, \ldots, z_p)$. Write the remaining variables as $\hat{y} = (y_1, \ldots, y_{\hat{r}})$, $\hat{z} = (z_1, \ldots, z_{\bar{r}})$. Applying the implicit function theorem to $g = 0$ around (y^*, z^*) write, as in (4.53),

$$\bar{y} = \psi_1(\hat{y}, \hat{z})$$
$$\bar{z} = \psi_2(\hat{y}, \hat{z}),$$

which yields a local parametrization of the manifold $\tilde{W}_1$ of the form $(u_1, u_2) \to \varphi(u_1, u_2) = ((u_1, \psi_1(u_1, u_2)), (u_2, \psi_2(u_1, u_2)))$. Inserting these relations into (4.65) and using $\tilde{P} = \begin{pmatrix} I_r & 0 \end{pmatrix}$ we are led to the singular reduction

$$C(\varphi(u_1, u_2)) \begin{pmatrix} I_{\hat{r}} & 0 \\ \psi_{1_{u_1}}(u_1, u_2) & \psi_{1_{u_2}}(u_1, u_2) \end{pmatrix} \begin{pmatrix} u_1' \\ u_2' \end{pmatrix} = h(\varphi(u_1, u_2)). \qquad (4.69)$$

Now, the leading matrix of (4.69) is nonsingular at a given (u_1, u_2) if and only if so are both $C(\varphi(u_1, u_2))$ and $\psi_{1_{u_2}}(u_1, u_2)$, and this can be

4.4. Singularities of autonomous quasilinear DAEs

proved in turn equivalent to the nonsingularity of $C(y,z)$ and $g_z(y,z)$, with $(y,z) = \varphi(u_1, u_2)$. Mind that, in particular, rank deficiencies in g_z yield 1-singularities of the differential-algebraic equation (4.65). The important point is that the reduction (4.69) can be performed in a way which accommodates both types of singular points; indeed, the leading matrix of the reduced equation reflects the singularities which arise in the original setting of the problem, i.e. in the matrix $C(y,z)$, but also after the first reduction step, that is, in $g_z(y,z)$.

Remark 4.12. The transversality assumption on the intersection of Σ and $\tilde{W}_1$ is generic. It implies in particular that $\overline{W_1} = \tilde{W}_1$, as depicted in (4.68), meaning that in this setting the singular reduction process is accommodated on the closure of W_1 which is hereby guaranteed to have a smooth manifold structure. In spite of this genericity, it is of interest to examine some cases in which (4.68) does not hold: see, specifically, the example considered in 4.4.6.6 below.

4.4.6.5 A DAE with 0- and 1-singularities

A simple illustration of the discussion above is given by the following modification of (4.61):

$$(x+z)x' = y \qquad (4.70a)$$
$$y' = 1 \qquad (4.70b)$$
$$0 = x - y^2 + z^2, \qquad (4.70c)$$

defined on $W_0 = \mathbb{R}^3$. Points $(x, y, z) \in W_0$ for which $x + z = 0$, $x = y^2 - z^2$ are now 0-singularities. The working assumptions of 4.4.6.4 hold in particular for this example, and a singular reduction is globally defined by $x = y^2 - z^2$ and $\tilde{P}_1 = \begin{pmatrix} I & 0 \end{pmatrix}$, yielding

$$\begin{pmatrix} y^2 - z^2 + z & 0 \\ 0 & 1 \end{pmatrix} \begin{pmatrix} 2y & -2z \\ 1 & 0 \end{pmatrix} \begin{pmatrix} y' \\ z' \end{pmatrix} = \begin{pmatrix} y \\ 1 \end{pmatrix}. \qquad (4.71)$$

The leading matrix of (4.71) is singular if $y^2 - z^2 + z = 0$, points which correspond to the above-mentioned 0-singularities, but also if $z = 0$; the latter correspond to 1-singular points in the parabola $x = y^2$, $z = 0$. This way the singularities of the leading matrix in (4.71) capture both 0- and 1-singular points of the original problem (4.70).

4.4.6.6 The differentiation-perturbation example of Campbell and Gear

Consider finally the DAE

$$y_1 N \begin{pmatrix} y' \\ z' \end{pmatrix} + \begin{pmatrix} y \\ z \end{pmatrix} = 0, \qquad (4.72)$$

where $y = (y_1, \ldots, y_{m-1}) \in \mathbb{R}^{m-1}$, $z \in \mathbb{R}$, and N is an index m nilpotent matrix in lower-triangular Jordan form. System (4.72) was constructed by Campbell and Gear in [45] in order to illustrate that the perturbation index [121, 122] may exceed the differentiation index by an arbitrarily large quantity. The reader can think e.g. of the two- and three-dimensional instances of (4.72), which read

$$y_1 = 0$$
$$y_1 y_1' + z = 0$$

and

$$y_1 = 0$$
$$y_1 y_1' + y_2 = 0$$
$$y_1 y_2' + z = 0,$$

respectively. These singular equations have the semi-implicit form depicted in (4.65) with $\det C(y, z) = y_1^{m-1}$ and $g(y, z) = y_1$, but in this case both sets $\Sigma = \{(y, z) \in \mathbb{R}^m \;/\; \det C(y, z) = 0\}$ and $\tilde{W}_1 = \{(y, z) \in \mathbb{R}^m \;/\; g(y, z) = 0\}$ are defined by the condition $y_1 = 0$; therefore Σ and $\tilde{W}_1$ do not intersect transversally, not even in the case $m = 2$ which makes $\det C(y, z)$ a submersion. Note that W_1 in (4.67) amounts to the origin and (4.68) does not hold.

However, this problem also falls in the working setting of subsection 4.4.3. Indeed, Assumptions S1 (with $L_0 = \mathbb{R}^{m-1} \times \{0\}$, $\tilde{r}_1 = m - 1$) and S2 are met in the first reduction step. A singular reduction is globally defined on $\tilde{W}_1$ by the parametrization $(0, y_2, \ldots, y_{m-1}, z)$. The reduced equation reads $y_2 = \ldots = y_{m-1} = z = 0$, so that $\tilde{r}_2 = 0$ and the origin is a harmless singularity which accommodates the unique solution of the DAE.

PART II
Semistate models of electrical circuits

Chapter 5

Nodal analysis

A major problem in electrical circuit analysis is how to set up the network equations. When time-varying and/or nonlinear effects are present, time-domain models replace those based on Laplace transforms or Fourier analysis, very common in the study of linear time-invariant circuits.

In the time-domain setting, much attention has been directed to the formulation of *state space* models based on explicit ODEs. This framework allows for the application of many analytical and numerical tools coming from the ODE context and dynamical systems theory. The systematic approach to state space circuit modeling emanates from the work of Bashkow, Bryant and other researchers in the late 1950s and early 1960s [18, 35, 36], and several issues on this topic have been the object of continuous interest since then; cf. [13, 32, 60, 63, 66, 128, 145, 164, 206, 235, 273, 274]. Note that this is just a sample of a huge amount of related literature.

However, the state space approach to circuit modeling displays some important limitations. For several circuit configurations an explicit state space equation may not exist, not even locally. Additionally, when state space descriptions do exist, their formulation may be hardly automatable. The latter is extremely important from the computational point of view, specially in very large scale integration systems. These limitations have led, in the last decades, to the formulation of *semistate* [82, 210] models, which use larger sets of network variables allowing some redundancy between them. Semistate models are currently framed in the differential-algebraic context.

The benefits of the semistate approach to circuit modeling can be roughly described along two directions. The first one concerns the automatic generation of circuit models. Modern schemes used to set up network equations are based on a differential-algebraic formalism; this is the case of Modified Nodal Analysis (MNA) [135], used in different circuit simulation programs such as SPICE (in its different versions) or TITAN [85, 87, 112–114, 198, 292, 293]. In the numerical simulation of circuit dynamics, a key aspect is the computation and monitorization of the index of these differential-algebraic models, a problem which has attracted much recent attention; see specifically [87, 292, 293], but also [84, 113, 114, 194, 234, 248, 253].

The second advantageous feature of the semistate framework is given by its chance to accommodate, in a comprehensive manner, different but tightly interrelated families of circuit models. A difference between model families is made by the set of semistate variables that they use. In particular, *nodal* methods are characterized by the use of node potentials as the fundamental model variables, in addition to some branch variables. This family includes the above-mentioned Modified Nodal Analysis, but also other techniques such as Node Tableau Analysis (NTA) [63, 66, 113, 114, 117] or Augmented Nodal Analysis (ANA) [84, 164, 248, 253]. The present Chapter presents a comprehensive discussion of different nodal methods, addressing several interrelations between NTA, ANA and MNA, together with an analysis of their indices; this analysis will include in particular some results involving non-passive circuits, out of the scope of [87, 292, 293].

Not only the formulation and analysis of nodal methods profit from the differential-algebraic framework. Other issues can be tackled advantageously in the DAE context; these include the above-mentioned state formulation problem and other related modeling issues, the analysis of singularities [61, 62, 233, 236], which in particular accommodate situations in which explicit ODE models do not exist, as well as qualitative studies [244, 252]. For reasons detailed later, these aspects will be better addressed through the models considered in Chapter 6. Although not considered in this book, loop and mesh analyses can also be framed naturally in the differential-algebraic context.

This Chapter is structured as follows: Section 5.1 presents a rather detailed background on graphs and electrical circuit theory. In particular, we detail the circuit families which will be considered throughout. A description of the differential-algebraic models resulting from nodal analysis methods can be found in Section 5.2; these include the above-mentioned NTA, ANA and MNA schemes. The rest of the Chapter is focused on index characterizations for these DAEs. In Section 5.3 we introduce the tools for index analysis, including the tractability index notion for the quasilinear DAEs arising from nodal methods. Note incidentally that, from the DAE

5.1. Background on graphs and electrical circuits

point of view, this Chapter illustrates the use of projector techniques in index analyses, whereas Chapter 6 will focus on reduction methods. Section 5.4 provides index characterizations for nodal models of passive circuits, whereas Section 5.5 illustrates how tree-based techniques provide a way to extend these results to non-passive systems. Several examples are discussed within Sections 5.4 and 5.5.

5.1 Background on graphs and electrical circuits

In this Section we compile some prerequisites for the circuit analyses performed in the rest of the book. We begin with a survey, in subsection 5.1.1, of elementary concepts coming from the theory of graphs and digraphs. As detailed later, the link between graphs and electrical circuits comes from the fact that every lumped circuit composed of two-terminal elements naturally has an associated graph, which results from identifying circuit branches and nodes with graph edges and vertices, respectively. *Topological* properties in circuit theory will be those which can be examined in terms of this graph, retaining only the electric (resistive, capacitive, etc.) nature of every branch; see 5.1.2.1 in this regard.

In subsection 5.1.2 we compile elementary aspects of circuit theory and detail the circuit devices allowed in later analyses. The attention will be restricted to *lumped* circuits in which the spatial dimensions do not play a role, in contrast to *distributed* systems where spatial variables must be taken into account; cf. [2, 25, 111, 264, 293] for *partial differential-algebraic equations* (PDAEs) modeling distributed systems. We will accommodate fully-coupled and nonlinear capacitive, inductive and resistive elements. For simplicity these elements will be assumed to be time-invariant. Also, for the sake of brevity we will focus on so-called *conventional* circuit models, which do not use the charge in capacitors or the flux in inductors as semistate variables; this will require global voltage- and current-control assumptions on capacitors and inductors, respectively. Many results can be extended to *charge-oriented* models (see [87, 253, 292]).

When the so-called incremental capacitance, inductance and conductance matrices are positive definite these devices will be called *passive*; in Section 5.5 we will also consider *non-passive* problems which remove this definiteness requirement. In order to emphasize the mathematical aspects of our analysis, the circuits here analyzed will exclude some devices which would introduce several technical complexities without making a real difference from the DAE point of view. This is the case of controlled sources; the reader is referred in particular to [87] for a detailed index analysis of MNA models including controlled sources. It is worth remarking, in any case, that many of our results can be extended to a broad family of circuits including controlled sources, along the lines discussed in subsection 6.2.6 for branch-oriented methods. Note also that the term 'active' (vs. 'non-passive' above) is usually reserved for circuits including controlled sources.

5.1.1 Graphs and digraphs

We refer the reader to [4, 5, 26, 89] for detailed introductions to graph theory. In particular, the proof of the properties compiled here can be found in these references. Many books on circuit theory (e.g. [13, 58, 63, 66]) also include some background on digraphs.

5.1.1.1 Graphs, digraphs, subgraphs

A *directed graph* or *digraph* is a triple (V, E, α), where $V \neq \emptyset$ and E are finite sets, and $\alpha : E \to V \times V$. The elements of E are called *edges* and will correspond to the *branches* of an electrical circuit, whereas the elements of V are *vertices* which stand for the *nodes* of the circuit. Denoting $\alpha(e) = (\alpha_1(e), \alpha_2(e)) \in V \times V$, we say that e is directed from $v_1 = \alpha_1(e)$ to $v_2 = \alpha_2(e)$, and call v_1 and v_2 the *initial* and *final* vertices of e, respectively; both are called the *terminal* vertices of e. The edge e is also said to be *incident* with both v_1 and v_2.

If we disregard directions in the edges we are led to a *graph*, which can be formally defined as a triple (V, E, γ) with V and E as above but where γ maps E into $V \times V/\sim$; here $\sim$ stands for the equivalence relation in $V \times V$ according to which $(v_1, v_2) \sim (v_1', v_2')$ if either $v_1 = v_1'$ and $v_2 = v_2'$, or $v_1 = v_2'$ and $v_2 = v_1'$. Thereby, in a graph the image through γ of a given edge is an unordered pair or vertices and we say that the edge is *incident* with these terminal vertices. Letting $\pi : V \times V \to V \times V/\sim$ stand for the canonical projection, the *underlying graph* of a given digraph (V, E, α) is defined as (V, E, γ) with $\gamma = \pi \circ \alpha$. Unless otherwise stated, the notions which are defined below for graphs must be understood to hold also for digraphs by applying them to the underlying graph.

Either for digraphs or graphs, note that we do not define an edge as a (ordered or unordered) pair of vertices, as it is done in many texts, since this rules out the description of multiple edges connecting the same pair

5.1. Background on graphs and electrical circuits

of vertices. This situation is often met in circuit theory, e.g. when two branches are connected in parallel. The definitions above also accommodate the presence of *self-loops* both in digraphs and graphs, that is, of edges for which both terminal vertices coincide. These self-loops have no interest in circuit theory and will need to be excluded when representing a digraph via the incidence matrix.

A graph (V', E', γ') is said to be a *subgraph* of (V, E, γ) if $V' \subseteq V$, $E' \subseteq E$ and $\gamma'(e) = \gamma(e)$ for any $e \in E'$. By assuming that (V', E', γ') is a graph, we are implicitly requiring that the incident vertices with any edge of E' are in V'. For a digraph, a directed subgraph is additionally assumed to inherit the direction of the edges; this means that the definition for digraphs is exactly the same as the one above provided that γ is replaced by α. Note that the removal of an arbitrary set of edges of a given (directed) graph results in a (directed) subgraph.

5.1.1.2 Loops, cutsets

Within a given graph, a *path* connecting two vertices v_0 and v_l is a sequence $(v_0, e_1, v_1, \ldots, v_{l-1}, e_l, v_l)$ in which the edge e_i is incident with the vertices v_{i-1} and v_i for $1 \leq i \leq l$. A path is said to be closed if $v_0 = v_l$. A closed path with $l \geq 1$ in which $e_i \neq e_j$ and $v_i \neq v_j$ for $1 \leq i < j \leq l$ is called a *loop*. We will normally use this term to mean just the set of edges $\{e_1, \ldots, e_l\}$. Note that a parallel connection of two branches in a circuit defines a loop. Self-loops (v_0, e_1, v_0) are allowed in the definition above, but since they will be precluded in later discussions we may understand from now on that $l \geq 2$.

A graph is said to be *connected* if for every pair of vertices there exists a path connecting them. A *connected component* is a maximal connected subgraph of a given graph. A subset K of the set of edges of a connected graph is a *cutset* if the removal of K results in a disconnected graph, and it is minimal with respect to this property, that is, the removal of any proper subset of K does not disconnect the graph.

5.1.1.3 Trees

Given a connected graph, a *tree* is a connected subgraph which contains all vertices and has no loops. Again, we shall often use this term to refer just to the set of edges within a given tree. In graph theory a tree is sometimes defined just as a connected subgraph without loops; when this is the case, the additional requirement that a given tree contains all vertices makes it a *spanning* tree. According to our definition, which is very often adopted in circuit theory, a 'tree' is implicitly assumed to be a 'spanning tree'. In a graph which is not connected, the choice of a tree in every connected component defines a *forest*.

Once a tree has been chosen in a given connected graph, the edges in the tree are called *twigs*, whereas the remaining ones are called *links* or *chords*. The set of links defines a *cotree*. If the (connected) graph has n vertices and b edges, any tree defines $n-1$ twigs and $b-n+1$ links; in a forest, the numbers of twigs and links are $n-k$ and $b-n+k$, respectively, k standing for the number of connected components. Note that because of the notion of a tree there cannot exist any loop just defined by twigs; similarly, the links may not include any cutset, since the set of twigs connects the whole graph and therefore no removal of any set of links may result in a disconnected graph. More is true, as displayed in the following statement. See also 5.1.1.8 below.

Lemma 5.1. *Let J, K be disjoint subsets of the set of edges of a given connected graph. Then there exists a tree which contains all edges from J and no edge from K if and only if J has no loops and K has no cutsets.*

5.1.1.4 *Incidence matrix*

Assume that the vertices and edges in a given digraph are numbered, so that V and E can be written as $\{1,\ldots,n\}$ and $\{1,\ldots,b\}$, respectively. We assume in the sequel that $b \geq 1$ and that the digraph has no self-loops, so that $n \geq 2$.

The incidence between edges and vertices can be described in terms of the so-called *incidence matrix* $\tilde{A} = (a_{ij}) \in \mathbb{R}^{n \times b}$, where

$$a_{ij} = \begin{cases} 1 & \text{if edge } j \text{ leaves vertex } i \\ -1 & \text{if edge } j \text{ enters vertex } i \\ 0 & \text{if edge } j \text{ is not incident with vertex } i. \end{cases}$$

Equivalently, $a_{ij} = 1$ (resp. -1) if and only if $\alpha_1(j) = i$ (resp. $\alpha_2(j) = i$), the other entries in the j-th column being null.

The results stated in Lemmas 5.2-5.4 below are well-known and widely used in the context of nodal analysis methods for circuits [87, 292, 293]; for the sake of brevity we therefore omit their proofs. In any case they follow from elementary properties of digraph theory: see for instance Chapter 3 in [5] or Chapters 6 and 7 in [89].

5.1. Background on graphs and electrical circuits

Lemma 5.2. *If n and k stand for the number of vertices and connected components of a digraph, then $\operatorname{rk} \tilde{A} = n - k$. In particular, for a connected digraph it is $\operatorname{rk} \tilde{A} = n - 1$.*

The quantity $r = n - k$ is called the *rank* of the digraph. In a connected digraph, any $n - 1$ rows of the incidence matrix $\tilde{A}$ are linearly independent; the removal of any row of $\tilde{A}$ yields a so-called *reduced incidence matrix* $A \in \mathbb{R}^{(n-1) \times b}$. From the point of view of circuit theory, this corresponds to the choice of a reference node. See (5.60) on p. 227 for an example. With terminological abuse, we will sometimes refer to A simply as the incidence matrix, $\tilde{A}$ being then termed the *complete* or *non-reduced* incidence matrix.

At several points we will need to check if a given subset of branches K within an electrical circuit contains a loop or a cutset. This can be performed in terms of the reduced incidence matrix via Lemmas 5.3 and 5.4 below. In these statements, if K is a subset of the set of edges of a connected digraph $\mathcal{G}$, we denote by A_K (resp. $A_{\mathcal{G}-K}$) the submatrix of A formed by the columns corresponding to edges in K (resp. not in K).

Lemma 5.3. *A subset K of the set of edges of a connected digraph $\mathcal{G}$ does not contain loops if and only if A_K has full column rank.*

Lemma 5.4. *A subset K of the set of edges of a connected digraph $\mathcal{G}$ does not contain cutsets if and only if $A_{\mathcal{G}-K}$ has full row rank.*

Equivalently, the absence of loops within the set K is characterized by $A_K y = 0 \Rightarrow y = 0$, and the absence of cutsets yields $x^\mathsf{T} A_{\mathcal{G}-K} = 0 \Rightarrow x = 0$.

Now, if T is a tree in a connected digraph, A_T is an $(n-1) \times (n-1)$ matrix and, since T contains no loops, it has full column rank and therefore is nonsingular. It is not difficult to prove the stronger statement depicted below (cf. Section 2.2 in [13]).

Lemma 5.5. *Let K be a set of $n-1$ edges of a connected digraph. Then A_K is nonsingular if and only if K defines a tree. In this case, $\det A_K = \pm 1$.*

5.1.1.5 Loop matrix

The definition of the loop matrix associated with a digraph requires introducing previously the notion of an *orientation* in a loop. With the notation used in 5.1.1.2, if a given loop is defined by the vertex-edge sequence $(v_0, e_1, v_1, \ldots, v_{l-1}, e_l, v_l)$ with $v_l = v_0$, one of the two possible orientations is defined by the sequence $(v_0, e_1, v_1, \ldots, v_{l-1}, e_l, v_l)$ itself and the other one by $(v_l, e_l, v_{l-1}, \ldots, v_1, e_1, v_0)$; this is not ambiguous, i.e., both orientations are actually different since self-loops are excluded and therefore $l \geq 2$. Furthermore, if the loop is orientated according (for instance) to the sequence $(v_0, e_1, v_1, \ldots, v_{l-1}, e_l, v_l)$, we say that the edge e_j has the same orientation as the loop if e_j is directed from v_{j-1} towards v_j, and that it has the opposite orientation if it leaves v_j and enters v_{j-1}.

Given an orientation in every loop, the *loop matrix* is then defined as $\tilde{B} = (b_{ij})$, with

$$b_{ij} = \begin{cases} 1 & \text{if edge } j \text{ is in loop } i \text{ with the same orientation} \\ -1 & \text{if edge } j \text{ is in loop } i \text{ with the opposite orientation} \\ 0 & \text{if edge } j \text{ is not in loop } i. \end{cases}$$

Lemma 5.6. *In a digraph with n vertices and k connected components, $\operatorname{rk} \tilde{B} = b - n + k$. In particular, for a connected digraph it is $\operatorname{rk} \tilde{B} = b - n + 1$.*

The quantity $s = b - n + k$ defines the *cyclomatic number* of the digraph. For a connected digraph, a submatrix $B \in \mathbb{R}^{(b-n+1) \times b}$ of $\tilde{B}$ with full row rank will be called a *reduced loop matrix*. Sometimes, with terminological abuse we will refer to this matrix simply as a 'loop matrix'. For later use note that, since the rows of any two reduced loop matrices B, $\hat{B}$ span the same space, it follows that the relation $\hat{B} = M_1 B$ holds for some nonsingular matrix M_1.

The following analogs of Lemmas 5.3 and 5.4 will be useful in Chapter 6. They can be actually derived from the above-mentioned Lemmas, as will be shown below.

Lemma 5.7. *A subset K of the set of edges of a connected digraph $\mathcal{G}$ does not contain cutsets if and only if B_K has full column rank.*

Lemma 5.8. *A subset K of the set of edges of a connected digraph $\mathcal{G}$ does not contain loops if and only if $B_{\mathcal{G}-K}$ has full row rank.*

For the sake of completeness, and regarding the property depicted in Lemma 5.5, it is worth mentioning that a set K of $b - n + 1$ edges of a given connected digraph yields a nonsingular submatrix B_K if and only if the edges in K define a cotree (see 3.21 in [5]). We will not make specific use of this property, though.

5.1.1.6 Cutset matrix

Finally, the *cutset matrix* $\tilde{Q} = (q_{ij})$ of a connected digraph is defined as detailed below; note that we restrict the definition to connected digraphs

5.1. Background on graphs and electrical circuits

only for simplicity. The removal of any cutset in a connected digraph results in a digraph with two connected components, to be denoted by C_1 and C_2. We can define two different orientations in this cutset, say *from C_1 to C_2* or *from C_2 to C_1*. Given an edge in the cutset, one terminal vertex must be in C_1 and the other one in C_2. Assume e.g. that the cutset is oriented from C_1 to C_2; an edge of the cutset is then said to have the same (resp. opposite) orientation as the cutset if the initial vertex of this edge is in C_1 (resp. C_2). The cutset matrix $\tilde{Q} = (q_{ij})$ is then defined by

$$q_{ij} = \begin{cases} 1 & \text{if edge } j \text{ is in cutset } i \text{ with the same orientation} \\ -1 & \text{if edge } j \text{ is in cutset } i \text{ with the opposite orientation} \\ 0 & \text{if edge } j \text{ is not in cutset } i. \end{cases}$$

The rank of $\tilde{Q}$ can be proved to be $n-1$ for a connected digraph; again, $n-1$ linearly independent rows of $\tilde{Q}$ define a *reduced cutset matrix* $Q \in \mathbb{R}^{(n-1)\times b}$.

5.1.1.7 Relations among digraph matrices

The proof of the following result can be found e.g. in [89] (Section 7.4).

Lemma 5.9. *If the columns of the reduced incidence, cutset, and loop matrices A, Q, B of a connected digraph are arranged according to the same order of edges, then $BA^\mathsf{T} = BQ^\mathsf{T} = 0$.*

From this statement and the conditions $\mathrm{rk}\, A = \mathrm{rk}\, Q = n-1$, $\mathrm{rk}\, B = b-n+1$ it follows that $\mathrm{im}\, A^\mathsf{T} = \mathrm{im}\, Q^\mathsf{T} = \ker B$, and $\mathrm{im}\, B^\mathsf{T} = \ker A = \ker Q$. In particular, the rows of A and those of Q span the same subspace of $\mathbb{R}^b$; this means that a relation of the form $Q = M_0 A$ holds for a nonsingular matrix M_0. Lemma 5.9 actually expresses an orthogonality relation between this so-called *cut space* $\mathrm{im}\, Q^\mathsf{T}$ and the *cycle space* $\mathrm{im}\, B^\mathsf{T}$ spanned by the rows of B; see [26] for details in this regard.

Proof of Lemmas 5.7 and 5.8. Lemma 5.9 makes it easy to derive the results stated in Lemmas 5.7 and 5.8 from the corresponding properties stated in terms of the incidence matrix in Lemmas 5.3 and 5.4. According to Lemma 5.3, the absence of loops within K will be characterized by the property

$$A_K y = 0 \Rightarrow y = 0, \tag{5.1}$$

whereas, following Lemma 5.4, for the absence of cutsets we will use

$$x^\mathsf{T} A_{\mathcal{G}-K} = 0 \Rightarrow x = 0. \tag{5.2}$$

Assume first that, as stated in Lemma 5.7, K does not contain cutsets. Suppose that B_K does not have full column rank, i.e., that there exists a non-vanishing v for which $B_K v = 0$. It would then follow that the vector $(v, 0)$ belongs to $\ker(B_K\ B_{\mathcal{G}-K}) = \ker B = \operatorname{im} A^\mathsf{T}$, so that $v = A_K^\mathsf{T} x$ and $0 = A_{\mathcal{G}-K}^\mathsf{T} x$ for some x. From (5.2) we derive the relation $x = 0$ and thereby the contradiction $v = 0$.

Conversely, suppose that B_K has full column rank, and assume that $x^\mathsf{T} A_{\mathcal{G}-K} = 0$. Premultiplying the relation $A_K B_K^\mathsf{T} + A_{\mathcal{G}-K} B_{\mathcal{G}-K}^\mathsf{T} = 0$ by x^T we obtain $x^\mathsf{T} A_K B_K^\mathsf{T} = 0$ which, by the full column rank of B_K, implies $x^\mathsf{T} A_K = 0$. This identity, together with $x^\mathsf{T} A_{\mathcal{G}-K} = 0$ and the fact that A has full row rank, implies that $x = 0$, so that (5.2) is met; this shows that K cannot contain cutsets.

The proof of Lemma 5.8 proceeds analogously. Assume first that K does not contain loops, so that (5.1) holds, and that $w^\mathsf{T} B_{\mathcal{G}-K} = 0$. The relation $B_K A_K^\mathsf{T} + B_{\mathcal{G}-K} A_{\mathcal{G}-K}^\mathsf{T} = 0$ then yields $w^\mathsf{T} B_K A_K^\mathsf{T} = 0$ and, from (5.1), we obtain $w^\mathsf{T} B_K = 0$. Together with $w^\mathsf{T} B_{\mathcal{G}-K} = 0$ this leads to $w^\mathsf{T} B = 0$, and the maximal row rank of B implies that $w = 0$. It then follows that $B_{\mathcal{G}-K}$ has full row rank.

Finally, let us assume that $B_{\mathcal{G}-K}$ has full row rank but K does include a loop. This means that there exists a non-vanishing y with $A_K y = 0$, and then the vector $(y, 0)$ belongs to $\ker A = \operatorname{im} B^\mathsf{T}$. Hence there must exist a vector w such that $y = B_K^\mathsf{T} w$ and $0 = B_{\mathcal{G}-K}^\mathsf{T} w$. The full row rank of $B_{\mathcal{G}-K}$ yields $w = 0$ and this in turn implies $y = 0$, against the hypothesis that y does not vanish. □

5.1.1.8 *Fundamental loops and cutsets and their associated matrices*

An important role will be played in Chapter 6 by certain loops and cutsets constructed from a given tree.

Lemma 5.10. *Assume that a tree has been chosen in a given connected graph. Then every link defines a unique loop together with some twigs, and every twig defines a unique cutset together with some links.*

Indeed (see e.g. [63]), given a link there must exist a unique path in the tree connecting its incident vertices; the link together with this path defines a loop. In turn, given a twig, its removal defines two connected components in the tree. Take the (maybe empty) set of links which connect both components; the twig together with these links defines a cutset. The loops (resp. cutsets) defined in this manner from the links (resp. twigs) associated with

5.1. Background on graphs and electrical circuits

the tree are sometimes called *fundamental* loops (resp. cutsets). Examples can be found on page 261.

Now, given a tree in a connected digraph, if we orientate every fundamental cutset and loop in a way such that it gets the same orientation as its defining twig or link, respectively, then Q and B take the form

$$Q = \begin{pmatrix} I_r & -F^\mathsf{T} \end{pmatrix} \quad (5.3\text{a})$$
$$B = \begin{pmatrix} F & I_s \end{pmatrix}, \quad (5.3\text{b})$$

for a certain matrix $F \in \mathbb{R}^{s \times r}$. We are using the notations $r = n - 1$ and $s = b - n + 1$ for the rank and the cyclomatic number, respectively. In (5.3), the first $r = n - 1$ columns of both matrices are assumed to be associated with the twigs, whereas the last $s = b - n + 1$ ones correspond to the links; the submatrices I_r and I_s then reflect the fact that exactly one twig (resp. link) enters each fundamental cutset (resp. loop), having additionally the same orientation as this cutset (resp. loop). The submatrices $-F^\mathsf{T}$ and F can be written in this form because of the relation $QB^\mathsf{T} = 0$ stated in Lemma 5.9.

The form depicted for Q in (5.3a) shows that the fundamental cutsets defined by a tree (or, more precisely, the corresponding vectors in $\mathbb{R}^b$) are linearly independent. The same happens with the fundamental loops, in the light of the expression given for B in (5.3b). The matrices Q and B constructed this way will be called the *fundamental matrices* associated with the tree.

5.1.2 Elementary aspects of circuit theory

An electrical circuit will be considered here as a set of interconnected *branches*, each one accommodating a two-terminal *circuit element* and being incident with two (distinct) *nodes*. Branches are interconnected by having (or, more precisely, being incident with) at least one node in common. Identifying branches and nodes with edges and vertices, a graph can be naturally associated with the circuit.

Every circuit branch has two associated variables: the branch *current* i and *voltage* v. When accommodating a so-called *reactive* element (a capacitor or an inductor), the branch has an additional variable, namely, the *charge* q for capacitors or the *flux* ϕ for inductors. Every branch is given an orientation which defines a *reference direction* for the current; thereby, if a branch incident with n_1, n_2 is directed away from the node n_1 towards the node n_2 then a current of, say, $+1$ mA means that this current flows out of n_1 into n_2, whereas -1 mA stands for 1 mA flowing out of n_2 into n_1. These orientations also induce a voltage reference direction in every circuit branch, so that a voltage of $+5$ (resp. -5) mV in the above-mentioned branch indicates that the electric potential at n_1 is 5 mV larger (resp. smaller) than the one at n_2. These reference directions make it possible to associate a digraph with a given circuit.

The equations governing circuit dynamics combine the graph-theoretic relations among currents and among voltages given by Kirchhoff laws with the electromagnetic relations characterizing circuit devices. As detailed below, Kirchhoff laws result in linear algebraic relations involving branch currents and voltages, whereas the electromagnetic relations include differential equations at reactive elements, namely capacitors and inductors, and algebraic (non-differential), possibly nonlinear relations coming from the devices' characteristics. The mixed nature of these equations explains the role of the DAE formalism in circuit modeling.

The reader without a background on electrical circuit theory may profit from taking a look at the examples discussed in Sections 5.4 and 5.5 while reading this material.

5.1.2.1 *Topological aspects*

We will focus our attention on connected circuits, denoting by b and n the number of branches and nodes, respectively. We assume that $b \geq 1$, and since self-loops are excluded, this yields $n \geq 2$. Nontrivial problems require $b \geq 2$. The circuit will contain b_r resistors, b_c capacitors, b_l inductors, b_u independent voltage sources, and b_j independent current sources, so that $b_r + b_c + b_l + b_u + b_j = b$. We use the subscripts u and j for voltage and current sources to avoid confusion with branch voltages and currents, denoted by v and i. Some of these quantities may of course vanish, meaning that there are no devices of the corresponding type. The characteristic equations of each type of circuit element are presented in 5.1.2.2 below.

Topological properties of a given circuit are those which can be assessed just in terms of the circuit graph and the electric nature of every branch, disregarding the specific characteristic equations of each circuit device. These topological aspects are often addressed for restricted circuit families; for instance, Theorems 5.1 and 5.2 in Section 5.4 will provide topological index characterizations of NTA, ANA and MNA models for *passive* circuits.

Kirchhoff laws. Kirchhoff's current law (KCL) states that *the sum of the currents leaving any circuit node is zero*. This must be understood

5.1. Background on graphs and electrical circuits

as follows: if the current in branch k is denoted by i_k and this branch is directed away from (resp. towards) a given node, then the current leaving this node is i_k (resp. $-i_k$). Using the reduced incidence matrix A introduced in 5.1.1.4, this law can be then checked to read

$$Ai = 0. \tag{5.4}$$

Here $i = (i_r, i_c, i_l, i_u, i_j) \in \mathbb{R}^b$ is the vector of branch currents, the subscripts r, c, l, u and j standing as above for resistors, capacitors, inductors, voltage sources and current sources, respectively. If the reduced incidence matrix A is split accordingly as $A = (A_r\ A_c\ A_l\ A_u\ A_j)$, (5.4) reads

$$A_r i_r + A_c i_c + A_l i_l + A_u i_u + A_j i_j = 0. \tag{5.5}$$

Note that e.g. $A_r \in \mathbb{R}^{(n-1) \times b_r}$ is the submatrix of A corresponding to resistor branches and all circuit nodes: this subgraph need not be connected and may well have isolated nodes.

Kirchhoff's current law can also be expressed in terms of any reduced cutset matrix as

$$Qi = 0. \tag{5.6}$$

Indeed, as indicated in 5.1.1.7, the fact that the rows of any reduced cutset matrix span the same space as the rows of a reduced incidence matrix yields a relation of the form $Q = M_0 A$ for some nonsingular matrix M_0. Equation (5.6) then follows from the premultiplication of (5.4) by M_0.

In turn, Kirchhoff's voltage law (KVL) states that *the sum of the voltage drops along the branches of any loop is zero*. In this statement, provided that an orientation is defined in every loop and denoting by v_k the voltage in branch k, the corresponding voltage drop must be understood as v_k if branch k has the same orientation as the loop, and $-v_k$ otherwise. In terms of the loop matrix, this can be expressed as

$$Bv = 0, \tag{5.7}$$

where $v = (v_r, v_c, v_l, v_u, v_j) \in \mathbb{R}^b$ denotes the vector of branch voltages. Splitting the matrix B as $(B_r\ B_c\ B_l\ B_u\ B_j)$, (5.7) reads

$$B_r v_r + B_c v_c + B_l v_l + B_u v_u + B_j v_j = 0. \tag{5.8}$$

Nodal analysis methods will be based on an alternative statement of Kirchhoff's voltage law. In a connected circuit with n nodes, we will denote by $e \in \mathbb{R}^{n-1}$ the vector of potentials of all nodes except for the reference one. The potential at the reference node is conventionally assumed to vanish. Node potentials are then unambiguously defined from branch voltages because of Kirchhoff's voltage law; indeed, the potential at any node can be defined as the sum of the voltage drops along some (hence any, because of KVL) path to the reference node. It is not difficult to check that this makes it possible to recast Kirchhoff's voltage law as

$$v = A^\mathsf{T} e. \tag{5.9}$$

Loops, cutsets and trees in electrical circuits. In the topological index analysis of different circuit models, certain loops and cutsets will play an important role. A *VC-loop* will be a loop formed exclusively by voltage sources and/or capacitors; an example can be found in Figure 5.2 (p. 229). Note that *C-loops* and *V-loops*, composed only of capacitors or voltage sources, respectively, are particular cases of a VC-loop. Similarly, an *IL-cutset* will be a cutset formed exclusively by current sources and/or inductors (cf. Figure 5.3 on p. 231), *L-cutsets* and *I-cutsets* being particular instances. Other configurations such as VL-loops, IC-cutsets, etc., are defined analogously.

In the light of Lemmas 5.3 and 5.4, a circuit does not include any VC-loop if and only if the matrix $(A_c \; A_u)$ has full column rank, and it does not include IL-cutsets if and only if $(A_r \; A_c \; A_u)$ has full row rank. A simultaneous characterization of both conditions is also possible in terms of trees via Lemma 5.1, as stated below. A *proper tree* [18] in a connected circuit is a tree which contains all voltage sources and all capacitors as well as (possibly) some resistors, but neither current sources nor inductors.

Proposition 5.1. *A given connected circuit has neither VC-loops nor IL-cutsets if and only if it contains a proper tree.*

All circuits will be hereafter assumed to be *well-posed*, meaning that they have neither V-loops nor I-cutsets; in the opposite case, the circuit is said to be *ill-posed*. A well-posed circuit need not have a proper tree, since it may certainly contain VC-loops (with at least one capacitor) and/or IL-cutsets (with at least one inductor). An important role will be played instead by the so-called normal trees introduced below [35, 36, 145].

A *normal tree* in a connected circuit is a tree which contains all voltage sources, no current sources, as many capacitors as possible and as few inductors as possible; it may also include some resistors. Note that this concept is well-defined, as shown in [32] (see Theorem 2 there), and that a well-posed circuit always contains at least one normal tree. In the absence of VC-loops and IL-cutsets, normal trees amount to the above-defined proper ones. By contrast, when VC-loops and/or IL-cutsets are present in

5.1. Background on graphs and electrical circuits

a well-posed circuit, normal trees are not proper. In this case, the number of capacitors in a normal tree is $b_c - x_c$, where x_c stands for the number of *excess* or *link* capacitors and equals the number of linearly independent VC-loops (including in particular C-loops); this linear independence notion relies on the description of loops as vectors of $\mathbb{R}^b$, as in 5.1.1.5. Analogously, the number x_l of *excess* or *twig* inductors in a normal tree equals the number of linearly independent IL-cutsets (including L-cutsets). Moreover, in a normal tree the fundamental loop defined by each link capacitor only involves twig capacitors and voltage sources and, similarly, the fundamental cutset defined by each twig inductor only has link inductors and current sources; cf. Theorem 3 in [32].

5.1.2.2 Circuit elements: dynamic relations and device characteristics

The different variables appearing in the below-considered circuit devices will be denoted as follows. The vector $(q, v_c, i_c) \in \mathbb{R}^{b_c} \times \mathbb{R}^{b_c} \times \mathbb{R}^{b_c}$ will stand for capacitors charges, voltages and currents; $(\phi, v_l, i_l) \in \mathbb{R}^{b_l} \times \mathbb{R}^{b_l} \times \mathbb{R}^{b_l}$ for inductors fluxes, voltages and currents; $(v_r, i_r) \in \mathbb{R}^{b_r} \times \mathbb{R}^{b_r}$ for resistors voltages and currents; $(v_u, i_u) \in \mathbb{R}^{b_u} \times \mathbb{R}^{b_u}$ for the voltages and currents in voltage source branches; and $(v_j, i_j) \in \mathbb{R}^{b_j} \times \mathbb{R}^{b_j}$ for the voltages and currents in the branches accommodating current sources.

Capacitors. The electromagnetic relations governing capacitors are defined by the differential equation

$$q' = i_c \quad (5.10)$$

relating charges and currents, together with the characteristic

$$g_c(q, v_c) = 0. \quad (5.11)$$

In the latter, $g_c \in C^1(\mathbb{R}^{b_c} \times \mathbb{R}^{b_c}, \mathbb{R}^{b_c})$ represents the capacitors' constitutive relations.

The fully implicit equation (5.11) defines a general form for time-invariant capacitors. In particular, it may accommodate coupling effects; capacitors are said to be *uncoupled* when, for $1 \leq k \leq b_c$, the k-th relation in (5.11) has the form $g_{c_k}(q_k, v_{c_k}) = 0$, that is, it involves only the charge and the voltage in the k-th capacitor.

Equation (5.11) does not comprise any assumption on the existence of a *controlling variable* for capacitors. A simplifying assumption which will be made in the sequel is that capacitors are globally *voltage-controlled*, i.e. that (5.11) has the form

$$q = \gamma_c(v_c) \quad (5.12)$$

for a C^1 mapping γ_c. Defining the *incremental capacitance matrix* as $C(v_c) = \gamma'_c(v_c)$, the relation (5.10) reads

$$C(v_c)v'_c = i_c, \qquad (5.13)$$

which comprises both the differential and the characteristic equations for capacitors.

In some cases, voltage- or charge-controlled descriptions of a capacitor only exist locally. In particular, if the matrix of partial derivatives $\frac{\partial g_c}{\partial q}$ is nonsingular at some (q^*, v_c^*) satisfying (5.11), a voltage-controlled description such as (5.12) exists locally around (q^*, v_c^*) by the implicit function theorem.

When the capacitors are linear and voltage-controlled, (5.12) can be written as $q = Cv_c$, where $C \in \mathbb{R}^{b_c \times b_c}$ is the *capacitance matrix*. In uncoupled linear cases, the capacitance matrix is diagonal.

Inductors. Analogously, inductors are characterized by the differential relation

$$\phi' = v_l \qquad (5.14)$$

between magnetic fluxes and voltages, together with

$$g_l(\phi, i_l) = 0, \qquad (5.15)$$

where $g_l \in C^1(\mathbb{R}^{b_l} \times \mathbb{R}^{b_l}, \mathbb{R}^{b_l})$. Coupling effects may be displayed also in this context.

For the sake of simplicity, unless otherwise stated inductors will be assumed to be globally current-controlled by a C^1 relation of the form

$$\phi = \gamma_l(i_l). \qquad (5.16)$$

The *incremental inductance matrix* is $L(i_l) = \gamma'_l(i_l)$ and, in this setting, (5.14) reads

$$L(i_l)i'_l = v_l. \qquad (5.17)$$

In some cases, current-controlled descriptions such as (5.16) will only exist locally. This is the case for instance in the *Josephson junction*. This device is composed of two superconductors separated by an oxide barrier [63]; it is characterized by the differential relation $\phi' = v$ together with a current-flux characteristic

$$i = I_0 \sin k\phi.$$

5.1. Background on graphs and electrical circuits

Here, $I_0 > 0$ is a device parameter, whereas $k = 4\pi e/h$, e and h standing for the electron charge and Planck's constant, respectively. The Josephson junction is an example of a nonlinear inductor.

In the linear case, (5.16) amounts to $\phi = Li_l$, $L \in \mathbb{R}^{b_l \times b_l}$ denoting the *inductance matrix*. In this context, if the inductance matrix is symmetric, its j-th diagonal entry is called the *self-inductance* of the j-th inductor, whereas the (j, k)-th entry (with $j \neq k$) is the *mutual inductance* of the j-th and k-th inductors. Certainly, the inductance matrix L is diagonal in the absence of coupling effects.

Resistors. We will use the term 'resistor' in a broad sense which accommodates any device characterized by an algebraic (non-differential) relation between its branch voltage and current. This includes not only linear resistors governed by Ohm's law but also diodes, for instance, whose i-v characteristic are typically defined by Shockley's equation $i = i_0(e^{v/v_0} - 1)$ for certain real constants i_0, v_0. We will assume resistors to be time-invariant. Coupling effects will be accommodated by letting the set of resistors be defined by a relation of the form

$$g_r(v_r, i_r) = 0, \tag{5.18}$$

with $g_r \in C^1(\mathbb{R}^{b_r} \times \mathbb{R}^{b_r}, \mathbb{R}^{b_r})$.

An instance of a pair of coupled resistors is given by a *gyrator* [63], defined by the linear relation

$$\begin{pmatrix} i_1 \\ i_2 \end{pmatrix} = \begin{pmatrix} 0 & k \\ -k & 0 \end{pmatrix} \begin{pmatrix} v_1 \\ v_2 \end{pmatrix},$$

where $k \in \mathbb{R}$ is called the *gyration conductance*. Analogously, an *ideal transformer* is defined by

$$\begin{pmatrix} v_1 \\ i_2 \end{pmatrix} = \begin{pmatrix} 0 & n \\ -n & 0 \end{pmatrix} \begin{pmatrix} i_1 \\ v_2 \end{pmatrix},$$

where $n \in \mathbb{R}$ is the *turns ratio* [63]. Controlled sources in which the controlling elements are resistors currents or voltages also fall in this setting.

In most cases, resistors will be assumed to be voltage-controlled through a C^1 mapping $\gamma_r \colon \mathbb{R}^{b_r} \to \mathbb{R}^{b_r}$; this means that resistors currents and voltages will be related by

$$i_r = \gamma_r(v_r). \tag{5.19}$$

In this case, the *incremental conductance matrix* reads

$$G(v_r) = \gamma_r'(v_r). \tag{5.20}$$

In the linear setting, the voltage-controlled description (5.19) amounts to $i_r = Gv_r$, where G stands for the *conductance matrix*.
The current-controlled counterpart is defined by a relation of the form

$$v_r = \rho(i_r), \qquad (5.21)$$

with incremental resistance matrix $R(i_r) = \rho'(i_r)$. In the uncoupled linear case, the resistance matrix R is a diagonal one, its entries modeling the individual resistances R_k defined by Ohm's law $v_{r_k} = R_k i_{r_k}$, $k = 1, \ldots, b_r$. Note finally that, when the conductance and resistance matrices are both well-defined (maybe in an incremental sense) and nonsingular, they are inverse to each other.

Voltage and current sources. Finally, in independent voltage sources the branch voltages are driven by a mapping of the form

$$v_u = v_s(t) \qquad (5.22)$$

and, analogously, for independent current sources we have

$$i_j = i_s(t). \qquad (5.23)$$

Both $v_s : \mathcal{J} \to \mathbb{R}^{b_u}$ and $i_s : \mathcal{J} \to \mathbb{R}^{b_j}$ are explicit functions of time defined on some working interval $\mathcal{J} \subseteq \mathbb{R}$, and will correspond to the excitation terms in the equations. When they are constant we speak of *DC sources*.

As indicated on page 206, well-posed circuits are those which have neither V-loops nor I-cutsets. Now it becomes clear that Kirchhoff's voltage law applied to a loop of voltage sources would result in a link between *a priori* independent time mappings, leading generically to an inconsistency. Similar considerations apply to I-cutsets; its exclusion can be understood as a consistency requirement for Kirchhoff's current law. Actually, as shown in Proposition 1 of [248], a V-loop or an I-cutset yields a singular matrix pencil in the Augmented Nodal Analysis models of linear time-invariant circuits discussed later; the same property may be easily shown to hold for Node Tableau Analysis and Modified Nodal Analysis.

Although excluded in most later analyses, controlled sources are defined by mappings of the form $v_u = g_u(v, i)$ for controlled voltage sources, and $i_j = g_j(v, i)$ for controlled current sources. These general expressions allow all branch variables in the circuit to control a given source, and may even result in an implicit form if the k-th component of g_u or g_j actually depends on the k-th component of v_u or i_j, respectively. It is often assumed that each source is controlled by just one type of variable, that is, either by branch voltages or by branch currents. In this setting, controlled sources can be

5.1. Background on graphs and electrical circuits

classified into *voltage-controlled voltage sources* (VCVSs), *current-controlled voltage sources* (CCVSs), *voltage-controlled current sources* (VCCSs) and *current-controlled current sources* (CCCSs). When the controlling element is a resistor or a set of resistors, controlled sources together with their controlling devices may also be modeled as coupled resistors. See subsection 6.2.6 for additional remarks concerning circuits with controlled sources.

A *short circuit*, which is a branch with vanishing voltage $v = 0$, may be modeled by a DC voltage source with vanishing voltage, whereas an *open circuit*, for which $i = 0$, can be seen as a DC current source with null current. Note that short and open circuits may be alternatively modeled by current- and voltage-controlled linear resistors with vanishing resistance and conductance, respectively.

Reciprocity and passivity. Symmetric capacitance or inductance matrices will be said to describe *reciprocal* devices [60]. At this point the reader should however be aware of the fact that the term 'reciprocal network' is usually reserved for linear circuits with uncoupled resistors and capacitors, symmetrically coupled inductors, and ideal transformers [37, 63].

Positive definite (resp. semidefinite) capacitance, inductance or conductance matrices yield *strictly passive* (resp. *passive*) elements [60]; positive definiteness (resp. semidefiniteness) of an $m \times m$ matrix M means that $u^\mathsf{T} M u > 0$ (resp. ≥ 0) for any $u \in \mathbb{R}^m - \{0\}$. Mind that in this definition we do not assume M to be symmetric. For nonlinear circuits, the local counterpart of these passivity notions are naturally defined in terms of incremental matrices to yield *strictly locally passive* and *locally passive* devices; more precisely, the set of capacitors, inductors or resistors is said to be strictly locally (resp. locally) passive on a given region Ω if the condition $u^\mathsf{T} M(x) u > 0$ (resp. ≥ 0) holds for all $u \in \mathbb{R}^m - \{0\}$ and all $x \in \Omega$, $M(x)$ standing for the incremental capacitance, inductance or conductance matrix, respectively. With terminological abuse, we will often use 'passive' to mean 'strictly locally passive'. Additional remarks on strictly locally passive devices can be found in subsection 6.2.2.

We will call the set of capacitors, inductors or resistors *non-passive* when this positive definiteness assumption does not hold. A circuit is non-passive if any one of these three groups of devices is non-passive. This term will be used most of the times in a local sense. Note that 'active' is not a synonym for 'non-passive' since an active circuit is usually understood as one including controlled sources.

5.2 Formulation of nodal models

The distinct feature of nodal analysis methods is the use of *node potentials* as some of the model variables. Broadly speaking, branch voltages will be written in terms of node potentials via the expression $v = A^T e$ depicted in (5.9) for Kirchhoff's voltage law. This makes it possible to write also the current of voltage-controlled devices in terms of node potentials. The chance to handle automatically the topological information in terms of the incidence matrix via (5.9) and the equation $Ai = 0$ (cf. (5.4)) makes these methods well-suited for simulation purposes; in particular, Modified Nodal Analysis (MNA) is actually used in many modern circuit simulation programs, because its compact form allows for efficient numerical computations. Index analyses for MNA models can be found in [87, 292, 293].

However, the term *nodal analysis* refers to a broad set of techniques, including MNA but also those based on Node Tableau Analysis (NTA) and Augmented Nodal Analysis (ANA), which use different sets of semistate variables. We present in this Section the differential-algebraic models resulting from all these techniques; their indices will be analyzed in Sections 5.4 and 5.5. Examples can be found at the end of those Sections.

Special attention will be paid to ANA systems. These models have received less attention than tableau or MNA equations, but certainly have a theoretical importance. On the one hand, at least in passive settings these models preserve the tractability index of tableau equations and can be therefore seen as the result of eliminating somehow "superfluous" variables from NTA, actually via a Schur reduction. On the other hand, keeping capacitor voltages and currents as model variables (in contrast to MNA) makes it possible to link these models with state formulations in terms of capacitor voltages and inductor currents; for instance, the state space model of [164] can be seen as a reduction of ANA systems, as shown in [253]. Note also that the semiexplicit form of ANA makes these models easier to handle than MNA-based ones.

For brevity, we will restrict the attention to problems with current-controlled inductors and voltage-controlled resistors and capacitors, yielding *conventional* models. So-called *charge-oriented* and *hybrid* models, which use also capacitor charges and inductor fluxes as semistate variables, display several similarities with conventional systems [113, 114, 253, 292]. To simplify later analyses the sources will be assumed to be independent, although many results can be extended to a broad class of circuits with controlled sources, as it is done in subsection 6.2.6 for branch-oriented methods.

5.2.1 Node Tableau Analysis

The node tableau approach to circuit analysis can be traced back to [117]. See also [63, 66, 112–114, 253]. It is characterized by the use of node potentials and *all* branch voltages and currents as model variables; as indicated in [117], basic to this approach "is the concept of a tableau which includes all network information in a nonreduced form." Later-considered models such as those arising in MNA and ANA can be seen as reductions of the Node Tableau Analysis (NTA) equations. This approach usually yields very large, sparse systems.

Under the voltage-control assumption on capacitors and resistors stated in (5.12) and (5.19), and the current-control one (5.16) for inductors, the relations introduced in subsection 5.1.2 yield the NTA model

$$\frac{d}{dt}\gamma_c(v_c) = i_c \tag{5.24a}$$

$$\frac{d}{dt}\gamma_l(i_l) = v_l \tag{5.24b}$$

$$0 = i_r - \gamma_r(v_r) \tag{5.24c}$$

$$0 = v_u - v_s(t) \tag{5.24d}$$

$$0 = i_j - i_s(t) \tag{5.24e}$$

$$0 = Ai \tag{5.24f}$$

$$0 = v - A^\mathsf{T} e. \tag{5.24g}$$

Using the incremental capacitance and inductance matrices, and expanding $i = (i_r, i_l, i_c, i_u, i_j)$, $v = (v_r, v_l, v_c, v_u, v_j)$ in (5.24f) and (5.24g), the node tableau equations (5.24) read

$$C(v_c)v_c' = i_c \tag{5.25a}$$

$$L(i_l)i_l' = v_l \tag{5.25b}$$

$$0 = i_r - \gamma_r(v_r) \tag{5.25c}$$

$$0 = v_u - v_s(t) \tag{5.25d}$$

$$0 = i_j - i_s(t) \tag{5.25e}$$

$$0 = A_r i_r + A_l i_l + A_c i_c + A_u i_u + A_j i_j \tag{5.25f}$$

$$0 = v_r - A_r^\mathsf{T} e \tag{5.25g}$$

$$0 = v_l - A_l^\mathsf{T} e \tag{5.25h}$$

$$0 = v_c - A_c^\mathsf{T} e \tag{5.25i}$$

$$0 = v_u - A_u^\mathsf{T} e \tag{5.25j}$$

$$0 = v_j - A_j^\mathsf{T} e, \tag{5.25k}$$

which can be trivially rewritten as a semiexplicit DAE if $C(v_c)$ and $L(i_l)$ in (5.25a) and (5.25b) are nonsingular.

5.2.2 Augmented Nodal Analysis

Several variables can be eliminated from the tableau equations (5.24) or (5.25) without substantially affecting the structure of the model. Via the substitutions detailed below, (5.24) can be rewritten as

$$\frac{d}{dt}\gamma_c(v_c) = i_c \tag{5.26a}$$

$$\frac{d}{dt}\gamma_l(i_l) = A_l^\mathsf{T} e \tag{5.26b}$$

$$0 = A_r \gamma_r(A_r^\mathsf{T} e) + A_l i_l + A_c i_c + A_u i_u + A_j i_s(t) \tag{5.26c}$$

$$0 = v_c - A_c^\mathsf{T} e \tag{5.26d}$$

$$0 = v_s(t) - A_u^\mathsf{T} e \tag{5.26e}$$

or, in terms of the incremental reactive matrices, as

$$C(v_c) v_c' = i_c \tag{5.27a}$$

$$L(i_l) i_l' = A_l^\mathsf{T} e \tag{5.27b}$$

$$0 = A_r \gamma_r(A_r^\mathsf{T} e) + A_l i_l + A_c i_c + A_u i_u + A_j i_s(t) \tag{5.27c}$$

$$0 = v_c - A_c^\mathsf{T} e \tag{5.27d}$$

$$0 = v_s(t) - A_u^\mathsf{T} e. \tag{5.27e}$$

In the NTA system (5.25), resistive currents and voltages have been eliminated using (5.25c) and (5.25g). Inductive voltages are substituted by means of (5.25h), whereas voltage and current variables in the corresponding sources have been eliminated using (5.25d) and (5.25e). Finally, (5.25k) can be considered as an output equation giving the voltages in current source branches and may therefore be removed from the model. System (5.26), or equivalently (5.27), defines the Augmented Nodal Analysis (ANA) model; see [84, 164, 248, 253].

Again, if the matrices $C(v_c)$ and $L(i_l)$ in (5.27a) and (5.27b) are nonsingular, this system can be rewritten in semiexplicit form; in this case, the transition from the tableau equations (5.25) to the ANA model (5.27) can be seen as a Schur reduction of semiexplicit DAEs.

5.2.3 Modified Nodal Analysis

The currents and voltages in capacitive branches can be removed from (5.26) using (5.26a) and (5.26d). This yields the Modified Nodal Analysis (MNA) model

$$A_c \frac{d}{dt}\gamma_c(A_c^\mathsf{T} e) = -A_r\gamma_r(A_r^\mathsf{T} e) - A_l i_l - A_u i_u - A_j i_s(t) \quad (5.28a)$$

$$\frac{d}{dt}\gamma_l(i_l) = A_l^\mathsf{T} e \quad (5.28b)$$

$$0 = v_s(t) - A_u^\mathsf{T} e. \quad (5.28c)$$

The seminal reference for this approach is the paper [135]. This method has attracted much attention in the differential-algebraic context; see specifically [85, 87, 112–114, 194, 292, 293]. System (5.28) can also be written, in terms of the incremental capacitance and inductance matrices, as

$$A_c C(A_c^\mathsf{T} e) A_c^\mathsf{T} e' = -A_r\gamma_r(A_r^\mathsf{T} e) - A_l i_l - A_u i_u - A_j i_s(t) \quad (5.29a)$$

$$L(i_l) i_l' = A_l^\mathsf{T} e \quad (5.29b)$$

$$0 = v_s(t) - A_u^\mathsf{T} e. \quad (5.29c)$$

The coefficient in front of e' in (5.29a) is called the *nodal capacitance matrix*. This matrix will be typically singular; indeed, from Lemma 5.5 (p. 199) it follows that the existence of at least one capacitive tree would be necessary for it to be nonsingular (cf. in this regard also subsections 5.4.2 and 5.5.2). This precludes a semiexplicit form for MNA systems which, on the other hand, usually display a more compact form than ANA models due to the absence of capacitive branch variables.

Remark 5.1. A key aspect in later analyses will be defined by the requirement that the nodal capacitance matrix verifies $\mathrm{rk}\, A_c C A_c^\mathsf{T} = \mathrm{rk}\, A_c$ pointwise or, equivalently, that $\ker A_c C A_c^\mathsf{T} = \ker A_c^\mathsf{T}$. This condition will be satisfied in the two settings considered in Sections 5.4 and 5.5. Indeed, for the coupled problems with positive definite C analyzed in Section 5.4, premultiplying the relation $A_c C A_c^\mathsf{T} w = 0$ by w^T we obtain $w^\mathsf{T} A_c C A_c^\mathsf{T} w = 0$ and, from the definiteness of C, it follows that $A_c^\mathsf{T} w = 0$, so that indeed $\ker A_c C A_c^\mathsf{T} = \ker A_c^\mathsf{T}$. In the uncoupled, non-passive problems considered in Section 5.5, this identity will be characterized by the condition (5.85) in Lemma 5.14 (see p. 245).

5.3 Index analysis: Fundamentals

Projector methods based on the tractability index have been proved in the last decade to be a valuable tool for the analysis of differential-algebraic models of electrical circuits [85, 87, 190, 198, 292, 293]. We present in this Section the index notion which will be used in later analyses, restricting the attention to the structural forms displayed by nodal equations. More details can be found in [190, 193, 293] and, specially, in the forthcoming title [157].

5.3.1 Structural form of nodal models

The nodal models (5.24), (5.26) and (5.28) discussed above have the form
$$E(d(x))' = f(x) + q(t), \qquad (5.30)$$
which is a quasilinear (time-invariant) analog of the properly stated linear DAE (2.3) considered in Chapter 2. In (5.30), $E \in \mathbb{R}^{m \times r}$ is a constant matrix, whereas $d \in C^1(W_0, \mathbb{R}^r)$, $f \in C^1(W_0, \mathbb{R}^m)$ and $q \in C(\mathcal{J}, \mathbb{R}^m)$; W_0 is open in $\mathbb{R}^m$ and $\mathcal{J} \subseteq \mathbb{R}$ is an open interval. The semistate space W_0, as well as its dimension m, will be different for each analysis method, whereas the dimension r of the space where the map $d(x)$ takes values will equal the number of reactances in all schemes, that is, $r = b_c + b_l$ in all cases.

Tableau Analysis

For the NTA system (5.24) the semistate dimension is $m = 2b + n - 1$, since the semistate vector is defined by $x = (v, i, e) \in \mathbb{R}^{2b+n-1}$; recall that b and n stand for the number of circuit branches and nodes, respectively. The leading term $E(d(x))'$ of these tableau equations is given by

$$E = \begin{pmatrix} I_c & 0 \\ 0 & I_l \\ 0 & 0 \end{pmatrix}, \quad d(x) = \begin{pmatrix} \gamma_c(v_c) \\ \gamma_l(i_l) \end{pmatrix}, \qquad (5.31)$$

I_c and I_l denoting the identity matrices of sizes b_c and b_l, whereas

$$f(x) = \begin{pmatrix} i_c \\ v_l \\ i_r - \gamma_r(v_r) \\ v_u \\ i_j \\ -Ai \\ v - A^\mathsf{T} e \end{pmatrix}, \quad q(t) = \begin{pmatrix} 0 \\ 0 \\ 0 \\ -v_s(t) \\ -i_s(t) \\ 0 \\ 0 \end{pmatrix}. \qquad (5.32)$$

5.3. Index analysis: Fundamentals

Mind that we have inserted a "$-$" sign in the sixth entry of $f(x)$ for later convenience.

Augmented Nodal Analysis

The ANA equations (5.26) have the same structure as the NTA model. Indeed, E and $d(x)$ have the form depicted in (5.31), the only difference relying on the smaller size of the vanishing blocks in the last rows of E. The semistate vector is $x = (v_c, i_c, e, i_l, i_u)$, with semistate dimension $m = 2b_c + b_l + b_u + n - 1$, and

$$f(x) = \begin{pmatrix} i_c \\ A_l^\mathsf{T} e \\ -A_r \gamma_r(A_r^\mathsf{T} e) - A_l i_l - A_c i_c - A_u i_u \\ v_c - A_c^\mathsf{T} e \\ -A_u^\mathsf{T} e \end{pmatrix}, \quad q(t) = \begin{pmatrix} 0 \\ 0 \\ -A_j i_s(t) \\ 0 \\ v_s(t) \end{pmatrix}, \quad (5.33)$$

the "$-$" sign in the third entry of f and q being also aimed at later convenience.

Modified Nodal Analysis

Finally, for the MNA system (5.28) the semistate dimension is given by $m = b_l + b_u + n - 1$. The semistate vector now reads $x = (e, i_l, i_u)$, whereas

$$E = \begin{pmatrix} A_c & 0 \\ 0 & I_l \\ 0 & 0 \end{pmatrix}, \quad d(x) = \begin{pmatrix} \gamma_c(A_c^\mathsf{T} e) \\ \gamma_l(i_l) \end{pmatrix}, \quad (5.34)$$

and

$$f(x) = \begin{pmatrix} -A_r \gamma_r(A_r^\mathsf{T} e) - A_l i_l - A_u i_u \\ A_l^\mathsf{T} e \\ -A_u^\mathsf{T} e \end{pmatrix}, \quad q(t) = \begin{pmatrix} -A_j i_s(t) \\ 0 \\ v_s(t) \end{pmatrix}. \quad (5.35)$$

Leading terms

From the C^1 assumption on $d(x)$, which holds for all nodal models because of the C^1 requirement on γ_c and γ_l in (5.12) and (5.16), respectively, and denoting by D the Jacobian matrix d_x, if we restrict the attention to C^1 solutions we can write $(d(x))' = D(x)x'$, equation (5.30) taking the form

$$A(x)x' = f(x) + q(t), \quad (5.36)$$

with $A(x) = ED(x)$. The leading matrix $A(x)$ in (5.36) should not be confused with the incidence matrix A. Equation (5.36) accommodates the forms (5.25) and (5.27) for NTA and ANA, for which

$$D(x) = \begin{pmatrix} C(v_c) & 0 & 0 \\ 0 & L(i_l) & 0 \end{pmatrix}, \quad A(x) = \begin{pmatrix} C(v_c) & 0 & 0 \\ 0 & L(i_l) & 0 \\ 0 & 0 & 0 \end{pmatrix} \quad (5.37)$$

with different sizes in the last null blocks of the tableau and augmented models. Assuming that the incremental capacitance and inductance matrices $C(v_c)$, $L(i_l)$ are nonsingular, we have $\operatorname{im} D(x) = \mathbb{R}^r$ and, together with the fact that the matrix E defined in (5.31) verifies $\ker E = \{0\}$, this yields the identity $\ker E \oplus \operatorname{im} D(x) = \mathbb{R}^r$ and makes the leading term of the DAE *properly stated* in the sense specified in [190, 293]. Note also that the nonsingularity of $C(v_c)$, $L(i_l)$ makes $\ker A(x)$ constant.

The form (5.36) also accounts for the MNA system (5.29), with

$$D(x) = \begin{pmatrix} C(A_c^\mathsf{T} e) A_c^\mathsf{T} & 0 & 0 \\ 0 & L(i_l) & 0 \end{pmatrix}, \quad A(x) = \begin{pmatrix} A_c C(A_c^\mathsf{T} e) A_c^\mathsf{T} & 0 & 0 \\ 0 & L(i_l) & 0 \\ 0 & 0 & 0 \end{pmatrix}. \quad (5.38)$$

In this case, the property $\ker E \oplus \operatorname{im} D(x) = \mathbb{R}^r$ (the matrix E being defined in (5.34)) will rely on the identity $\operatorname{rk} A_c = \operatorname{rk} A_c C(A_c^\mathsf{T} e) A_c^\mathsf{T}$ holding in the problems considered in this Chapter (cf. Remark 5.1, p. 215). Indeed, from the relation $\operatorname{rk} A_c = \operatorname{rk} A_c C(A_c^\mathsf{T} e) A_c^\mathsf{T}$ we derive $\ker A_c \cap \operatorname{im} C(A_c^\mathsf{T} e) A_c^\mathsf{T} = \{0\}$ and $\operatorname{rk} C(A_c^\mathsf{T} e) A_c^\mathsf{T} = \operatorname{rk} A_c$, so that $\mathbb{R}^{b_c} = \ker A_c \oplus \operatorname{im} C(A_c^\mathsf{T} e) A_c^\mathsf{T}$. Together with the nonsingularity of $L(i_l)$, this yields $\ker E \oplus \operatorname{im} D(x) = \mathbb{R}^r$. Note that also in this setting $\ker A(x) = \ker A_c^\mathsf{T} \times \{0\} \times \mathbb{R}^{b_u}$ is constant.

5.3.2 On the tractability index of quasilinear DAEs

A detailed discussion concerning the tractability index for properly stated nonlinear DAEs, together with solvability results supported on it, can be found in [157]. We just compile here, from [190, 293], the notions of a tractability index zero, one and two quasilinear DAE, aimed at the analysis of nodal circuit models. For the sake of simplicity we restrict the attention to C^1 solutions and therefore direct the index notions below to the DAE (5.36), which accounts for the nodal models (5.25), (5.27) and (5.29), as explained in subsection 5.3.1. In (5.36), we assume that $A \in C(W_0, \mathbb{R}^{m \times m})$, $f \in C^1(W_0, \mathbb{R}^m)$ and $q \in C(\mathcal{J}, \mathbb{R}^m)$, W_0 being an open subset of $\mathbb{R}^m$ and $\mathcal{J} \subseteq \mathbb{R}$ an open interval.

As a cautionary remark, it is worth indicating that the matrices $A_1(x)$ and $A_2(x)$ in (5.39) and (5.40) below will extend the construction carried out for linear constant coefficient problems in Chapter 2 (see specifically (2.21) on page 34), under the requirement that $A(x)$ and $A_1(x)$ have constant kernel and constant rank, respectively. Note however that the notion of an index ν DAE introduced in [157, 190, 293] is a more general one, not restricted to problems with $\nu \leq 2$, and that the general tractability index definition is more involved; in other words, the reader should not try to parallelize the linear constant coefficient case, as we will do here, in the definition of index $\nu \geq 3$ quasilinear problems, since the simplifications allowing one to use the expressions (5.39) and (5.40) are only valid for $\nu \leq 2$; find details in this regard in [190, 293].

The DAE (5.36) will be said to have tractability index zero on W_0 if $A(x)$ is nonsingular for all $x \in W_0$. Even though index zero problems can be displayed in MNA models (cf. item (1) in Theorems 5.2 and 5.4), they are somehow exceptional in nodal analysis. Therefore, assume in the sequel that $A(x)$ is everywhere singular.

As shown in subsection 5.3.1, in the circuit models here considered the kernel of the leading matrix $A(x)$ is constant. In this situation we may write $B(x) = -f'(x)$ (cf. [190, 293]) and, letting Q_0 be a constant projector onto $\ker A(x)$, state the tractability index one condition for (5.36) as the nonsingularity of

$$A_1(x) = A(x) + B(x)Q_0 \tag{5.39}$$

on the open set W_0.

Let us now focus on cases displaying a singular matrix $A_1(x)$. Suppose that $A_1(x)$ has constant rank on W_0 and, furthermore, that there exists a continuous projector $Q_1(x)$ onto $\ker A_1(x)$ defined on the whole of W_0. Letting $B_1(x) = B(x)P_0$, with $P_0 = I - Q_0$, the tractability index two notion on W_0 can be formulated as the nonsingularity of the matrix

$$A_2(x) = A_1(x) + B_1(x)Q_1(x) \tag{5.40}$$

for all $x \in W_0$; see specifically Remark A.18 in [293].

5.4 Index analysis: Passive circuits

The above-introduced notions make it possible to characterize the index of the DAEs (5.25), (5.27) and (5.29), modeling nodal analysis methods. Following the seminal ideas of [292], the goal is to analyze the index in topological terms (cf. 5.1.2.1) or, more precisely, in terms of the existence or absence of certain types of loops and cutsets.

A key aspect in this regard is to figure out conditions on the incremental conductance, capacitance and inductance matrices which allow for such a topological characterization. As detailed in this Section, strict local passivity assumptions on these devices (that is, positive definiteness assumptions on the corresponding incremental matrices) support a topological index characterization in all cases. Depending on the model, these passivity requirements can be partially relaxed; for instance, the index one topological characterization of NTA and ANA stated in Theorem 5.1 only requires nonsingular reactances, whereas the corresponding index one statement for MNA in Theorem 5.2 is based on the use of positive definite capacitance. The positive definiteness requirements will be relaxed further in Section 5.5. Full coupling within all these devices is allowed in this Section.

5.4.1 Tableau equations and Augmented Nodal Analysis

Recall that a circuit is well-posed if it does not include V-loops nor I-cutsets. To simplify notation, C and L stand below for the incremental matrices $C(v_c)$ and $L(i_l)$, whereas G stands for $\gamma'_r(v_r)$ and $\gamma'_r(A_r^T e)$ in the Node Tableau Analysis and Augmented Nodal Analysis models (5.25) and (5.27), respectively; note that v_r is one of the variables eliminated from NTA, via $v_r = A_r^T e$, in the Schur reduction which leads to ANA.

The nonsingularity and definiteness conditions on the incremental matrices are assumed to hold pointwise. From a mathematical point of view positive definiteness assumptions, describing strictly locally passive devices, might be replaced everywhere by negative definiteness ones, although this has no practical meaning in circuit theory.

Theorem 5.1. *Consider a well-posed, connected circuit with nonsingular capacitance and inductance matrices C, L, and positive definite conductance matrix G.*

(1) The Node Tableau Analysis (NTA) equations (5.25) and the Augmented Nodal Analysis (ANA) system (5.27) have tractability index one if and only if the circuit has neither VC-loops nor IL-cutsets.

(2) Assuming additionally that the matrices C and L are positive definite, the existence of VC-loops and/or IL-cutsets makes the above-mentioned DAEs (5.25) and (5.27) index two.

5.4. Index analysis: Passive circuits

Proof. Let us fist prove both assertions for the Augmented Nodal Analysis model (5.27). Ordering the semistate variables as v_c, i_l, e, i_c, i_u, from subsection 5.3.1 we have

$$A = \begin{pmatrix} C & 0 & 0 & 0 & 0 \\ 0 & L & 0 & 0 & 0 \\ 0 & 0 & 0 & 0 & 0 \\ 0 & 0 & 0 & 0 & 0 \\ 0 & 0 & 0 & 0 & 0 \end{pmatrix}, \quad B = \begin{pmatrix} 0 & 0 & 0 & -I & 0 \\ 0 & 0 & -A_l^T & 0 & 0 \\ 0 & A_l & A_r G A_r^T & A_c & A_u \\ -I & 0 & A_c^T & 0 & 0 \\ 0 & 0 & A_u^T & 0 & 0 \end{pmatrix}.$$

Note that B stands for the derivative $-f'$, the mapping $f(x)$ being defined in (5.33). From the nonsingularity of C and L, we can take

$$Q_0 = \begin{pmatrix} 0 & 0 & 0 & 0 & 0 \\ 0 & 0 & 0 & 0 & 0 \\ 0 & 0 & I & 0 & 0 \\ 0 & 0 & 0 & I & 0 \\ 0 & 0 & 0 & 0 & I \end{pmatrix}, \tag{5.41}$$

which yields for the matrix A_1 introduced in (5.39) the expression

$$A_1 = A + BQ_0 = \begin{pmatrix} C & 0 & 0 & -I & 0 \\ 0 & L & -A_l^T & 0 & 0 \\ 0 & 0 & A_r G A_r^T & A_c & A_u \\ 0 & 0 & A_c^T & 0 & 0 \\ 0 & 0 & A_u^T & 0 & 0 \end{pmatrix}. \tag{5.42}$$

Since C and L are nonsingular, the matrix A_1 in (5.42) is nonsingular if and only if so it is

$$J_1 = \begin{pmatrix} A_r G A_r^T & A_c & A_u \\ A_c^T & 0 & 0 \\ A_u^T & 0 & 0 \end{pmatrix}. \tag{5.43}$$

This nonsingularity condition holds if and only if the homogeneous linear system

$$A_r G A_r^T w + A_c y + A_u z = 0 \tag{5.44a}$$
$$A_c^T w = 0 \tag{5.44b}$$
$$A_u^T w = 0 \tag{5.44c}$$

only has the zero solution. Premultiply (5.44a) by w^T and use (5.44b) and (5.44c) to derive $w^T A_r G A_r^T w = 0$ which, because of the positive definiteness on G, implies

$$A_r^T w = 0. \tag{5.45}$$

This reduces (5.44a) to

$$A_c y + A_u z = 0. \tag{5.46}$$

The existence of a non-vanishing solution holds simultaneously for (5.44) and for (5.44b), (5.44c), (5.45), (5.46) altogether. Following Lemmas 5.3 and 5.4, the index one condition in the ANA model (5.27) is then equivalent to the absence of VC-loops and IL-cutsets in the circuit; indeed, a non-vanishing vector (y, z) satisfying (5.46) indicates the existence of a VC-loop, whereas a non-trivial solution w of (5.44b), (5.44c) and (5.45) expresses the existence of an IL-cutset. This completes the proof of item (1) for the ANA model (5.27).

Assume now that there exists at least one VC-loop or IL-cutset. From the positive definiteness of G, proceeding as above it is easy to check that $\ker A_1$ is defined by the vectors (p, q, w, y, z) which satisfy

$$p = C^{-1} y \tag{5.47a}$$

$$q = L^{-1} A_l^\mathsf{T} w \tag{5.47b}$$

$$w \in \ker (A_r \ A_c \ A_u)^\mathsf{T} \tag{5.47c}$$

$$(y, z) \in \ker (A_c \ A_u). \tag{5.47d}$$

The expressions depicted in (5.47) imply that the dimension of $\ker A_1$ is constant, so that A_1 has constant rank in the whole semistate space.

We show below that, under the assumed positive definiteness on C and L in item (2) and the exclusion of V-loops and I-cutsets in well-posed circuits, the ANA system (5.27) is actually index two. To achieve this we need to prove that there exists a globally defined, continuous projector Q_1 onto $\ker A_1$ making the matrix $A_2 = A_1 + B_1 Q_1$ in (5.40) nonsingular, with

$$B_1 = B P_0 = \begin{pmatrix} 0 & 0 & 0 & 0 & 0 \\ 0 & 0 & 0 & 0 & 0 \\ 0 & A_l & 0 & 0 & 0 \\ -I & 0 & 0 & 0 & 0 \\ 0 & 0 & 0 & 0 & 0 \end{pmatrix}$$

provided that $P_0 = I - Q_0$.

Let $\bar{Q}$ be a projector onto $\ker (A_r \ A_c \ A_u)^\mathsf{T}$, and $\hat{Q}$ a projector onto $\ker (A_c \ A_u)$. Splitting the latter as

$$\hat{Q} = \begin{pmatrix} \hat{Q}_{11} & \hat{Q}_{12} \\ \hat{Q}_{21} & \hat{Q}_{22} \end{pmatrix}, \tag{5.48}$$

5.4. Index analysis: Passive circuits

a globally defined projector onto $\ker A_1$ is

$$Q_1 = \begin{pmatrix} 0 & 0 & 0 & C^{-1}\hat{Q}_{11} & C^{-1}\hat{Q}_{12} \\ 0 & 0 & L^{-1}A_l^\mathsf{T}\bar{Q} & 0 & 0 \\ 0 & 0 & \bar{Q} & 0 & 0 \\ 0 & 0 & 0 & \hat{Q}_{11} & \hat{Q}_{12} \\ 0 & 0 & 0 & \hat{Q}_{21} & \hat{Q}_{22} \end{pmatrix}. \tag{5.49}$$

Some elementary computations lead to

$$A_2 = A_1 + B_1 Q_1 = \begin{pmatrix} C & 0 & 0 & -I & 0 \\ 0 & L & -A_l^\mathsf{T} & 0 & 0 \\ 0 & 0 & A_r G A_r^\mathsf{T} + A_l L^{-1} A_l^\mathsf{T} \bar{Q} & A_c & A_u \\ 0 & 0 & A_c^\mathsf{T} & -C^{-1}\hat{Q}_{11} & -C^{-1}\hat{Q}_{12} \\ 0 & 0 & A_u^\mathsf{T} & 0 & 0 \end{pmatrix}$$

and, since C and L are nonsingular, the invertibility of A_2 relies on that of the matrix

$$J_2 = \begin{pmatrix} A_r G A_r^\mathsf{T} + A_l L^{-1} A_l^\mathsf{T} \bar{Q} & A_c & A_u \\ A_c^\mathsf{T} & -C^{-1}\hat{Q}_{11} & -C^{-1}\hat{Q}_{12} \\ A_u^\mathsf{T} & 0 & 0 \end{pmatrix}. \tag{5.50}$$

In order to prove the nonsingularity of J_2 we only need to show that the unique solution to

$$A_r G A_r^\mathsf{T} w + A_l L^{-1} A_l^\mathsf{T} \bar{Q} w + A_c y + A_u z = 0 \tag{5.51a}$$
$$A_c^\mathsf{T} w - C^{-1}\hat{Q}_{11} y - C^{-1}\hat{Q}_{12} z = 0 \tag{5.51b}$$
$$A_u^\mathsf{T} w = 0 \tag{5.51c}$$

is the trivial one. Since $\hat{Q}$ is a projector onto $\ker(A_c \; A_u)$, we have $A_c\hat{Q}_{11} = -A_u\hat{Q}_{21}$ and $A_c\hat{Q}_{12} = -A_u\hat{Q}_{22}$. Using these identities and (5.51c), the premultiplication of (5.51b) by $w^\mathsf{T} A_c C$ leads to $w^\mathsf{T} A_c C A_c^\mathsf{T} w = 0$ which, by the positive definiteness of C, yields

$$A_c^\mathsf{T} w = 0, \tag{5.52}$$

and transforms (5.51b) into

$$\hat{Q}_{11} y + \hat{Q}_{12} z = 0. \tag{5.53}$$

On the other hand, $\bar{Q}$ being a projector onto $\ker(A_r \; A_c \; A_u)^\mathsf{T}$ implies that $\bar{Q}^\mathsf{T} A_r = \bar{Q}^\mathsf{T} A_c = \bar{Q}^\mathsf{T} A_u = 0$. Multiplying (5.51a) by $w^\mathsf{T} \bar{Q}^\mathsf{T}$ then yields $w^\mathsf{T} \bar{Q}^\mathsf{T} A_l L^{-1} A_l^\mathsf{T} \bar{Q} w = 0$, and this in turn leads to

$$A_l^\mathsf{T} \bar{Q} w = 0 \tag{5.54}$$

in virtue of the positive definiteness of L, which implies that of L^{-1}. Therefore, (5.51a) reads
$$A_r G A_r^\mathsf{T} w + A_c y + A_u z = 0. \tag{5.55}$$
Together with (5.51c) and (5.52), the premultiplication of (5.55) by w^T yields $w^\mathsf{T} A_r G A_r^\mathsf{T} w = 0$, and then
$$A_r^\mathsf{T} w = 0, \tag{5.56}$$
because of the positive definiteness of G. Equation (5.55) can be then simplified to
$$A_c y + A_u z = 0, \tag{5.57}$$
showing that $(y, z) \in \ker(A_c\ A_u)$. This means that $y = \hat{Q}_{11} y + \hat{Q}_{12} z$ and, according to (5.53), $y = 0$. Additionally, (5.57) can be simplified to $A_u z = 0$ and, due to the absence of V-loops in well-posed circuits, Lemma 5.3 implies that $z = 0$.

Finally, (5.51c), (5.52) and (5.56) show that $w \in \ker(A_r\ A_c\ A_u)^\mathsf{T}$, which implies $\bar{Q} w = w$. This transforms (5.54) into
$$A_l^\mathsf{T} w = 0. \tag{5.58}$$
Equations (5.51c), (5.52), (5.56) and (5.58), together with the assumed exclusion of I-cutsets and Lemma 5.4, yield $w = 0$. Together with $y = 0$, $z = 0$, this shows that indeed (5.51) only has the trivial solution, meaning that A_2 is nonsingular and, therefore, that the ANA system (5.27) has tractability index two under the assumptions stated in item (2).

The statements for the tableau equations (5.25) are proved in a similar manner, and hence several details can be omitted. Ordering now the semistate variables as v_c, i_l, e, i_c, i_u, v_l, v_j, i_r, v_u, i_j, v_r, and taking the constant projector Q_0 as in (5.41) with the obvious modifications in the dimensions, the matrix $A_1 = A + BQ_0$ (cf. (5.39)) can be checked to read

$$A_1 = \begin{pmatrix} C & 0 & 0 & -I & 0 & 0 & 0 & 0 & 0 & 0 & 0 \\ 0 & L & 0 & 0 & 0 & -I & 0 & 0 & 0 & 0 & 0 \\ 0 & 0 & 0 & 0 & 0 & 0 & 0 & -I & 0 & 0 & G \\ 0 & 0 & 0 & 0 & 0 & 0 & 0 & 0 & -I & 0 & 0 \\ 0 & 0 & 0 & 0 & 0 & 0 & 0 & 0 & 0 & -I & 0 \\ 0 & 0 & 0 & A_c & A_u & 0 & 0 & A_r & 0 & A_j & 0 \\ 0 & 0 & A_r^\mathsf{T} & 0 & 0 & 0 & 0 & 0 & 0 & 0 & -I \\ 0 & 0 & A_l^\mathsf{T} & 0 & 0 & -I & 0 & 0 & 0 & 0 & 0 \\ 0 & 0 & A_c^\mathsf{T} & 0 & 0 & 0 & 0 & 0 & 0 & 0 & 0 \\ 0 & 0 & A_u^\mathsf{T} & 0 & 0 & 0 & 0 & 0 & -I & 0 & 0 \\ 0 & 0 & A_j^\mathsf{T} & 0 & 0 & 0 & -I & 0 & 0 & 0 & 0 \end{pmatrix}, \tag{5.59}$$

5.4. Index analysis: Passive circuits

where G now stands for $\gamma'_r(v_r)$. Mind that $B = -f'$ comes from the map $f(x)$ defined in (5.32). Since G is positive definite, proceeding as above the nonsingularity of A_1 can be proved equivalent to the absence of VC-loops and IL-cutsets. The index one claim in item (1) for NTA follows.

Assuming now that there is at least one VC-loop or IL-cutset, the matrix A_1 in (5.59) can be checked to have constant rank and to admit

$$Q_1 = \begin{pmatrix} 0 & 0 & 0 & C^{-1}\hat{Q}_{11} & C^{-1}\hat{Q}_{12} & 0 & 0 & 0 & 0 & 0 & 0 \\ 0 & 0 & L^{-1}A_l^\mathsf{T}\bar{Q} & 0 & 0 & 0 & 0 & 0 & 0 & 0 & 0 \\ 0 & 0 & \bar{Q} & 0 & 0 & 0 & 0 & 0 & 0 & 0 & 0 \\ 0 & 0 & 0 & \hat{Q}_{11} & \hat{Q}_{12} & 0 & 0 & 0 & 0 & 0 & 0 \\ 0 & 0 & 0 & \hat{Q}_{21} & \hat{Q}_{22} & 0 & 0 & 0 & 0 & 0 & 0 \\ 0 & 0 & A_l^\mathsf{T}\bar{Q} & 0 & 0 & 0 & 0 & 0 & 0 & 0 & 0 \\ 0 & 0 & A_j^\mathsf{T}\bar{Q} & 0 & 0 & 0 & 0 & 0 & 0 & 0 & 0 \\ 0 & 0 & 0 & 0 & 0 & 0 & 0 & 0 & 0 & 0 & 0 \\ 0 & 0 & 0 & 0 & 0 & 0 & 0 & 0 & 0 & 0 & 0 \\ 0 & 0 & 0 & 0 & 0 & 0 & 0 & 0 & 0 & 0 & 0 \\ 0 & 0 & 0 & 0 & 0 & 0 & 0 & 0 & 0 & 0 & 0 \end{pmatrix}$$

as a globally defined projector onto $\ker A_1$; as above, $\bar{Q}$ and $\hat{Q}$ are projectors onto $\ker(A_r \; A_c \; A_u)^\mathsf{T}$ and $\ker(A_c \; A_u)$, respectively. The matrix A_2 reads

$$A_2 = \begin{pmatrix} C & 0 & 0 & -I & 0 & 0 & 0 & 0 & 0 & 0 & 0 \\ 0 & L & 0 & 0 & 0 & -I & 0 & 0 & 0 & 0 & 0 \\ 0 & 0 & 0 & 0 & 0 & 0 & 0 & -I & 0 & 0 & G \\ 0 & 0 & 0 & 0 & 0 & 0 & 0 & 0 & -I & 0 & 0 \\ 0 & 0 & 0 & 0 & 0 & 0 & 0 & 0 & 0 & -I & 0 \\ 0 & 0 & A_l L^{-1} A_l^\mathsf{T} \bar{Q} & A_c & A_u & 0 & 0 & A_r & 0 & A_j & 0 \\ 0 & 0 & A_r^\mathsf{T} & 0 & 0 & 0 & 0 & 0 & 0 & 0 & -I \\ 0 & 0 & A_l^\mathsf{T} & 0 & 0 & -I & 0 & 0 & 0 & 0 & 0 \\ 0 & 0 & A_c^\mathsf{T} & -C^{-1}\hat{Q}_{11} & -C^{-1}\hat{Q}_{12} & 0 & 0 & 0 & 0 & 0 & 0 \\ 0 & 0 & A_u^\mathsf{T} & 0 & 0 & 0 & 0 & 0 & -I & 0 & 0 \\ 0 & 0 & A_j^\mathsf{T} & 0 & 0 & 0 & -I & 0 & 0 & 0 & 0 \end{pmatrix}.$$

The nonsingularity of this matrix is easily seen equivalent, via a Schur reduction, to that of

$$\begin{pmatrix} C & 0 & 0 & -I & 0 \\ 0 & L & -A_l^\mathsf{T} & 0 & 0 \\ 0 & 0 & A_l L^{-1} A_l^\mathsf{T} \bar{Q} + A_r G A_r^\mathsf{T} & A_c & A_u \\ 0 & 0 & A_c^\mathsf{T} & -C^{-1}\hat{Q}_{11} & -C^{-1}\hat{Q}_{12} \\ 0 & 0 & A_u^\mathsf{T} & 0 & 0 \end{pmatrix},$$

which has been proved above to be nonsingular under positive definiteness assumptions on G, C and L. This completes the proof of the index two assertion in item (2) for the node tableau system (5.25). □

Remark 5.2. Some readers might try to derive the tractability index characterization above for NTA from that of ANA, which is a Schur reduction of the tableau equations, somehow parallelizing the result stated in Proposition 3.7 (p. 128) for reduction methods. However, for nonlinear problems this is not possible in the tractability index framework, for the reasons indicated below.

In the terms of Proposition 3.7, the results can be transferred from the Schur reduction to the original DAE only on the manifold $g_2 = 0$. This is acceptable in the context of reduction methods for semiexplicit DAEs because, roughly, the geometric index is only defined at points which satisfy the constraints. In contrast, the tractability index is based on continuous projectors which are assumed to be defined on the whole semistate space. Note, incidentally, that Theorem 5.1 is stated globally. Broadly speaking, the verification of certain working assumptions on the semistate space of ANA does not imply that they hold in the (higher dimensional) semistate space of NTA. This idea obstructs the extension of Proposition 3.7 to the tractability index framework in nonlinear cases. Not even locally is it possibly to guarantee that certain constant (not maximal) rank assumptions holding on $g_2 = 0$ are satisfied around this manifold.

Remark 5.3. Following Proposition 5.1 (p. 206), the absence of VC-loops and IL-cutsets in the index one statement within item (1) of Theorem 5.1 can be equivalently expressed as the existence of at least one proper tree in the circuit. Proper trees will actually play a key role in the analysis of ANA in the non-passive framework of Section 5.5; see specifically Theorem 5.3 on page 238 which, for circuits without resistive coupling, can be seen as a generalization of item (1) in Theorem 5.1 above for ANA models.

Example 1

The circuit depicted in Figure 5.1 was introduced in Bashkow's seminal paper on tree methods for the state formulation problem [18]. We will use this circuit, as well as certain modifications of it, as a running example to illustrate the ideas discussed in this Chapter. This circuit can be modified without difficulties in order to illustrate different aspects while, at the same time, being simple enough as to allow for an easily readable discussion.

5.4. Index analysis: Passive circuits

As in [18], we assume that resistors, capacitors and inductors are linear, although the results hold locally for nonlinear problems if the resistance, capacitance and inductance are understood in an incremental sense. For the sake of simplicity we restrict the discussion to uncoupled problems. We also assume that the positive definiteness and nonsingularity hypotheses on the devices arising in Theorem 5.1 are met; in this uncoupled setting, they will amount to positiveness and non-vanishing requirements on the individual resistances, capacitances and inductances. The analysis will show how to relax these hypotheses, introducing some ideas which will be addressed in general in Section 5.5.

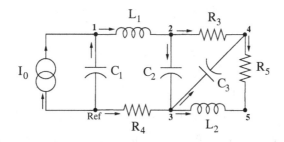

Fig. 5.1 Example 1: Bashkow's circuit.

Circuit nodes apart from the reference one are numbered in Figure 5.1, and every branch is given a reference direction indicated by an arrow. Ordering the branches according to the sequence C_1, C_2, C_3, L_1, L_2, R_3, R_4, R_5, I_0, the reduced incidence matrix reads

$$A = \begin{pmatrix} -1 & 0 & 0 & 1 & 0 & 0 & 0 & 0 & -1 \\ 0 & 1 & 0 & -1 & 0 & 1 & 0 & 0 & 0 \\ 0 & -1 & 1 & 0 & 1 & 0 & -1 & 0 & 0 \\ 0 & 0 & -1 & 0 & 0 & -1 & 0 & 1 & 0 \\ 0 & 0 & 0 & 0 & -1 & 0 & 0 & -1 & 0 \end{pmatrix}. \quad (5.60)$$

Since the circuit has no voltage sources, A can be split into the submatrices

$$A_c = \begin{pmatrix} -1 & 0 & 0 \\ 0 & 1 & 0 \\ 0 & -1 & 1 \\ 0 & 0 & -1 \\ 0 & 0 & 0 \end{pmatrix}, \; A_l = \begin{pmatrix} 1 & 0 \\ -1 & 0 \\ 0 & 1 \\ 0 & 0 \\ 0 & -1 \end{pmatrix}, \; A_r = \begin{pmatrix} 0 & 0 & 0 \\ 1 & 0 & 0 \\ 0 & -1 & 0 \\ -1 & 0 & 1 \\ 0 & 0 & -1 \end{pmatrix}, \; A_j = \begin{pmatrix} -1 \\ 0 \\ 0 \\ 0 \\ 0 \end{pmatrix}.$$

(5.61)

Assume that all conductances $G_i = 1/R_i$ are positive. The Augmented Nodal Analysis model (5.27) for the circuit in Figure 5.1 is defined by

$$C_1 v'_{c_1} = i_{c_1} \quad (5.62\text{a})$$
$$C_2 v'_{c_2} = i_{c_2} \quad (5.62\text{b})$$
$$C_3 v'_{c_3} = i_{c_3} \quad (5.62\text{c})$$
$$L_1 i'_{l_1} = e_1 - e_2 \quad (5.62\text{d})$$
$$L_2 i'_{l_2} = e_3 - e_5 \quad (5.62\text{e})$$
$$0 = i_{l_1} - i_{c_1} - i_0(t) \quad (5.62\text{f})$$
$$0 = G_3(e_2 - e_4) - i_{l_1} + i_{c_2} \quad (5.62\text{g})$$
$$0 = G_4 e_3 + i_{l_2} - i_{c_2} + i_{c_3} \quad (5.62\text{h})$$
$$0 = -G_3(e_2 - e_4) + G_5(e_4 - e_5) - i_{c_3} \quad (5.62\text{i})$$
$$0 = -G_5(e_4 - e_5) - i_{l_2} \quad (5.62\text{j})$$
$$0 = v_{c_1} + e_1 \quad (5.62\text{k})$$
$$0 = v_{c_2} - e_2 + e_3 \quad (5.62\text{l})$$
$$0 = v_{c_3} - e_3 + e_4, \quad (5.62\text{m})$$

where $i_0(t)$ is the current injected by the current source. The matrix J_1 in (5.43), which characterizes index one configurations in ANA models, reads for this problem

$$J_1 = \begin{pmatrix} 0 & 0 & 0 & 0 & 0 & -1 & 0 & 0 \\ 0 & G_3 & 0 & -G_3 & 0 & 0 & 1 & 0 \\ 0 & 0 & G_4 & 0 & 0 & 0 & -1 & 1 \\ 0 & -G_3 & 0 & G_3 + G_5 & -G_5 & 0 & 0 & -1 \\ 0 & 0 & 0 & -G_5 & G_5 & 0 & 0 & 0 \\ -1 & 0 & 0 & 0 & 0 & 0 & 0 & 0 \\ 0 & 1 & -1 & 0 & 0 & 0 & 0 & 0 \\ 0 & 0 & 1 & -1 & 0 & 0 & 0 & 0 \end{pmatrix},$$

and its determinant can be shown to be given by

$$\det J_1 = -G_4 G_5. \quad (5.63)$$

The strict passivity conditions $G_3 > 0$, $G_4 > 0$, and $G_5 > 0$ obviously make the determinant in (5.63) non-null, so that the ANA model (5.62) is index one provided that capacitances and inductances do not vanish. This is consistent with Theorem 5.1, since the circuit displays neither VC-loops nor IL-cutsets. By Theorem 5.1, the tableau equations for this circuit would also yield an index one DAE. The same index one property may be shown

5.4. Index analysis: Passive circuits

to hold if we assume arbitrary coupling among resistors, as long as the conductance matrix is positive definite.

Note, however, that in the light of (5.63) the mere non-vanishing of G_4 and G_5 would define a more accurate index one requirement for this circuit; that is, it is not necessary that G_4 and G_5 are actually positive, and the value of G_3 does not play a role in the index one characterization. These facts will be explained using tree-based techniques in Section 5.5.

Example 2: VC-loops

Insert a voltage source in parallel to the resistor R_3 in Figure 5.1, to obtain the circuit shown in Figure 5.2. Together with the capacitors C_2 and C_3, the voltage source defines a VC-loop which, according to Theorem 5.1, should raise the tractability index of the ANA model of this circuit to two, under the assumption that all conductances, capacitances and inductances are positive. We show below that this is indeed the case, illustrating the form of the different projectors and matrices arising in the proof of the index two case in Theorem 5.1.

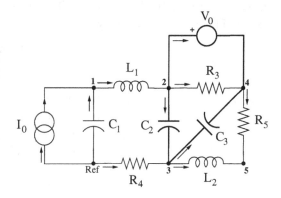

Fig. 5.2 A VC-loop resulting from the insertion of a voltage source.

Note that the modified circuit has the same nodes as the one in Example 1 (Figure 5.1), with an additional branch for which

$$A_u = \begin{pmatrix} 0 \\ 1 \\ 0 \\ -1 \\ 0 \end{pmatrix}.$$

The ANA model is obtained by replacing (5.62g) and (5.62i) by

$$0 = G_3(e_2 - e_4) - i_{l_1} + i_{c_2} + i_u$$
$$0 = -G_3(e_2 - e_4) + G_5(e_4 - e_5) - i_{c_3} - i_u,$$

respectively, and adding an equation for the source of the form

$$0 = v_0(t) - e_2 + e_4.$$

The reader can easily check that the VC-loop avoids the matrix $(A_c\ A_u)$ from having full column rank. This makes the J_1 matrix in (5.43) singular, so that the problem is not index one. Following the proof of item (2) in Theorem 5.1, the projector $\hat{Q}$ onto $\ker(A_c\ A_u)$ can be taken as

$$\hat{Q} = \begin{pmatrix} 0 & 0 & 0 & 0 \\ 0 & 0 & 0 & -1 \\ 0 & 0 & 0 & -1 \\ 0 & 0 & 0 & 1 \end{pmatrix},\ \text{with}\ \hat{Q}_{12} = \begin{pmatrix} 0 \\ -1 \\ -1 \end{pmatrix}.$$

Mind that the block $\hat{Q}_{11}$ in (5.48) vanishes.

On the other hand, the matrix $(A_r\ A_c\ A_u)^\mathsf{T}$ is easily seen to have full column rank, consistently with the absence of IL-cutsets, so that the projector $\overline{Q}$ vanishes. This yields the expression

$$J_2 = \begin{pmatrix} 0 & 0 & 0 & 0 & 0 & -1 & 0 & 0 & 0 \\ 0 & G_3 & 0 & -G_3 & 0 & 0 & 1 & 0 & 1 \\ 0 & 0 & G_4 & 0 & 0 & 0 & -1 & 1 & 0 \\ 0 & -G_3 & 0 & G_3+G_5 & -G_5 & 0 & 0 & -1 & -1 \\ 0 & 0 & 0 & -G_5 & G_5 & 0 & 0 & 0 & 0 \\ -1 & 0 & 0 & 0 & 0 & 0 & 0 & 0 & 0 \\ 0 & 1 & -1 & 0 & 0 & 0 & 0 & 0 & C_2^{-1} \\ 0 & 0 & 1 & -1 & 0 & 0 & 0 & 0 & C_3^{-1} \\ 0 & 1 & 0 & -1 & 0 & 0 & 0 & 0 & 0 \end{pmatrix} \quad (5.64)$$

for the matrix depicted in (5.50), which characterizes index two configurations in ANA models. The determinant of J_2 in (5.64) reads

$$\det J_2 = G_4 G_5 \left(\frac{1}{C_2} + \frac{1}{C_3} \right),$$

which is indeed non-null in the light of the positiveness of the conductances and capacitances, thus making the ANA model of this circuit index two. Again, the same assertion would hold true for arbitrary coupling among resistors, capacitors and inductors, as long as the corresponding matrices are positive definite, and also for the tableau model (5.25).

5.4. Index analysis: Passive circuits

Example 3: IL-cutsets

The circuit displayed in Figure 5.3 is obtained from the one in Figure 5.1 by replacing the capacitor C_1 by the inductor labeled as L_3. This generates an IL-cutset defined by the current source I_0 together with the inductors L_1 and the newly introduced L_3.

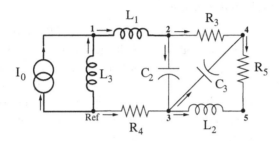

Fig. 5.3 An IL-cutset resulting from the replacement of C_1 by the inductor L_3.

Now the projector $\hat{Q}$ vanishes since $(A_c \ A_u) = A_c$ can be checked to have full column rank. In contrast, $(A_r \ A_c \ A_u)^\mathsf{T}$ has a non-trivial kernel due to the presence of an IL-cutset, which makes J_1 in (5.43) a singular matrix. A projector onto this kernel is

$$\overline{Q} = \begin{pmatrix} 1 & 0 & 0 & 0 & 0 \\ 0 & 0 & 0 & 0 & 0 \\ 0 & 0 & 0 & 0 & 0 \\ 0 & 0 & 0 & 0 & 0 \\ 0 & 0 & 0 & 0 & 0 \end{pmatrix},$$

yielding for the J_2 matrix in (5.50) the expression

$$J_2 = \begin{pmatrix} L_1^{-1} + L_3^{-1} & 0 & 0 & 0 & 0 & 0 & 0 \\ -L_1^{-1} & G_3 & 0 & -G_3 & 0 & 1 & 0 \\ 0 & 0 & G_4 & 0 & 0 & -1 & 1 \\ 0 & -G_3 & 0 & G_3 + G_5 & -G_5 & 0 & -1 \\ 0 & 0 & 0 & -G_5 & G_5 & 0 & 0 \\ 0 & 1 & -1 & 0 & 0 & 0 & 0 \\ 0 & 0 & 1 & -1 & 0 & 0 & 0 \end{pmatrix}. \quad (5.65)$$

The determinant of J_2 in (5.65) now reads

$$\det J_2 = G_4 G_5 \left(\frac{1}{L_1} + \frac{1}{L_3} \right),$$

which again does not vanish due to the positiveness of the conductances and inductances. This means that the tractability index of the ANA model for this circuit is two. As in Example 2 above, the same would be true in the presence of coupling, and the NTA model (5.25) can also be checked to be index two for this circuit.

5.4.2 Modified Nodal Analysis

The ideas leading to the index characterization of NTA and ANA in Theorem 5.1 are based on the ones originally introduced in [292] for Modified Nodal Analysis (MNA). We compile in Theorem 5.2 below a slightly improved form of the topological characterizations of index one and two configurations in MNA discussed in [292], accommodating also the index zero analysis of [293]. The improvement concerns the relaxation of some passivity requirements, since the original statements in [292, 293] assume the positive definiteness of the incremental conductance, capacitance and inductance matrices G, C, L. The proof of these results in the above-mentioned references [292, 293] apply identically in this slightly broader setting. Note also that positive definiteness of the incremental capacitance matrix C (standing here for $C(A_c^T e)$) is required already for index one problems in MNA, in contrast to the NTA/ANA case.

Theorem 5.2. *Consider a well-posed, connected circuit with nonsingular inductance matrix L and positive definite capacitance matrix C.*

(1) The MNA system (5.29) is index zero if and only if

(a) there are no voltage sources; and
(b) there exists a capacitive tree.

Assume in the sequel that at least one of the conditions (a) or (b) fails.

(2) Suppose additionally that the conductance matrix G is positive definite. Then the MNA model (5.29) has tractability index one if and only if the network contains neither VC-loops (except for C-loops) nor IL-cutsets.

(3) Let also L be positive definite. If the network contains VC-loops (with at least one voltage source) and/or IL-cutsets, then (5.29) has tractability index two.

The reader is referred to [292, 293] for the proofs of these statements. Let us just point out the key ideas and some facts which will be useful later.

5.4. Index analysis: Passive circuits

First, in the light of the expression (5.38) for the leading matrix $A(x)$ of MNA, the absence of voltage sources is clearly necessary for system (5.29) to be index zero; in this situation, the index zero condition amounts to the nonsingularity of the matrix

$$A(x) = \begin{pmatrix} A_c C(A_c^T e) A_c^T & 0 \\ 0 & L(i_l) \end{pmatrix}. \tag{5.66}$$

Because of the nonsingularity of L, the characterization relies on the nodal capacitance matrix $A_c C A_c^T \in \mathbb{R}^{(n-1) \times (n-1)}$; for it to be nonsingular the requirement $\mathrm{rk}\, A_c = n - 1$ must be met, meaning that there must exist a nonsingular $(n-1) \times (n-1)$-submatrix of A_c which accounts for a capacitive tree (cf. Lemma 5.5 on page 199). Conversely, if there exists a capacitive tree and $A_c C A_c^T w = 0$, we derive $w^T A_c C A_c^T w = 0$ and, from the positive definiteness of C, the identity $A_c^T w = 0$ follows. Since the columns of A_c^T are linearly independent due to the condition $\mathrm{rk}\, A_c = n - 1$, which owes to the existence of a capacitive tree, we obtain $w = 0$. This reasoning supports the index zero characterization depicted in item (1) above.

The index one analysis is based on the matrix

$$B = \begin{pmatrix} A_r G A_r^T & A_l & A_u \\ -A_l^T & 0 & 0 \\ A_u^T & 0 & 0 \end{pmatrix}, \tag{5.67}$$

constructed as $B = -f'$ with $f(x)$ defined in (5.35). This yields for $A_1(x) = A(x) + B(x)Q_0$ the expression

$$A_1 = \begin{pmatrix} A_c C A_c^T + A_r G A_r^T Q_c & 0 & A_u \\ -A_l^T Q_c & L & 0 \\ A_u^T Q_c & 0 & 0 \end{pmatrix}, \tag{5.68}$$

where Q_c is a projector onto $\ker A_c^T$, so that

$$Q_0 = \begin{pmatrix} Q_c & 0 & 0 \\ 0 & 0 & 0 \\ 0 & 0 & I \end{pmatrix} \tag{5.69}$$

is a projector onto $\ker A(x)$, with $A(x)$ defined in (5.38). Here we are making use of the assumption that $L(i_l)$ is nonsingular and the property $\ker A_c C A_c^T = \ker A_c^T$ (cf. Remark 5.1 on page 215). In problems without capacitors set $Q_c = I_{n-1}$.

The nonsingularity of A_1 is proved in [292] to be equivalent to the topological conditions depicted in item (2). We only emphasize here that, with respect to Theorem 5.1, C-loops yield an index one configuration in

MNA models; see Example 5 on p. 252. The key element in this regard is the $A_u^\mathsf{T} Q_c$ block in (5.68), which has full row rank if all VC-loops are actually C-loops. This is a consequence of Theorem 2.2 (item 4) in [87], which for later use we recast here in the following slightly generalized form.

Lemma 5.11. *Assume that J and K are two disjoint sets of edges within a connected digraph, and let Q_J be a projector onto $\ker A_J^\mathsf{T}$. Then $Q_J^\mathsf{T} A_K$ has full column rank if and only if all JK-loops are actually J-loops.*

In particular, $Q_c^\mathsf{T} A_u$ has full column rank (or, equivalently, $A_u^\mathsf{T} Q_c$ has full row rank) if and only if all VC-loops are C-loops. Find additional details, specially regarding index two configurations, in the above-mentioned references [87, 292, 293]. Again, the reader is referred to Section 5.5 for index analyses of MNA models without positive definiteness assumptions.

Index of the MNA model for Example 1

As in item (2) of Theorem 5.2, let us now assume that in the circuit depicted in Figure 5.1 the conductances and capacitances are positive and the inductances do not vanish. The MNA model (5.29) reads for this circuit

$$C_1 e_1' = -i_{l_1} + i_0(t) \tag{5.70a}$$
$$C_2 e_2' - C_2 e_3' = -G_3(e_2 - e_4) + i_{l_1} \tag{5.70b}$$
$$-C_2 e_2' + (C_2 + C_3) e_3' - C_3 e_4' = -G_4 e_3 - i_{l_2} \tag{5.70c}$$
$$-C_3 e_3' + C_3 e_4' = G_3(e_2 - e_4) - G_5(e_4 - e_5) \tag{5.70d}$$
$$0 = G_5(e_4 - e_5) + i_{l_2} \tag{5.70e}$$
$$L_1 i_{l_1}' = e_1 - e_2 \tag{5.70f}$$
$$L_2 i_{l_2}' = e_3 - e_5. \tag{5.70g}$$

Since the circuit has no voltage sources, the leading matrix of this MNA model has the form

$$A(x) = \begin{pmatrix} A_c C A_c^\mathsf{T} & 0 \\ 0 & L \end{pmatrix}$$

5.4. Index analysis: Passive circuits

where $L = \mathrm{diag}\,(L_1, L_2)$, and the nodal capacitance matrix is given by

$$A_c C A_c^\mathsf{T} = \begin{pmatrix} -1 & 0 & 0 \\ 0 & 1 & 0 \\ 0 & -1 & 1 \\ 0 & 0 & -1 \\ 0 & 0 & 0 \end{pmatrix} \begin{pmatrix} C_1 & 0 & 0 \\ 0 & C_2 & 0 \\ 0 & 0 & C_3 \end{pmatrix} \begin{pmatrix} -1 & 0 & 0 & 0 & 0 \\ 0 & 1 & -1 & 0 & 0 \\ 0 & 0 & 1 & -1 & 0 \end{pmatrix}$$

$$= \begin{pmatrix} C_1 & 0 & 0 & 0 & 0 \\ 0 & C_2 & -C_2 & 0 & 0 \\ 0 & -C_2 & C_2 + C_3 & -C_3 & 0 \\ 0 & 0 & -C_3 & C_3 & 0 \\ 0 & 0 & 0 & 0 & 0 \end{pmatrix}. \qquad (5.71)$$

The B matrix in (5.67) is defined in this case by

$$B = \begin{pmatrix} 0 & 0 & 0 & 0 & 0 & 1 & 0 \\ 0 & G_3 & 0 & -G_3 & 0 & -1 & 0 \\ 0 & 0 & G_4 & 0 & 0 & 0 & 1 \\ 0 & -G_3 & 0 & G_3 + G_5 & -G_5 & 0 & 0 \\ 0 & 0 & 0 & -G_5 & G_5 & 0 & -1 \\ -1 & 1 & 0 & 0 & 0 & 0 & 0 \\ 0 & 0 & -1 & 0 & 1 & 0 & 0 \end{pmatrix}.$$

Under the assumption that the capacitances are positive, the identity $\ker A_c C A_c^\mathsf{T} = \ker A_c^\mathsf{T}$ can be easily checked to hold. A projector Q_0 onto the kernel of the leading matrix $A(x)$ with the form displayed in (5.69) is

$$Q_0 = \begin{pmatrix} 0 & 0 & 0 & 0 & 0 & 0 & 0 \\ 0 & 0 & 0 & 1 & 0 & 0 & 0 \\ 0 & 0 & 0 & 1 & 0 & 0 & 0 \\ 0 & 0 & 0 & 1 & 0 & 0 & 0 \\ 0 & 0 & 0 & 0 & 1 & 0 & 0 \\ 0 & 0 & 0 & 0 & 0 & 0 & 0 \\ 0 & 0 & 0 & 0 & 0 & 0 & 0 \end{pmatrix}.$$

This yields (cf. (5.68))

$$A_1 = A + B Q_0 = \begin{pmatrix} C_1 & 0 & 0 & 0 & 0 & 0 & 0 \\ 0 & C_2 & -C_2 & 0 & 0 & 0 & 0 \\ 0 & -C_2 & C_2 + C_3 & -C_3 + G_4 & 0 & 0 & 0 \\ 0 & 0 & -C_3 & C_3 + G_5 & -G_5 & 0 & 0 \\ 0 & 0 & 0 & -G_5 & G_5 & 0 & 0 \\ 0 & 0 & 0 & 1 & 0 & L_1 & 0 \\ 0 & 0 & 0 & -1 & 1 & 0 & L_2 \end{pmatrix}.$$

The determinant of this matrix equals
$$\det A_1 = L_1 L_2 G_4 G_5 C_1 C_2 C_3, \qquad (5.72)$$
which does not vanish provided that the conductances and capacitances are positive and the inductances are non-null. In this situation the MNA model (5.70) is index one, consistently with item (2) of Theorem 5.2 since the circuit has neither VC-loops at all nor IL-cutsets.

Note however that the non-vanishing of the parameters appearing in (5.72), where it is worth remarking that G_3 is not involved, is enough to yield an index one MNA model; again, this will be explained in terms of certain circuit trees in Section 5.5.

5.5 Index analysis: Tree methods for non-passive circuits

Most index characterizations for DAE models of electrical circuits are directed to (strictly locally) passive systems [87, 234, 248, 253, 292, 293]. Incidentally, a key role in the proof of Theorem 5.1, focused on Node Tableau and Augmented Nodal Analysis models, is played by the positive definiteness of certain incremental matrices; the same happens in the index characterization of Modified Nodal Analysis discussed in Theorem 5.2.

Nevertheless, Example 1 above and, in particular, the expressions depicted in (5.63) and (5.72) for the determinants characterizing index one configurations in ANA and MNA, suggest that these positive definiteness assumptions are far from necessary, at least in uncoupled problems. We show below that this type of expressions reflects certain properties that can be stated in terms of the structure of certain *trees* within the circuit; these properties will make it possible to formulate more accurate requirements on the circuit devices for index characterizations.

Specifically, following [84] we relax the positive definiteness assumption on the incremental conductance matrix G for index one characterizations of ANA systems. The definiteness requirement on G will be replaced by an algebraic assumption on the set of *proper* trees (p. 206) in the circuit. Similarly, we weaken the definiteness requirements on both G and C for index one configurations in MNA; in this case the characterization will be stated in terms of *normal* trees. The same approach makes it possible to relax the definiteness assumption on the capacitance C for index zero MNA models. The attention is restricted to circuits without coupling in these devices. Fully-coupled problems and index two configurations are the subject of undergoing research. For the sake of brevity we do not consider NTA models, although the results can be extended to this setting along the lines presented in subsection 5.5.1.

5.5.1 Augmented Nodal Analysis

In the tractability index one characterization of ANA, an important role is played by the so-called *nodal conductance matrix* $A_r G A_r^\mathsf{T}$; cf. the matrix A_1 depicted in (5.42). Similar considerations apply to the nodal capacitance matrix $A_c C A_c^\mathsf{T}$ in MNA, as illustrated in subsection 5.4.2 above. The appearance of these nodal matrices is certainly a distinct feature of nodal analysis methods.

When trying to eliminate the definiteness assumption on G or C in index analyses, we will be faced with the computation of the determinant of certain matrix products related with the above-mentioned nodal matrices. The matrices within these products will not be square. For the computation of these determinants we will therefore make use of the below-stated Cauchy-Binet formula; see e.g. [137].

Within Lemma 5.12, ω will represent the set $\{1, \ldots, p\}$, whereas α and β will stand for subsets of $\{1, \ldots, r\}$ with cardinality $p \leq r$. These sets will be used as indices to specify $p \times p$ submatrices of certain matrices. Specifically, we denote by $D^{\omega, \alpha}$ the $p \times p$ submatrix of $D \in \mathbb{R}^{p \times r}$ defined by the columns indexed by α, whereas $F^{\beta, \omega}$ describes the $p \times p$ submatrix of $F \in \mathbb{R}^{r \times p}$ defined by the rows indexed by β. Additionally, $B^{\alpha, \beta}$ will be the $p \times p$ submatrix of $B \in \mathbb{R}^{r \times r}$ that lies in the rows indexed by α and the columns indexed by β.

Lemma 5.12. *Let* $D \in \mathbb{R}^{p \times r}$, $E \in \mathbb{R}^{r \times r}$, $F \in \mathbb{R}^{r \times p}$, *with* $p \leq r$. *Then*

$$\det DEF = \sum_{\alpha, \beta} \det D^{\omega, \alpha} \det E^{\alpha, \beta} \det F^{\beta, \omega}, \qquad (5.73)$$

the sum ranging over all index sets α, $\beta \subseteq \{1, \ldots, r\}$ *with cardinality* p.

This result is easily derived from the one corresponding to the product of *two* matrices (cf. item 0.8.7 in [137]). Indeed, let H be a matrix in $\mathbb{R}^{r \times p}$. The Cauchy-Binet formula for the product DH reads

$$\det DH = \sum_{\alpha} \det D^{\omega, \alpha} \det H^{\alpha, \omega}, \qquad (5.74)$$

where the sum is taken over all index sets $\alpha \subseteq \{1, \ldots, r\}$ with cardinality p. Now, if H stands for the product EF above, then the submatrix $H^{\alpha, \omega}$

equals the product $E^{\alpha,\zeta}F$ with $\zeta = \{1,\ldots,r\}$, and its determinant can be written as

$$\det H^{\alpha,\omega} = \sum_\beta \det E^{\alpha,\beta} \det F^{\beta,\omega}, \qquad (5.75)$$

the sum ranging again over all index sets $\beta \subseteq \{1,\ldots,r\}$ with cardinality p. The expressions depicted in (5.74) and (5.75) yield (5.73).

Theorem 5.3. *Consider a well-posed, connected circuit with nonsingular capacitance and inductance matrices C, L, and no coupling among resistors. Then the Augmented Nodal Analysis system (5.27) has tractability index one if and only if either there exists a proper VC-tree or*

(a) there are neither VC-loops nor IL-cutsets; and
(b) the sum of conductance products in proper trees does not vanish.

Proof. We follow essentially the proof presented in [84]. As shown in the proof of Theorem 5.1, the index one condition on the ANA system (5.27) with nonsingular C, L amounts to the nonsingularity of J_1 in (5.43). This matrix reads

$$J_1 = \begin{pmatrix} A_r G A_r^\mathsf{T} & A_{cu} \\ A_{cu}^\mathsf{T} & 0 \end{pmatrix}, \qquad (5.76)$$

where A_{cu} stands for $(A_c \; A_u)$ to simplify notation. Note that the existence of a proper VC-tree, which comprises all capacitors and voltage sources, implies, in the light of Lemma 5.5, that A_{cu} is a square nonsingular matrix, and this makes J_1 in (5.76) nonsingular because of its block structure. Focus in the sequel on cases in which there exists no proper VC-tree.

Let us now check that an index one condition on the ANA system (5.27) rules out the existence of VC-loops and IL-cutsets. Following the idea introduced in [84], we will separate the relevant topological circuit information from the device characteristics by factorizing J_1 in the form

$$J_1 = \underbrace{\begin{pmatrix} A_r & A_{cu} & 0 \\ 0 & 0 & I_{cu} \end{pmatrix}}_{D} \underbrace{\begin{pmatrix} G & 0 & 0 \\ 0 & 0 & I_{cu} \\ 0 & I_{cu} & 0 \end{pmatrix}}_{E} \underbrace{\begin{pmatrix} A_r^\mathsf{T} & 0 \\ A_{cu}^\mathsf{T} & 0 \\ 0 & I_{cu} \end{pmatrix}}_{F}, \qquad (5.77)$$

the block I_{cu} standing for the identity matrix of size $b_{cu} = b_c + b_u$; recall that b_c and b_u define the number of capacitors and voltage sources, respectively. Here, $D \in \mathbb{R}^{p \times r}$, $E \in \mathbb{R}^{r \times r}$, and $F \in \mathbb{R}^{r \times p}$ with $p = n - 1 + b_{cu}$ and $r = b_r + 2b_{cu}$; n and b_r stand for the number of nodes and resistors.

5.5. Index analysis: Tree methods for non-passive circuits 239

The existence of a VC-loop avoids A_{cu} from having full column rank, according to Lemma 5.3; it follows that J_1 in (5.76) is a singular matrix, against the index one hypothesis. If there exists an IL-cutset, then following Lemma 5.4 the matrix $(A_r \ A_{cu})$ cannot have full row rank; this means that D in (5.77) does not have full row rank, making J_1 a singular matrix, again in contradiction to the index one hypothesis. This means that VC-loops and IL-cutsets are precluded in index one configurations.

The absence of VC-loops and IL-cutsets guarantees the existence of at least one proper tree, as stated in Proposition 5.1 (page 206). This implies in particular that the dimensions p, r of the matrices D, E and F above verify $p \leq r$ since $n - 1 \leq b_r + b_{cu}$ because of the existence of at least one proper tree. It is also worth remarking that the exclusion of proper VC-trees mentioned earlier in the proof forces all proper trees to have at least one resistor.

It then remains to show that the nonsingularity of J_1 defining the index one condition for system (5.27) is equivalent, in the absence of VC-loops, IL-cutsets and proper VC-trees, to the non-vanishing of the sum of conductance products in proper trees. This relies on the Cauchy-Binet formula stated in Lemma 5.12, that is,

$$\det J_1 = \det DEF = \sum_{\alpha,\beta} \det D^{\omega,\alpha} \det E^{\alpha,\beta} \det F^{\beta,\omega}, \qquad (5.78)$$

for $p \times p$ submatrices $D^{\omega,\alpha}$, $E^{\alpha,\beta}$ and $F^{\beta,\omega}$ of D, E and F, respectively.

Let us examine, in terms of (5.77), the structure of the matrices $D^{\omega,\alpha}$, $E^{\alpha,\beta}$ and $F^{\beta,\omega}$ yielding non-vanishing determinants in (5.78). Note first that the last b_{cu} columns of D must be in $D^{\omega,\alpha}$ since otherwise the last b_{cu} rows of this matrix would not have full rank. Similarly, the last b_{cu} rows of F need to be fully present in $F^{\beta,\omega}$, because on the contrary the last b_{cu} columns of this matrix could not have full rank. This means that the blocks I_{cu} from D and F must be fully present in $D^{\omega,\alpha}$ and $F^{\beta,\omega}$, respectively, for the corresponding determinants not to vanish.

In turn, the full presence of the block I_{cu} from D in $D^{\omega,\alpha}$ implies that in $E^{\alpha,\beta}$ there will be entries coming from each one of the last b_{cu} rows of E; for the corresponding block in $E^{\alpha,\beta}$ to have full row rank, the block I_{cu} coming from the last rows of E must be fully present in $E^{\alpha,\beta}$. Additionally, this implies that the whole block A_{cu}^{T} from F must be present in $F^{\beta,\omega}$. Similarly, the presence of the block I_{cu} from F in $F^{\beta,\omega}$ forces the full block I_{cu} from the last columns of E to be present in $E^{\alpha,\beta}$, and this in turn implies that A_{cu} from D must be fully present in $D^{\omega,\alpha}$.

Now, remark that the absence of coupling among resistors gives the conductance matrix a diagonal form, namely $G = \text{diag}(G_1, G_2, \ldots, G_{b_r})$, where G_i is the conductance of the i-th resistor. Letting α_1 and β_1 stand for the indices of the rows and columns defining the submatrix $\tilde{G}$ of G present in $E^{\alpha,\beta}$, it is easy to check that

$$\det \tilde{G} = \begin{cases} 0 & \text{if } \alpha_1 \neq \beta_1, \\ \prod_{i \in \alpha_1} G_i & \text{if } \alpha_1 = \beta_1, \end{cases} \qquad (5.79)$$

meaning that α_1 must equal β_1 for $\det E^{\alpha,\beta}$ not to vanish.

So far, we have shown that any non-zero term of (5.78) must be defined by submatrices of the form

$$D^{\omega,\alpha} = \begin{pmatrix} A_{\tilde{r}} & A_{cu} & 0 \\ 0 & 0 & I_{cu} \end{pmatrix}, \quad E^{\alpha,\beta} = \begin{pmatrix} \tilde{G} & 0 & 0 \\ 0 & 0 & I_{cu} \\ 0 & I_{cu} & 0 \end{pmatrix}, \quad F^{\beta,\omega} = \begin{pmatrix} A_{\tilde{r}}^\mathsf{T} & 0 \\ A_{cu}^\mathsf{T} & 0 \\ 0 & I_{cu} \end{pmatrix},$$

where the tilde $\tilde{}$ indicates that only the entries corresponding to some resistors must be present in each non-vanishing term. The above-mentioned identity $\alpha_1 = \beta_1$ explains that the branches corresponding to *the same* set of resistors enter $A_{\tilde{r}}$ and $A_{\tilde{r}}^\mathsf{T}$ in $D^{\omega,\alpha}$ and $F^{\beta,\omega}$, respectively; that is, the columns of the submatrix of A_r entering $D^{\omega,\alpha}$ must match the rows of the submatrix of A_r^T within $F^{\beta,\omega}$, and therefore $F^{\beta,\omega} = (D^{\omega,\alpha})^\mathsf{T}$.

Finally, this set of resistors must be such that the square matrix $(A_{\tilde{r}} \ A_{cu})$ is nonsingular for the corresponding determinant not to vanish. Letting T be the set defined by these resistors, all the capacitors and all the voltage sources, Lemma 5.5 implies that T must be a tree; this implies in particular that $\det D^{\omega,\alpha} = \det F^{\beta,\omega} = \pm 1$ and then $\det D^{\omega,\alpha} \cdot \det F^{\beta,\omega} = 1$ since $F^{\beta,\omega} = (D^{\omega,\alpha})^\mathsf{T}$. Furthermore, T will actually be a *proper* tree since it includes all capacitors and voltage sources and neither inductors nor current sources.

Altogether, the non-vanishing terms in the sum (5.78) have been shown to be the ones corresponding to proper trees, so that $\det J_1$ can be expressed as

$$\det J_1 = (-1)^{b_{cu}} \sum_{T \in \mathcal{T}_p} \prod_{G_i \in T} G_i, \qquad (5.80)$$

where $\mathcal{T}_p$ is the set of proper trees in the circuit, and the notational abuse $G_i \in T$ is used to denote the resistors within the tree T. Note that the $(-1)^{b_{cu}}$ sign in (5.80) comes from the block structure of $E^{\alpha,\beta}$. This shows that the non-vanishing requirement on the sum of conductance products within proper trees is equivalent to the index one condition in the differential-algebraic system (5.27), thus completing the proof. □

5.5. Index analysis: Tree methods for non-passive circuits

Remark 5.4. The existence of a proper VC-tree, that is, of a tree which includes all voltage sources and capacitors, and no resistor, inductor or current source, can be equivalently stated as the absence of VC-loops and ILR-cutsets, in the light of Lemma 5.1. The proof of Theorem 5.3 shows that the existence of a proper VC-tree actually makes the ANA system (5.27) index one also for circuits with resistive coupling.

Remark 5.5. For circuits without resistive coupling, the index one characterization for problems with strictly locally passive resistors depicted in Theorem 5.1 can be derived from Theorem 5.3. Indeed, as indicated above, in this setting G has a diagonal structure diag$(G_1, G_2, \ldots, G_{b_r})$ with $G_i > 0$ for $i \in \{1, \ldots, b_r\}$; therefore, all conductance products have positive sign in (5.80) and the sum in item (b) of Theorem 5.3 does not vanish.

Example 1 revisited: ANA model and proper trees

Theorem 5.3 shows that, by means of a tree-based approach and in the absence of VC-loops and IL-cutsets, the positive definite requirement on the conductance matrix G can be relaxed in the index one characterization of ANA models. We illustrate this result by explaining the properties of the circuit depicted in Figure 5.1 in terms of trees. Additional insight will be obtained from the modified circuit considered in Example 4 below.

Indeed, the expression displayed for the determinant of J_1 in (5.63), namely det $J_1 = -G_4 G_5$, becomes clear if we realize that the circuit in Figure 5.1 has just one proper tree in which the resistors are R_4 and R_5 (cf. Figure 5.4).

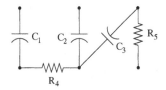

Fig. 5.4 Proper tree in Bashkow's circuit.

Consistently with Theorem 5.3, the minus sign in det J_1 equals $(-1)^{b_{cu}}$ since $b_{cu} = b_c = 3$. This approach yields the exact index one requirement $G_4 \neq 0 \neq G_5$, instead of the (sufficient but not necessary) one arising in the strictly passive setting, which would require $G_3 > 0$, $G_4 > 0$ and $G_5 > 0$. Note in particular that the resistor R_3 forms a loop together with some capacitors and hence it does not enter any proper tree; for this reason the value of the conductance G_3 does not play a role in the index one characterization. See also, in this regard, Corollary 4.3 in [87].

Example 4

The circuit obtained by replacing the inductor L_1 in Figure 5.1 by a resistor R_1 is shown in Figure 5.5.

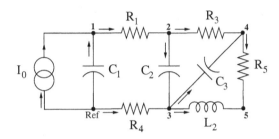

Fig. 5.5 Change L1 by R1 in Bashkow's circuit.

According to Theorem 5.3, the determinant of the matrix J_1, characterizing index one configurations in ANA, can be written by direct inspection of the proper trees in the circuit. These are displayed in Figure 5.6.

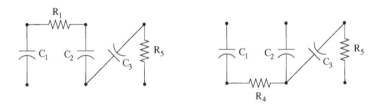

Fig. 5.6 Proper trees in the circuit of Figure 5.5.

In this case there are two proper trees. The resistor R_5 belongs to both, whereas R_1 and R_4 enter one proper tree each. Since $b_{cu} = b_c = 3$, in the light of (5.80) the determinant of det J_1 must read

$$\det J_1 = -G_1 G_5 - G_4 G_5 = -(G_1 + G_4)G_5. \qquad (5.81)$$

From (5.81) we derive the more precise index one requirement $G_1 + G_4 \neq 0$, which certainly holds in particular in a strictly passive setting since in

5.5. Index analysis: Tree methods for non-passive circuits   243

this case it would be $G_1 > 0$, $G_4 > 0$. The fact that the resistor R_5 belongs to both proper trees makes G_5 a common factor in (5.81), yielding the requirement $G_5 \neq 0$ instead of the more restrictive condition $G_5 > 0$ holding in strictly passive problems. Again R_3 does not appear in any of the proper trees and, as before, G_3 does not play a role in the index one characterization.

The expression depicted in (5.81) is confirmed via the explicit computation of the matrix J_1, which in this case can be checked to read

$$J_1 = \begin{pmatrix} G_1 & -G_1 & 0 & 0 & 0 & -1 & 0 & 0 \\ -G_1 & G_1+G_3 & 0 & -G_3 & 0 & 0 & 1 & 0 \\ 0 & 0 & G_4 & 0 & 0 & 0 & -1 & 1 \\ 0 & -G_3 & 0 & G_3+G_5 & -G_5 & 0 & 0 & -1 \\ 0 & 0 & 0 & -G_5 & G_5 & 0 & 0 & 0 \\ -1 & 0 & 0 & 0 & 0 & 0 & 0 & 0 \\ 0 & 1 & -1 & 0 & 0 & 0 & 0 & 0 \\ 0 & 0 & 1 & -1 & 0 & 0 & 0 & 0 \end{pmatrix},$$

its determinant being given indeed by (5.81).

5.5.2 Modified Nodal Analysis

The ideas presented above can be refined in order to relax positive definiteness assumptions on the incremental capacitance and conductance matrices $C(A_c^\mathsf{T} e)$, $G(A_r^\mathsf{T} e)$ for the MNA model (5.29). Briefly, the goal is to characterize index zero and index one configurations for MNA by assessing the nonsingularity of the matrices A and A_1 depicted in (5.66) and (5.68), respectively, without such definiteness requirements. In this discussion, we follow again the ideas introduced in [84].

As indicated in subsection 5.4.2, a key condition in this regard is

$$\ker A_c C A_c^\mathsf{T} = \ker A_c^\mathsf{T}, \tag{5.82}$$

which is equivalent to $\operatorname{rk} A_c C A_c^\mathsf{T} = \operatorname{rk} A_c$ and holds under a positive definiteness assumption on C (cf. Remark 5.1 on page 215). Lemma 5.14 below provides a characterization of (5.82) for uncoupled capacitors without the need for a definiteness assumption on C. The proof of Lemma 5.14 will be based on Lemma 5.13, which will be of interest also in the analysis of index zero configurations. Recall that, according to our definition (p. 197), trees are implicitly understood to be spanning.

Lemma 5.13. *Consider a connected circuit composed only of uncoupled capacitors, with reduced incidence matrix A_c and (diagonal) incremental capacitance matrix $C = \mathrm{diag}\,(C_1, \ldots, C_b)$. Then the nodal capacitance matrix $A_c C A_c^\mathsf{T}$ is nonsingular if and only if*

$$\sum_{T \in \mathcal{T}_c} \prod_{C_i \in T} C_i \neq 0, \tag{5.83}$$

where $\mathcal{T}_c$ stands for the set of trees in the capacitive circuit.

Proof. This is a consequence of the Cauchy-Binet formula (5.73). With the notation introduced for the index sets ω, α, β before Lemma 5.12, we can expand the determinant of $A_c C A_c^\mathsf{T}$ as

$$\det A_c C A_c^\mathsf{T} = \sum_{\alpha, \beta} \det A_c^{\omega, \alpha} \det C^{\alpha, \beta} \det (A_c^\mathsf{T})^{\beta, \omega}. \tag{5.84}$$

Note that for $\det A_c^{\omega, \alpha}$ not to vanish, in the sum we can restrict the values of the index set α to those which define a tree (cf. Lemma 5.5); in this case it is $\det A_c^{\omega, \alpha} = \pm 1$. The same happens with the capacitors indexed by β, since $\det(A_c^\mathsf{T})^{\beta, \omega}$ must not vanish; we then have $\det(A_c^\mathsf{T})^{\beta, \omega} = \pm 1$. Additionally, as we did for conductances in the proof of Theorem 5.3, from the diagonal structure of the incremental capacitance matrix we get

$$\det C^{\alpha, \beta} = \begin{cases} 0 & \text{if } \alpha \neq \beta, \\ \prod_{i \in \alpha} C_i & \text{if } \alpha = \beta, \end{cases}$$

so that $\det A_c^{\omega, \alpha} = \det(A_c^\mathsf{T})^{\beta, \omega} = \pm 1$ and then $\det A_c^{\omega, \alpha} \det(A_c^\mathsf{T})^{\beta, \omega} = 1$.

This means that the sum in (5.84) can be written as

$$\det A_c C A_c^\mathsf{T} = \sum_{\alpha} \det C^{\alpha, \alpha}$$

provided that α takes values within the family of index sets defining trees. This yields

$$\det A_c C A_c^\mathsf{T} = \sum_{T \in \mathcal{T}} \prod_{C_i \in T} C_i,$$

showing that the nonsingularity of the nodal capacitance matrix is indeed equivalent to (5.83). □

Lemma 5.14 below is no longer restricted to capacitive circuits. Recall that A_c represents the (reduced) incidence matrix of the subgraph defined by all capacitors and *all* circuit nodes. We will also consider in the sequel the subgraph defined by all capacitors together with their incident nodes,

5.5. Index analysis: Tree methods for non-passive circuits

and call a *capacitive block* a connected component of this subgraph. The number of capacitive blocks will be denoted by k_c, and $\mathcal{T}_j$ will stand for the set of trees within the j-th block; trees are here assumed to be spanning in each capacitive block.

Lemma 5.14. *Consider a connected circuit in which the capacitors are uncoupled, with incremental capacitance matrix C. Then* $\operatorname{rk} A_c C A_c^\mathsf{T} = \operatorname{rk} A_c$ *if and only if*

$$\prod_{j=1}^{k_c} \left(\sum_{T \in \mathcal{T}_j} \prod_{C_i \in T} C_i \right) \neq 0, \qquad (5.85)$$

that is, iff within each capacitive block the sum of capacitance products in trees does not vanish.

Proof. Under an appropriate ordering of the branches and nodes, the reduced incidence matrix A_c can be written as

$$A_c = \begin{pmatrix} \tilde{A}_{c^{(1)}} & 0 & \cdots & 0 \\ 0 & \tilde{A}_{c^{(2)}} & \cdots & 0 \\ \vdots & \vdots & \ddots & \vdots \\ 0 & 0 & \cdots & \tilde{A}_{c^{(k_c)}} \\ 0 & 0 & 0 & 0 \end{pmatrix}. \qquad (5.86)$$

This results from the fact that different connected components of a given digraph do not share nodes or branches. The last vanishing rows correspond to nodes which are not incident with any capacitor. Note that $\tilde{A}_{c^{(j)}}$ stands above for the (non-reduced) incidence matrix of the j-th capacitive block, with just one possible exception: if the reference node is incident with some capacitor belonging, say, to the j_0-th block, then $\tilde{A}_{c^{(j_0)}}$ is a reduced incidence matrix for that block. It may also happen that the reference node is not incident with any capacitor; in this case all the matrices $\tilde{A}_{c^{(j)}}$ are non-reduced incidence matrices.

In turn, due to the uncoupled assumption on capacitors, the nodal capacitance matrix reads

$$A_c C A_c^\mathsf{T} = \begin{pmatrix} \tilde{A}_{c^{(1)}} C^{(1)} \tilde{A}_{c^{(1)}}^\mathsf{T} & 0 & \cdots & 0 & 0 \\ 0 & \tilde{A}_{c^{(2)}} C^{(2)} \tilde{A}_{c^{(2)}}^\mathsf{T} & \cdots & 0 & 0 \\ \vdots & \vdots & \ddots & \vdots & \vdots \\ 0 & 0 & \cdots & \tilde{A}_{c^{(k_c)}} C^{(k_c)} \tilde{A}_{c^{(k_c)}}^\mathsf{T} & 0 \\ 0 & 0 & 0 & 0 & 0 \end{pmatrix},$$

where $C^{(j)}$ stands for the (diagonal) incremental capacitance matrix of the j-th capacitive block.

The assertion will then follow in a straightforward manner from the identity $\operatorname{rk} \tilde{A}_{c^{(j)}} = \operatorname{rk} \tilde{A}_{c^{(j)}} C^{(j)} \tilde{A}_{c^{(j)}}^\mathsf{T}$ proved in the sequel. Let us first remark that, if the j-th block does not include the circuit reference node, then any row of $\tilde{A}_{c^{(j)}}$ can be written as a linear combination of the remaining ones, which are linearly independent. If we denote by $A_{c^{(j)}}$ the submatrix which results from the removal of any row of $\tilde{A}_{c^{(j)}}$, then $\operatorname{rk} A_{c^{(j)}} = \operatorname{rk} \tilde{A}_{c^{(j)}} = n_j - 1$, being n_j the number of nodes within the j-th capacitive block. It is easy to check additionally that

$$\operatorname{rk} \tilde{A}_{c^{(j)}} C^{(j)} \tilde{A}_{c^{(j)}}^\mathsf{T} = \operatorname{rk} A_{c^{(j)}} C^{(j)} \tilde{A}_{c^{(j)}}^\mathsf{T} = \operatorname{rk} A_{c^{(j)}} C^{(j)} A_{c^{(j)}}^\mathsf{T}.$$

It then remains to show that the condition

$$\sum_{T \in \mathcal{T}_j} \prod_{C_i \in T} C_i \neq 0 \qquad (5.87)$$

for the j-th block is equivalent to the identity $\operatorname{rk} A_{c^{(j)}} C^{(j)} A_{c^{(j)}}^\mathsf{T} = \operatorname{rk} A_{c^{(j)}}$. Since $\operatorname{rk} A_{c^{(j)}} = n_j - 1$ and $A_{c^{(j)}} C^{(j)} A_{c^{(j)}}^\mathsf{T}$ is an $(n_j - 1) \times (n_j - 1)$ matrix, this amounts to saying that $A_{c^{(j)}} C^{(j)} A_{c^{(j)}}^\mathsf{T}$ is nonsingular. But noting that $A_{c^{(j)}}$ is a reduced incidence matrix for the j-th capacitive block, which is connected, Lemma 5.13 applies. Because of the block-diagonal structure of the matrix $A_c C A_c^\mathsf{T}$, the condition $\operatorname{rk} A_c C A_c^\mathsf{T} = \operatorname{rk} A_c$ is then equivalent to (5.85). □

A capacitive block for which the condition (5.87) holds is said in [84] to be *non-degenerate*.

We may now weaken the positive definiteness requirements on C, G in a low index characterization of MNA models with uncoupled capacitors and resistors, as done below. For the index zero statement in item (1), the condition (5.83) referred to in (c) makes sense since the existence of a capacitive tree makes the subgraph defined by the capacitors connected, and the nodal capacitance matrix of this subgraph is $A_c C A_c^\mathsf{T}$; mind that Lemma 5.13 is stated for connected capacitive circuits. In the index one setting of item (2), the condition depicted in (5.85) replaces the above-mentioned (5.83). Note that, in contrast to the corresponding result for ANA stated in Theorem 5.3, which is based on proper trees, the MNA setting naturally leads to *normal* trees; this is due to the fact that C-loops may be displayed in index one configurations in MNA.

5.5. Index analysis: Tree methods for non-passive circuits   247

Theorem 5.4. *Consider a well-posed, connected circuit with nonsingular inductance matrix L and no coupling among capacitors.*

(1) The MNA system (5.29) is index zero if and only if

  (a) there are no voltage sources;
  (b) there exists at least one capacitive tree; and
  (c) the condition on capacitive trees depicted in (5.83) holds.

Assume below that at least one of the conditions (a) or (b) fails.

(2) Suppose that the nondegeneracy condition on capacitors (5.85) holds, and that there is no coupling either among resistors. Then the MNA model (5.29) has tractability index one if and only if either there exists a proper V-tree or

  (d) there are neither VC-loops (except for C-loops) nor IL-cutsets; and
  (e) the sum of conductance products in normal trees does not vanish.

Proof. The index zero characterization stated in item (1) follows immediately from the form of the matrix A in (5.66), together with Lemma 5.13 and the nonsingularity of L.

For the index one statement in item (2), be aware of the fact that the expression (5.68) for the A_1 matrix is based on the nondegeneracy condition (5.85). Let us first analyze the case in which the circuit has a proper V-tree. This precludes the existence of capacitors, since otherwise such a tree would not be proper. In this situation the projector Q_c amounts to I_{n-1}. Note that other voltage sources apart from those in the tree are ruled out by the well-posedness of the circuit. This means that A_u and $A_u^\mathsf{T} Q_c = A_u^\mathsf{T}$ within A_1 in (5.68) are square nonsingular matrices; the nonsingularity of L together with the block structure of A_1 then makes this matrix nonsingular.

Assume in the sequel that there is not such a tree. Next we show that an index one assumption implies that the topological conditions stated in (d) are met. In this direction, following again [84] we will make use of the factorization

$$A_1 = \underbrace{\begin{pmatrix} A_c & A_r & A_u & 0 & 0 \\ 0 & 0 & 0 & I_l & 0 \\ 0 & 0 & 0 & 0 & I_u \end{pmatrix}}_{D} \underbrace{\begin{pmatrix} C & 0 & 0 & 0 & 0 \\ 0 & G & 0 & 0 & 0 \\ 0 & 0 & 0 & 0 & I_u \\ 0 & 0 & 0 & I_l & 0 \\ 0 & 0 & I_u & 0 & 0 \end{pmatrix}}_{E} \underbrace{\begin{pmatrix} A_c^\mathsf{T} & 0 & 0 \\ A_r^\mathsf{T} Q_c & 0 & 0 \\ A_u^\mathsf{T} Q_c & 0 & 0 \\ -A_l^\mathsf{T} Q_c & L & 0 \\ 0 & 0 & I_u \end{pmatrix}}_{F}. \quad (5.88)$$

In an index one setting, the nonsingularity of A_1 implies that the matrix $A_u^\mathsf{T} Q_c$ within the last rows of (5.68) must have full row rank or, equivalently, $Q_c^\mathsf{T} A_u$ must have full column rank. According to Lemma 5.11, this means that all VC-loops must be C-loops. Additionally, the factorization depicted in (5.88) shows that D must have full row rank if A_1 is nonsingular; this amounts to saying that $(A_c\ A_r\ A_u)$ must have full row rank, ruling out IL-cutsets in the light of Lemma 5.4.

We then need to show that, in the absence of proper V-trees and provided that the topological requirements depicted in (d) hold, the nonsingularity of A_1 is equivalent to the non-vanishing of the sum of conductance products in normal trees, as stated in (e). To achieve this, as in Theorem 5.3 we make use of the Cauchy-Binet formula (5.73) to write

$$\det A_1 = \det DEF = \sum_{\alpha,\beta} \det D^{\omega,\alpha} \det E^{\alpha,\beta} \det F^{\beta,\omega}. \quad (5.89)$$

Proceeding analogously to Theorem 5.3, the reader can check that all the blocks I_l and I_u, as well as the blocks A_u, $A_u^\mathsf{T} Q_c$, $-A_l^\mathsf{T} Q_c$ and L in (5.88) must be fully present in the submatrices $D^{\omega,\alpha}$, $E^{\alpha,\beta}$ and $F^{\beta,\omega}$ for the corresponding determinants not to vanish. This, together with the diagonal structure on the capacitance and conductance matrices C and G which results from the absence of coupling within these elements, shows that the matrices resulting in non-vanishing products in (5.89) have the form

$$D^{\omega,\alpha} = \begin{pmatrix} A_{\tilde{c}} & A_{\tilde{r}} & A_u & 0 & 0 \\ 0 & 0 & 0 & I_l & 0 \\ 0 & 0 & 0 & 0 & I_u \end{pmatrix},\ E^{\alpha,\beta} = \begin{pmatrix} \tilde{C} & 0 & 0 & 0 & 0 \\ 0 & \tilde{G} & 0 & 0 & 0 \\ 0 & 0 & 0 & 0 & I_u \\ 0 & 0 & 0 & I_l & 0 \\ 0 & 0 & I_u & 0 & 0 \end{pmatrix},\ F^{\beta,\omega} = \begin{pmatrix} A_{\tilde{c}}^\mathsf{T} & 0 & 0 \\ A_{\tilde{r}}^\mathsf{T} Q_c & 0 & 0 \\ A_u^\mathsf{T} Q_c & 0 & 0 \\ -A_l^\mathsf{T} Q_c & L & 0 \\ 0 & 0 & I_u \end{pmatrix},$$

where a tilde ~ is used again to indicate that only the branches which correspond to *some* capacitors and resistors must be present for the determinants not to vanish. Note that all voltage sources must be present in all cases.

In the analysis of the matrices $D^{\omega,\alpha}$ and $F^{\beta,\omega}$ depicted above, let us denote by T the set of branches defined by all voltage sources together with the capacitors and resistors specified by $A_{\tilde{c}}$ and $A_{\tilde{r}}$. To yield nontrivial terms, the set T must necessarily define a tree since otherwise $A_T = (A_{\tilde{c}}\ A_{\tilde{r}}\ A_u)$ would be a singular matrix (cf. Lemma 5.5). Additionally,

5.5. Index analysis: Tree methods for non-passive circuits

denote by K the upper left block from $F^{\beta,\omega}$, namely

$$K = \begin{pmatrix} A_{\tilde{c}}^{\mathsf{T}} \\ A_{\tilde{r}}^{\mathsf{T}} Q_c \\ A_u^{\mathsf{T}} Q_c \end{pmatrix}. \tag{5.90}$$

We show below that, in the present framework,

$$\det K = \begin{cases} \det A_T & \text{if } T \text{ is a normal tree} \\ 0 & \text{otherwise.} \end{cases} \tag{5.91}$$

Indeed, if T is a normal tree it must include no inductor, since IL-cutsets and in particular L-cutsets are excluded, and all capacitors away from T must define a C-loop with capacitors within T, because all VC-loops are C-loops. The latter implies $\ker A_c^{\mathsf{T}} = \ker A_{\tilde{c}}^{\mathsf{T}}$, since every row in A_c^{T} corresponding to a capacitor which is not in T must be linearly dependent with those of $A_{\tilde{c}}^{\mathsf{T}}$. Additionally, since T is a tree we may write

$$\mathbb{R}^{n-1} = \ker A_{\tilde{c}}^{\mathsf{T}} \oplus \ker \begin{pmatrix} A_{\tilde{r}}^{\mathsf{T}} \\ A_u^{\mathsf{T}} \end{pmatrix} = \ker A_c^{\mathsf{T}} \oplus \ker \begin{pmatrix} A_{\tilde{r}}^{\mathsf{T}} \\ A_u^{\mathsf{T}} \end{pmatrix}.$$

Let $\tilde{P}_c$ stand for the projector along $\ker A_c^{\mathsf{T}}$ onto $\ker (A_{\tilde{r}} \ A_u)^{\mathsf{T}}$, and set $\tilde{Q}_c = I - \tilde{P}_c$. In the light of the relations $A_{\tilde{c}}^{\mathsf{T}} \tilde{Q}_c = A_{\tilde{c}}^{\mathsf{T}} Q_c = 0$ $A_{\tilde{c}}^{\mathsf{T}} \tilde{P}_c = A_{\tilde{c}}^{\mathsf{T}} (I - \tilde{Q}_c) = A_{\tilde{c}}^{\mathsf{T}}$, $A_{\tilde{r}}^{\mathsf{T}} \tilde{P}_c = A_u^{\mathsf{T}} \tilde{P}_c = 0$, we derive

$$K = \begin{pmatrix} A_{\tilde{c}}^{\mathsf{T}} \\ A_{\tilde{r}}^{\mathsf{T}} \\ A_u^{\mathsf{T}} \end{pmatrix} (\tilde{P}_c + Q_c) = A_T^{\mathsf{T}}(\tilde{P}_c + Q_c). \tag{5.92}$$

Write $\tilde{P}_c + Q_c = I - (\tilde{Q}_c - Q_c)$ and remark that $\tilde{Q}_c - Q_c$ is a nilpotent index two matrix since $(\tilde{Q}_c - Q_c)(\tilde{Q}_c - Q_c) = \tilde{Q}_c - Q_c - \tilde{Q}_c + Q_c = 0$, where we have used the identities $\tilde{Q}_c Q_c = Q_c$, $Q_c \tilde{Q}_c = \tilde{Q}_c$ for projectors onto the same space (cf. Proposition 2.5 on page 49). We then derive $\det(I - (\tilde{Q}_c - Q_c)) = 1$, equation (5.92) yielding $\det K = \det A_T$ for normal trees.

Assume now that T is not a normal tree. Since it contains all voltage sources there must necessarily exist a capacitor $C_i \notin T$ which does not form any VC-loop with elements of T; in particular it does not form any C-loop with elements of T. On the other hand, since T is a tree, C_i defines a loop with certain elements of this set. This means that there exists a loop in $T \cup \{C_i\}$ which is not a C-loop, showing that $Q_c^{\mathsf{T}} (A_{\tilde{r}} \ A_u)$ does not have full column rank by Lemma 5.11. This means that

$$\begin{pmatrix} A_{\tilde{r}}^{\mathsf{T}} Q_c \\ A_u^{\mathsf{T}} Q_c \end{pmatrix}$$

does not have full row rank and hence K in (5.90) is a singular matrix, so that $\det K = 0$. This completes the proof of the identity (5.91).

In the light of (5.91), the non-trivial terms in (5.89) must correspond to normal trees. For them we have

$$\det D^{\omega,\alpha} = \det A_T, \quad \det F^{\beta,\omega} = \det K \det L = \det A_T \det L,$$

using again (5.91). Since $\det A_T = \pm 1$, (5.89) reads

$$\det A_1 = (-1)^{b_u} \det L \sum_{T \in \mathcal{T}_n} \left(\prod_{G_i \in T} G_i \prod_{C_i \in T} C_i \right), \quad (5.93)$$

where $\mathcal{T}_n$ is the set of normal trees within the circuit. The $(-1)^{b_u}$ factor owes to the block structure of $E^{\alpha,\beta}$; recall that b_u is the number of voltage sources in the circuit.

Note finally that the hypothesis that all VC-loops are actually C-loops implies that all normal trees include exactly one capacitive tree from within each capacitive block, and all possible combinations of these capacitive trees enter the sum (5.93); see e.g. Example 5 below. This means that the expression depicted in (5.93) can be rewritten as

$$\det A_1 = (-1)^{b_u} \det L \left(\sum_{T \in \mathcal{T}_N} \prod_{G_i \in T} G_i \right) \cdot \prod_{j=1}^{k_c} \left(\sum_{T \in \mathcal{T}_j} \prod_{C_i \in T} C_i \right). \quad (5.94)$$

The factors defined by sums of capacitive products do not vanish because of the non-degeneracy hypothesis (5.85) on the capacitive blocks. Since L is nonsingular, the nonsingularity of the matrix A_1 characterizing index one configurations in MNA relies indeed on the requirement that the sum of conductance products in normal trees does not vanish. □

Within Theorem 5.4 it must be understood that, if normal trees do not include resistors, then the requirement depicted in item (e) may be disregarded. Similarly, the non-degeneracy condition (5.85) should be ignored in problems without capacitors. Note that, excluding cases with proper V-trees (which are treated separately in Theorem 5.4), at least one capacitor or resistor must be present in every normal tree; otherwise, the existence of a V-tree and the assumption that VC-loops are C-loops preclude the existence of capacitors and this would make this V-tree a proper one.

Remark 5.6. Under the existence of a proper V-tree, the assumption that resistors are uncoupled can be removed from the index one characterization in item (2), since the nonsingularity of A_1 in (5.68) relies in this situation only on the nonsingularity of A_u, L and $A_u^\mathsf{T} Q_c = A_u^\mathsf{T}$. In this setting capacitors cannot be present, as indicated above, so that the hypothesis that capacitors are uncoupled can be obviously dropped.

Remark 5.7. For circuits without resistive and capacitive coupling, the index zero and index one characterizations stated in items (1) and (2) of Theorem 5.2 under passivity assumptions, can be derived from the corresponding statements in Theorem 5.4. Actually, the positive definiteness of C guarantees that the conditions (5.83) or (5.85), respectively, are met, since capacitive products are positive; analogously, the positive definiteness of G makes all the conductance products positive, so that the sum in item (e) of Theorem 5.4 is not null.

Example 1 again: MNA model, capacitive blocks and normal trees

In virtue of Theorem 5.4, the requirements of strict passivity on capacitors and resistors in the MNA context can be relaxed using normal trees. Focus the attention again on Bashkow's circuit depicted in Figure 5.1, for which the matrices A_c and $A_c C A_c^\mathsf{T}$ can be found in (5.61) and (5.71), respectively. Removing the positiveness assumption on the capacitances, the condition $\ker A_c C A_c^\mathsf{T} = \ker A_c^\mathsf{T}$ or, equivalently, $\mathrm{rk}\, A_c C A_c^\mathsf{T} = \mathrm{rk}\, A_c^\mathsf{T}$, may be easily proved equivalent to the requirements $C_1 \neq 0$, $C_2 \neq 0$, $C_3 \neq 0$.

This property relies on the fact that the circuit has two capacitive blocks, the first one defined by C_1 alone and the second one by C_2 and C_3, with one tree each; cf. Figure 5.7. This drives the non-vanishing requirements to the capacitance C_1 and the product $C_2 C_3$, respectively.

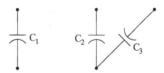

Fig. 5.7 Capacitive blocks C_1 and C_2-C_3 in Bashkow's circuit.

Additionally, since the circuit has neither VC-loops nor IL-cutsets, the normal trees are just the proper ones. Therefore, the circuit has just the normal tree displayed in Figure 5.4, in which the resistors are R_4 and R_5, and therefore the product (5.94) amounts in this example to the expression $L_1 L_2 G_4 G_5 C_1 C_2 C_3$ computed in (5.72). Mind the positive sign in this expression, which owes to the absence of voltage sources.

Example 5

Similar considerations apply to the circuit obtained after replacing the resistor R_3 by a capacitor C_4 in Figure 5.5 (p. 242), as shown in Figure 5.8. We will use this example to illustrate how the conditions arising in Theorem 5.4 can be checked by examining the capacitive blocks and the structure of the normal trees in the circuit.

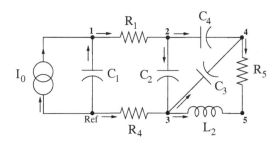

Fig. 5.8 Change R3 by C4 in Figure 5.5.

The above-mentioned replacement generates a C-loop defined by C_2, C_3 and C_4. The reader can check that, similarly to the case defined by the VC-loop in Example 2 (Figure 5.2), and consistently with Theorem 5.1, this raises the index of the ANA model of this circuit to two, at least in a positive definite setting. In contrast, a C-loop will typically result in an index one MNA model, as shown below.

The reduced incidence matrix for capacitive branches is now

$$A_c = \begin{pmatrix} -1 & 0 & 0 & 0 \\ 0 & 1 & 0 & 1 \\ 0 & -1 & 1 & 0 \\ 0 & 0 & -1 & -1 \\ 0 & 0 & 0 & 0 \end{pmatrix},$$

and the nodal capacitance matrix reads

$$A_c C A_c^\mathsf{T} = \begin{pmatrix} C_1 & 0 & 0 & 0 & 0 \\ 0 & C_2 + C_4 & -C_2 & -C_4 & 0 \\ 0 & -C_2 & C_2 + C_3 & -C_3 & 0 \\ 0 & -C_4 & -C_3 & C_3 + C_4 & 0 \\ 0 & 0 & 0 & 0 & 0 \end{pmatrix}. \qquad (5.95)$$

5.5. Index analysis: Tree methods for non-passive circuits

The reader can check that the nodal capacitance matrix has rank three (and therefore the identity $\ker A_c C A_c^\mathsf{T} = \ker A_c^\mathsf{T}$ holds) if and only if

$$C_1 \neq 0, \quad C_2 C_3 + C_2 C_4 + C_3 C_4 \neq 0. \tag{5.96}$$

These conditions can also be derived directly from the fact that the first capacitive block is just defined by the C_1 capacitor, whereas the second one contains the three capacitive trees shown in Figure 5.9.

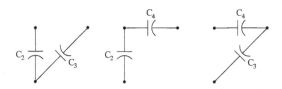

Fig. 5.9 Trees within the capacitive block C_2-C_3-C_4 in Figure 5.8.

If the requirements stated in (5.96) are met, set

$$Q_c = \begin{pmatrix} 0 & 0 & 0 & 0 & 0 \\ 0 & 0 & 0 & 1 & 0 \\ 0 & 0 & 0 & 1 & 0 \\ 0 & 0 & 0 & 1 & 0 \\ 0 & 0 & 0 & 0 & 1 \end{pmatrix},$$

for the projector onto the kernel of the nodal capacitance matrix (5.95). The matrix A_1 in (5.68) may then be shown to read

$$A_1 = \begin{pmatrix} C_1 & 0 & 0 & -G_1 & 0 & 0 \\ 0 & C_2+C_4 & -C_2 & -C_4+G_1 & 0 & 0 \\ 0 & -C_2 & C_2+C_3 & -C_3+G_4 & 0 & 0 \\ 0 & -C_4 & -C_3 & C_3+C_4+G_5 & -G_5 & 0 \\ 0 & 0 & 0 & -G_5 & G_5 & 0 \\ 0 & 0 & 0 & -1 & 1 & L_2 \end{pmatrix},$$

with

$$\det A_1 = L_2(G_1 + G_4)G_5 C_1 (C_2 C_3 + C_2 C_4 + C_3 C_4). \tag{5.97}$$

This yields, in addition to (5.96) and $L_2 \neq 0$, the index one requirements $G_1 + G_4 \neq 0$ and $G_5 \neq 0$. As expected, these conditions are less restrictive than the ones holding in a strictly passive setting, for which all conductances, capacitances and inductances would be required to be positive.

Again, these requirements could be directly obtained by examining the form of the normal trees in the circuit. Indeed, the expression (5.97) reflects, besides the above-detailed structure of capacitive blocks, the fact that R_5 belongs to all normal trees, whereas R_1 and R_4 are split among them as detailed in Figure 5.10.

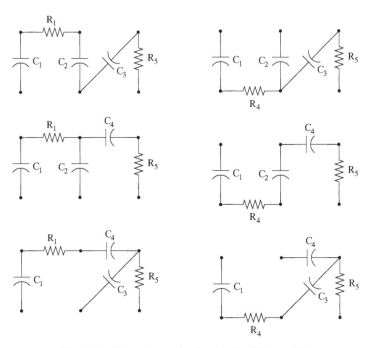

Fig. 5.10 Normal trees in the circuit of Figure 5.8.

The six normal trees in Figure 5.10 correspond to the different terms arising in the expansion

$$\det A_1 = L_2(G_1G_5C_1C_2C_3 + G_4G_5C_1C_2C_3 + G_1G_5C_1C_2C_4$$
$$+ G_4G_5C_1C_2C_4 + G_1G_5C_1C_3C_4 + G_4G_5C_1C_3C_4)$$

of (5.97). All trees coming from the capacitive block defined by C_2, C_3 and C_4 enter the normal trees which include R_1 and R_5 (left column of Figure 5.10) as well as those including R_4 and R_5 (right column); this explains the fact that the conductances and capacitances can be grouped as indicated in (5.97). Remark that this is a general property which in Theorem 5.4 made it possible to rewrite (5.93) in the form depicted in (5.94).

Chapter 6

Branch-oriented methods

Differential-algebraic equations play nowadays an important role in circuit modeling and simulation, as discussed in Chapter 5. Nevertheless, some important topics in nonlinear circuit theory have been somehow overlooked in the differential-algebraic context. Among others, these aspects include the state formulation problem and the analysis of different qualitative properties of circuit dynamics. In this Chapter we show that semistate formulations based on DAEs are certainly of interest regarding these topics, which will be here addressed via reduction methods for so-called *branch-oriented* circuit models.

The state formulation problem arises naturally in the time-domain analysis of electrical circuits. It can be simply defined as the problem of setting up an explicit ODE modeling the circuit dynamics, either in a local or a global sense. The goal is often to formulate state space models not for individual problems but for general circuit families, in terms of their topology (cf. 5.1.2.1) and the electrical features of the devices. As detailed in subsection 6.1.3, following the seminal ideas of Bashkow and Bryant [18, 35, 36] this problem can be tackled by means of tree-based techniques.

Qualitative properties of nonlinear circuits have been mainly addressed within the framework of state space models. Many qualitative studies are directed to the analysis of restricted circuit families displaying specific dynamical features, such as nonlinear oscillations or chaotic effects. Van der Pol's and Chua's circuits are paradigmatic examples in these directions; see e.g. [69, 110, 265, 266]. Explicit ODEs describing the dynamics of these circuits are given and therefore modeling is not an issue.

However, many other works address qualitative properties of general circuits or broad circuit families, and in these settings modeling is certainly a key aspect. See for instance [60, 63–65, 103–105, 286, 287]. Incidentally, in many cases the topological conditions imposed to guarantee the existence of a state space model are somehow merged with those involving qualitative aspects and, sometimes, this makes it difficult to distinguish the aim of the different topological hypotheses appearing in a given statement. The assumptions needed for the formulation of a state model are often unduly restrictive from the qualitative point of view. Moreover, in several cases the results are bound to certain (say, index one) circuit configurations. A more detailed discussion can be found in Section 6.3 below.

The differential-algebraic formalism is of help in both the state formulation problem and the analysis of qualitative properties. Roughly speaking, DAE theory defines a sufficiently broad framework as to accommodate general circuit models and, at the same time, has reached a stage in which efficient mathematical tools are available to tackle these problems in terms of the topological and electrical features of the circuit. We elaborate on this point of view in the next two paragraphs.

The state formulation problem can be naturally addressed as a *reduction* problem on semistate (DAE) models. The reduction methods discussed in Chapter 3 make it possible to formulate local state equations in terms of the original problem variables. This approach clarifies the role of the different equations and variables within the tree-based methods mentioned above, as well as the meaning of the topological and electrical assumptions arising in the analysis. Our discussion will also accommodate intermediate formulations such as *multiport* models in a comprehensive, interrelated framework. The "elimination of unwanted variables" already referred to by Bashkow [18] will be seen as a reduction process which preserves the geometric index of the system.

Regarding qualitative aspects, the DAE framework allows us to separate clearly the topological conditions arising in qualitative studies from those involved in the state formulation problem and index analyses; certain results from graph theory will be needed to achieve this. Additionally, linear stability properties of equilibria such as e.g. their hyperbolicity or exponential stability can be assessed directly in the DAE setting via matrix pencil theory. Specifically, the above-mentioned geometric index analysis, together with the results in Section 3.5, makes it possible to examine linear stability properties of equilibria in terms of the pencil coming from the linearization of the DAE at equilibrium. These ideas may also be of interest in future analyses of other qualitative aspects involving for instance bifurcations or oscillations.

The gap between the above-mentioned state space and qualitative problems and the DAE literature is quite apparent, in spite of several connections investigated in [39, 40, 74, 118, 235]. This gap may be partially due to the fact that the formulation of state space models is difficult to automatize. The numerical simulation of circuit dynamics is usually focused on *nodal* methods, which allow for an automatic generation of network equations in semistate form. The DAE literature on circuits has mainly focused on index computations for such nodal models, with some exceptions [234]. But node potentials, which are the key variables for setting up automatically circuit equations, will not appear in state space reductions, typically formulated in terms of capacitor voltages (or charges) and inductor currents (or fluxes). Branch-oriented models, which do not use node potentials as semistate variables, seem to be more appropriate for these purposes. Additionally, the results in Section 3.5 make the framework based on the geometric index better suited for qualitative studies than the tractability index supporting the analysis of nodal methods (cf. Chapter 5).

This Chapter is structured as follows. Branch-oriented models are introduced in Section 6.1. Note that this approach has received different names in the circuit literature; in particular, the analysis method underlying these models is sometimes referred to as 'hybrid analysis' or 'mixed analysis'. Once the models have been introduced, the state formulation problem can be presented in detail, as done in subsection 6.1.3. The geometric index of these models is analyzed in Section 6.2, where the attention is focused on problems with strictly passive resistors. Multiport and state space models are also discussed in this Section. The results are mainly directed to circuits without controlled sources, although subsection 6.2.6 extends the analysis to problems including certain controlled sources. Finally, Section 6.3 illustrates how linear stability properties of equilibria in DC circuits can be addressed using branch-oriented models.

6.1 Branch-oriented semistate models

Branch-oriented models are characterized by the statement of Kirchhoff laws in the form depicted in (5.4) and (5.7), that is, $Ai = 0$ and $Bv = 0$. Here A and B are the reduced incidence and loop matrices introduced in subsection 5.1.1. This will make it possible to formulate the circuit equations just in terms of the branch currents i and voltages v, which explains the 'branch-oriented' label for these methods. The main difference with nodal analysis stems from the fact that, in the latter, Kirchhoff's voltage law is formulated in terms of the node potentials e as $v = A^\mathsf{T} e$.

6.1.1 The basic model

The reader is referred to subsection 5.1.2 in Chapter 5 for background on elementary aspects of electrical circuit theory. Henceforth all circuits will be assumed to be connected. As in Chapter 5, they may be nonlinear. Full coupling within the sets of capacitors, inductors and resistors is allowed; all these devices are assumed to be time-invariant.

Although most results can be also applied to charge-oriented models, for simplicity we will assume throughout that capacitors are globally voltage-controlled by the C^1 relation $q = \gamma_c(v_c)$ depicted in (5.12), with incremental capacitance matrix $C(v_c) = \gamma'_c(v_c)$. Analogously, as displayed in (5.16), inductors will be globally current-controlled by a C^1 relation of the form $\phi = \gamma_l(i_l)$, with incremental inductance matrix $L(i_l) = \gamma'_l(i_l)$.

Contrary to Chapter 5, we do not make any assumption on controlling variables for resistors. For the moment we just let the resistive branch currents and voltages be related by the implicit equation depicted in (6.1e) below, with $g_r \in C^1(\mathbb{R}^{2b_r}, \mathbb{R}^{b_r})$. Note that for instance gyrators or ideal transformers (p. 209) can be considered as coupled resistors.

Voltage and current sources will be for the moment assumed to be independent; cf. subsection 6.2.6 for problems with controlled sources. Splitting in Kirchhoff current and voltage laws the vectors of currents and voltages as $i = (i_c, i_l, i_r, i_j, i_u)$ and $v = (v_c, v_l, v_r, v_j, v_u)$, where the subscripts c, l, r, j and u correspond to capacitors, inductors, resistors, current sources and voltage sources, respectively, we are led to the branch-oriented model

$$C(v_c)v'_c = i_c \tag{6.1a}$$
$$L(i_l)i'_l = v_l \tag{6.1b}$$
$$0 = A_c i_c + A_l i_l + A_r i_r + A_j i_j + A_u i_u \tag{6.1c}$$
$$0 = B_c v_c + B_l v_l + B_r v_r + B_j v_j + B_u v_u \tag{6.1d}$$
$$0 = g_r(v_r, i_r) \tag{6.1e}$$
$$0 = i_j - i_s(t) \tag{6.1f}$$
$$0 = v_u - v_s(t). \tag{6.1g}$$

Below we will also use the formulation of Kirchhoff's current law (6.1c) in terms of a reduced cutset matrix, that is, in the form $Qi = 0$ (cf. (5.6)). Recall, for later use, that b_c, b_l, b_r, b_j and b_u represent the number of capacitors, inductors, resistors, and current and voltage sources in the circuit.

6.1.2 Tree-based formulations

In the analysis of branch-oriented models, at several points we will profit from the choice of certain trees in a connected circuit. All circuits will be assumed hereafter to be well-posed (see p. 206) and, allowed by Lemma 5.1, we will choose all trees with the restriction that voltage sources correspond to twigs and current sources are located within the links.

As detailed in 5.1.1.8, the choice of a tree in a connected digraph yields two sets of so-called fundamental cutsets and loops, each one uniquely defined by a twig or a link, respectively. Choosing the orientation of the cutsets and loops coherently with that of the corresponding twigs or links, and using the orthogonality property $QB^\mathsf{T} = 0$ following from Lemma 5.9, the corresponding reduced cutset and loop matrices have the expressions depicted in (5.3a)-(5.3b), namely

$$Q = \begin{pmatrix} I_r & -F^\mathsf{T} \end{pmatrix}$$
$$B = \begin{pmatrix} F & I_s \end{pmatrix},$$

for a certain matrix $F \in \mathbb{R}^{s \times r}$; we are denoting $r = n-1$ and $s = b-n+1$, n and b being the number of nodes and branches, respectively. Recall that the first r columns of both matrices correspond to twigs, whereas the last s ones are associated with the links.

The use of these matrices makes it possible to recast Kirchhoff laws $Qi = 0$ and $Bv = 0$ as

$$i_1 = F^\mathsf{T} i_2 \qquad (6.2a)$$
$$v_2 = -Fv_1. \qquad (6.2b)$$

Here and in the sequel the boldface subscripts '1' and '2' refer to tree and cotree elements, respectively, that is, twigs and links. For instance, the vector of tree voltages v_1 can be split in terms of twig capacitors, inductors, resistors and voltage sources as $v_1 = (v_{c_1}, v_{l_1}, v_{r_1}, v_u)$; analogously, the vector of cotree voltages v_2 can be written in terms of link capacitors, inductors, resistors and current sources as $v_2 = (v_{c_2}, v_{l_2}, v_{r_2}, v_j)$. The same holds for the tree and cotree current vectors i_1 and i_2. The relations depicted in (6.2) for Kirchhoff laws express the widely used fact that tree currents can be written in terms of cotree currents, whereas cotree voltages can be expressed in terms of tree voltages [283].

Once a tree has been chosen, the model (6.1) can be rewritten in the form

$$C(v_c)v_c' = i_c \quad (6.3a)$$
$$L(i_l)i_l' = v_l \quad (6.3b)$$
$$0 = i_1 - F^\mathsf{T} i_2 \quad (6.3c)$$
$$0 = Fv_1 + v_2 \quad (6.3d)$$
$$0 = g_r(v_r, i_r) \quad (6.3e)$$
$$0 = i_j - i_s(t) \quad (6.3f)$$
$$0 = v_u - v_s(t). \quad (6.3g)$$

Remark 6.1. We know from 5.1.1.7 that the rows of any two reduced cutset and incidence matrices Q, A span the same space, meaning that the relation $Q = M_0 A$ holds for a nonsingular matrix M_0. The same happens with any two reduced loop matrices B, $\hat{B}$ so that $\hat{B} = M_1 B$ for some nonsingular matrix M_1. This means that (6.3) results from the premultiplication of (6.1) by a nonsingular matrix with a block diagonal structure defined by the blocks (I_c, I_l, M_0, M_1, I_r, I_j, I_u), the identities having sizes b_c, b_l, b_r, b_j and b_u, respectively. This transformation does not change the geometric index of the system, as indicated in 3.4.7.1, so that the circuit model (6.1) has the same index as the tree-based one (6.3). Later analyses will often use models of the form depicted in (6.3), the results holding as well for (6.1). Note that, for the same reason, the choice of *any* two trees yields the same index.

Example

The use of fundamental cutsets and loops can be easily illustrated by means of Bashkow's circuit, already considered in Example 1 (pp. 226-229). This circuit is displayed in Figure 6.1 below, together with the tree already introduced in Figure 5.4.

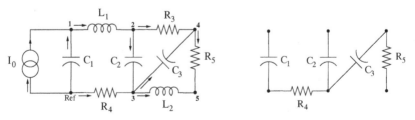

Fig. 6.1 Bashkow's circuit and tree.

6.1. Branch-oriented semistate models

Within this tree, the three capacitors C_1, C_2 and C_3, as well as the resistors R_4 and R_5 are twigs, whereas the two inductors L_1 and L_2 as well as the current source I_0 and the resistor R_3 are links. The fundamental cutset defined by the twig capacitor C_1 includes also the links accommodating L_1 and the current source. According to the chosen reference directions, Kirchhoff's current law yields for this cutset

$$0 = i_{c_1} - i_{l_1} + i_j, \qquad (6.4)$$

the current in the branch accommodating the current source being denoted by i_j. The reader can easily check that the fundamental cutsets defined by the other twigs lead to the relations

$$0 = i_{c_2} - i_{l_1} + i_{r_3} \qquad (6.5a)$$
$$0 = i_{c_3} + i_{l_2} + i_{r_3} \qquad (6.5b)$$
$$0 = i_{r_4} + i_{l_1} \qquad (6.5c)$$
$$0 = i_{r_5} + i_{l_2}. \qquad (6.5d)$$

In turn, the fundamental loop defined by the link inductor L_1 includes the twig capacitors C_1 and C_2, as well as the twig resistor R_4. Kirchhoff's voltage law yields

$$0 = v_{l_1} + v_{c_1} + v_{c_2} - v_{r_4} \qquad (6.6)$$

and, for the remaining fundamental loops,

$$0 = v_{l_2} - v_{c_3} - v_{r_5} \qquad (6.7a)$$
$$0 = v_j - v_{c_1} \qquad (6.7b)$$
$$0 = v_{r_3} - v_{c_2} - v_{c_3}, \qquad (6.7c)$$

where v_j stands for the voltage across the current source I_0. Equations (6.4), (6.5), (6.6) and (6.7), together with the dynamic relations $C_k v'_{c_k} = i_{c_k}$ ($k = 1, 2, 3$), $L_k i'_{l_k} = v_{l_k}$ ($k = 1, 2$), the characteristic relations for resistors and the equation $i_j = i_0(t)$ for the current source would define the branch-oriented model (6.3) for Bashkow's circuit, based on the tree depicted above. For later convenience, the equations for resistors will be written as

$$0 = v_{r_4} - R_4 i_{r_4} \qquad (6.8a)$$
$$0 = v_{r_5} - R_5 i_{r_5} \qquad (6.8b)$$
$$0 = i_{r_3} - G_3 v_{r_3}, \qquad (6.8c)$$

where we use a current-controlled description for the twig resistors R_4, R_5, and a voltage-controlled one, based on the conductance $G_3 = 1/R_3$, for the link resistor R_3. The chance to use this *hybrid* description will be extensively discussed in subsection 6.2.2 for coupled, strictly passive resistors.

6.1.3 The state formulation problem

The state space formulation problem can be now detailed in terms of the semistate circuit models (6.1) and (6.3) introduced above. We assume throughout that the incremental capacitance and inductance matrices $C(v_c)$, $L(i_l)$ are nonsingular for all values of v_c and i_l.

In this setting, the problem can be stated as the formulation of an explicit ODE in terms of certain state variables defined by the voltages of some capacitors and the currents of some inductors which, together with an output mapping expressing the remaining currents and voltages in terms of these state variables and the sources, fully characterizes the solutions of (6.1), at least in a local sense. The state dimension of this ODE, that is, the number of capacitor voltages and inductor currents appearing in this state space model, is called the *order of complexity* of the circuit [35, 36]. The problem can be analogously formulated in terms of capacitor charges and inductor fluxes.

In order to elaborate on this problem, let us first assume that the circuit has neither VC-loops nor IL-cutsets or, equivalently, that there exists a *proper tree*, that is, a tree containing all voltage sources and capacitors, and no current sources or inductors (cf. 5.1.2.1, p. 206). In this case, Kirchhoff laws yield no constraints involving voltages coming just from capacitors or voltage sources, or currents coming only from inductors or current sources. Following the ideas of Bashkow [18], in this setting the state formulation problem requires expressing the variables i_c, i_r, i_j, i_u, v_l, v_r, v_j and v_u in terms of v_c, i_l, $v_s(t)$ and $i_s(t)$, using equations (6.1c)-(6.1g). Note that all capacitor currents i_c in a proper tree will be twig currents, belonging to i_{1} within the tree-based model (6.3), and all inductor voltages v_l will be link voltages, entering v_{2} in (6.3). If the above-mentioned procedure is feasible, the insertion of the expressions for i_c and v_l into equations (6.1a), (6.1b) would yield an explicit ODE describing the dynamics of the circuit; this state space model would have the form

$$v'_c = f_1(v_c, i_l, q(t)) \qquad (6.9a)$$

$$i'_l = f_2(v_c, i_l, q(t)), \qquad (6.9b)$$

with all capacitor voltages and all inductor currents entering the equation. Here $q(t)$ comprises the excitation terms $v_s(t)$ and $i_s(t)$. The relations defining i_c, i_r, i_j, i_u, v_l, v_r, v_j and v_u in terms of v_c, i_l, $v_s(t)$ and $i_s(t)$ would yield the above-mentioned output mapping.

In the presence of VC-loops in a well-posed circuit, Kirchhoff's voltage law would include some constraints involving voltages coming only from

6.1. Branch-oriented semistate models

some capacitors and (maybe) some voltage sources. Analogously, the existence of IL-cutsets would lead to constraints involving currents coming just from some inductors and (possibly) some current sources. In these situations, the order of complexity will be actually smaller than the number of reactances. According to the work of Bryant [35, 36], this problem can be tackled using a *normal tree* (cf. again p. 206), the variables entering the state space model corresponding to the voltages v_{c_1} of twig capacitors and the currents i_{l_2} of link inductors. The state space model sought will then have the form

$$v'_{c_1} = f_1(v_{c_1}, i_{l_2}, q(t), q'(t)) \qquad (6.10a)$$
$$i'_{l_2} = f_2(v_{c_1}, i_{l_2}, q(t), q'(t)). \qquad (6.10b)$$

The reasons for the dependence on the derivatives of the excitation terms will become clear later.

The goal in this context is of course to figure out general conditions supporting the elimination procedure sketched above. The related literature assumes different conditions on the circuit to allow for such a derivation; besides the above-mentioned works of Bashkow and Bryant, see [13, 32, 58, 60, 63, 66, 75, 76, 145, 206, 209, 213, 280, 294, 299] and also the more recent papers [164, 235, 253, 271–274]. A nice survey of the literature can be found in particular in [273]. The working conditions can be roughly grouped along three directions, defined by the kind of coupling allowed among resistors, capacitors and inductors; the possible requirement of (strict) passivity on these devices, either in a local or in a global sense; and the presence or absence of controlled sources.

As will be detailed later, in the derivation of state models along the above-defined lines, one encounters two technical difficulties (cf. [13, 273, 274]). The first one concerns the compatibility between the choice of a normal tree and the existence of coupled resistors; the derivation of a state space model often assumes that the coupled resistors admit a *hybrid* description in terms of twig currents and link voltages, that is, a description of the form

$$v_{r_1} = \eta_1(i_{r_1}, v_{r_2}) \qquad (6.11a)$$
$$i_{r_2} = \eta_2(i_{r_1}, v_{r_2}), \qquad (6.11b)$$

where, as indicated above, the subscripts '1' and '2' refer to tree and cotree elements, respectively. The second requirement is that certain matrices arising in the analysis are nonsingular; see specifically pp. 289-290 in [13], as well as 6.2.3.2 below. In nonlinear problems, these requirements may be understood in a local sense around an operating point (cf. subsection 6.2.1).

The following Section tackles this state formulation problem as a *reduction problem* for the branch-oriented circuit model (6.3). In fully coupled, strictly locally passive problems the above-mentioned procedure will be shown to be feasible. This will be done by showing that (6.3) (and then (6.1), following Remark 6.1) has a well-defined geometric index, which will be one or two depending on the absence or presence of VC-loops and/or IL-cutsets. State space models will be naturally obtained as a consequence of the reduction process; see specifically (6.61) and (6.62) in subsection 6.2.5. Although the state space derivation in the passive, coupled setting is not new (cf. in particular [213]), the DAE formalism is of help in clarifying the interrelations among different models, as well as the role of the different assumptions and, in particular, of the above-mentioned compatibility and nonsingularity requirements.

6.2 Geometric index analysis and reduction of branch models

In this Section we undertake the geometric index analysis of the branch-oriented models (6.1) and (6.3), using the framework introduced in subsection 3.4.7. A distinction must be made between linear and nonlinear problems, since for the latter the results will hold locally around an operating point, as discussed in subsection 6.2.1. The attention will be then focused on problems with strictly (locally) passive resistors; cf. subsection 6.2.2. As we did in Chapter 5 for nodal models, we will perform in the branch equations (6.1) several variable eliminations which lead to "intermediate" formulations, in particular to the multiport model considered in subsection 6.2.3. These eliminations will actually be Schur reductions, which were shown in 3.4.7.4 to preserve the geometric index of semiexplicit DAEs; note that, if the incremental capacitance and inductance matrices $C(v_c)$, $L(i_l)$ are everywhere nonsingular, the models (6.1) and (6.3) can be trivially rewritten in semiexplicit form. This will pave the way for the index characterization presented in 6.2.4 and for the state space reductions of subsection 6.2.5. Finally, in subsection 6.2.6 we indicate how to extend these results to a large family of circuits including controlled sources.

6.2. Geometric index analysis and reduction of branch models

6.2.1 Operating points

A vector $(v, i) = (v_c, v_l, v_r, v_j, v_u, i_c, i_l, i_r, i_j, i_u)$ satisfying equations (6.1c)-(6.1g) (or, equivalently, (6.3c)-(6.3g)) at a given time t is said to be an *operating point* at t. Mind that the system of equations (6.1c)-(6.1g) defines $2b - (b_c + b_l)$ relations among $2b$ variables and, therefore, in circuits including reactances the operating points at a given t will typically be non-unique. The set of operating points is often called the *configuration space* of the circuit [128].

In linear circuits (or in circuits with linear resistors) the configuration space at a given time t is, if non-empty, a linear space. In this context, the results discussed below can be understand to hold globally; for illustrative purposes this will be detailed, in particular, for the properties involving strictly passive resistors in subsection 6.2.2.

The results for nonlinear problems will be stated in a local way; fixing a time t^* and certain vectors v_r^*, i_r^* of resistors' voltages and currents, we will assume in the sequel that there exists at least one operating point (v^*, i^*) at t^* with $v_r = v_r^*$ and $i_r = i_r^*$. Under this assumption, we will call the pair (v_r^*, i_r^*) a *resistive operating point* at t^*. It is worth emphasizing that this requires not only that $g_r(v_r^*, i_r^*) = 0$ but also the existence of a solution to the linear system which results from the substitutions $i_r = i_r^*$, $i_j = i_s(t^*)$, $v_r = v_r^*$ and $v_u = v_s(t^*)$ in (6.1c)-(6.1d).

6.2.2 Implicitly described resistors and strict passivity

As indicated in subsection 6.1.3, the compatibility between the choice of a tree and the existence of a description of the form (6.11) for coupled resistors is often a key requirement in the derivation of a state space circuit model. In this subsection we show that, under a strict passivity assumption on resistors, such a description is always well-defined, that is, it exists for any choice of a tree. More precisely, all possible assignments of current and voltage variables to the resistive branches lead to well-defined descriptions of the set of resistors and, moreover, yield positive definite hybrid matrices. This property will hold locally for nonlinear problems and globally for linear circuits. This result will also show that the voltage-controlled assumption on resistors (5.19) performed for nodal analysis methods in Chapter 5 is not unduly restrictive. The ideas here presented, which in essence can be traced back at least to [213], may be also applied to the fully-implicit description of capacitors and inductors displayed in (5.11) and (5.15), respectively.

In the models (6.1) and (6.3), no assumption has been made on the resistors' characteristic

$$g_r(v_r, i_r) = 0, \qquad (6.12)$$

except that $g_r \in C^1(\mathbb{R}^{2b_r}, \mathbb{R}^{b_r})$; recall that b_r is the number of resistors. In particular, full coupling among resistors is allowed. In this nonlinear context, the analysis will be performed locally around a resistive operating point (v_r^*, i_r^*) (cf. subsection 6.2.1 above); for the results discussed in this subsection it is actually enough to assume just that $g_r(v_r^*, i_r^*) = 0$.

In the linear setting, $g_r(v_r, i_r)$ reads $Dv_r + Ei_r$ for certain $b_r \times b_r$ real matrices D and E, and (6.12) has the form

$$Dv_r + Ei_r = 0. \qquad (6.13)$$

The results will be discussed here in both nonlinear and linear contexts, holding globally for the latter. In later subsections the results will be restricted to the nonlinear framework, although they can be applied globally to linear problems in a way similar to the one here presented.

We provide below a strict (local, in nonlinear cases) passivity definition directed to these fully implicit descriptions of resistors. Proposition 6.1 will prove that this notion is well-defined and, in particular, that it amounts to the positive definiteness of the (incremental) conductance and resistance matrices G, R in (local) voltage- and current-controlled descriptions, respectively. Essentially, we will assume that there exists some explicit voltage-current description of the resistors which uses exactly one variable (that is, the current or the voltage) from each circuit branch; the appearance of the current or the voltage variable may differ from one branch to another, that is, we certainly do not assume this description to use either all voltages or all currents.

In the nonlinear setting defined by (6.12), this property can be derived from the existence of a disjoint union $\alpha \cup \beta = \{1, \ldots, b_r\}$ such that the derivative

$$\left. \left(\frac{\partial g_r}{\partial v_{r_\alpha}} \; \frac{\partial g_r}{\partial i_{r_\beta}} \right) \right|_{(v_r^*, i_r^*)} \qquad (6.14)$$

is nonsingular. Then, by the implicit function theorem there exists a local C^1 map $\eta = (\eta_1, \eta_2)$ yielding a local description of $g_r(v_r, i_r) = 0$ of the form

$$v_{r_\alpha} = \eta_1(i_{r_\alpha}, v_{r_\beta}) \qquad (6.15a)$$
$$i_{r_\beta} = \eta_2(i_{r_\alpha}, v_{r_\beta}). \qquad (6.15b)$$

6.2. Geometric index analysis and reduction of branch models

The Jacobian matrix $H(i_{r_\alpha}, v_{r_\beta}) = \eta'(i_{r_\alpha}, v_{r_\beta})$ will be called the *incremental hybrid matrix* associated with the partition defined by α, β. The implicit function theorem locally yields

$$H(i_{r_\alpha}, v_{r_\beta}) = \left(\frac{\partial \eta}{\partial i_{r_\alpha}} \ \frac{\partial \eta}{\partial v_{r_\beta}}\right) = -\left(\frac{\partial g_r}{\partial v_{r_\alpha}} \ \frac{\partial g_r}{\partial i_{r_\beta}}\right)^{-1} \left(\frac{\partial g_r}{\partial i_{r_\alpha}} \ \frac{\partial g_r}{\partial v_{r_\beta}}\right), \quad (6.16)$$

the partial derivatives of η being evaluated at $(i_{r_\alpha}, v_{r_\beta})$ and those of g_r at (v_r, i_r).

In particular, if $\alpha = \{1, \ldots, b_r\}$ and $\beta = \emptyset$ then (6.15) corresponds to the current-controlled representation $v_r = \eta_1(i_r)$ (cf. (5.21)); the dual case $\alpha = \emptyset$, $\beta = \{1, \ldots, b_r\}$ yields the voltage-controlled description $i_r = \eta_2(v_r)$, as in (5.19). The incremental hybrid matrix would then amount to the incremental resistance or conductance matrices, respectively.

For the linear case, in which the implicit description $g_r(v_r, i_r) = 0$ reads $Dv_r + Ei_r = 0$ as depicted in (6.13), we use a notation for submatrices analogous to the one introduced in subsection 5.5.1 (see p. 237); specifically, ω will represent the set $\{1, \ldots, b_r\}$, whereas α and β stand, as above, for disjoint subsets of $\{1, \ldots, b_r\}$ with $\alpha \cup \beta = \{1, \ldots, b_r\}$. We then denote by $D^{\omega, \alpha}$ and $E^{\omega, \beta}$ the submatrices of D and E defined by the columns indexed by α and β, respectively. The nonsingularity of (6.14) can be then expressed as that of

$$(D^{\omega, \alpha} \ E^{\omega, \beta}). \quad (6.17)$$

Now the map η in (6.15) is globally defined and reads

$$\begin{pmatrix} v_{r_\alpha} \\ i_{r_\beta} \end{pmatrix} = -(D^{\omega, \alpha} \ E^{\omega, \beta})^{-1}(E^{\omega, \alpha} \ D^{\omega, \beta}) \begin{pmatrix} i_{r_\alpha} \\ v_{r_\beta} \end{pmatrix} \equiv H \begin{pmatrix} i_{r_\alpha} \\ v_{r_\beta} \end{pmatrix}, \quad (6.18)$$

where the $b_r \times b_r$ real matrix

$$H = -(D^{\omega, \alpha} \ E^{\omega, \beta})^{-1}(E^{\omega, \alpha} \ D^{\omega, \beta}) \quad (6.19)$$

is called the *hybrid matrix* associated with the above-indicated partition.

Definition 6.1. Assume that the set of resistors is characterized by the implicit description (6.12), and let the matrix in (6.14) be nonsingular at a given $(v_r^*, i_r^*) \in \mathbb{R}^{2b_r}$ such that $g_r(v_r^*, i_r^*) = 0$. The resistors are then said to be *strictly locally passive* at (v_r^*, i_r^*) if the incremental hybrid matrix $H(i_{r_\alpha}^*, v_{r_\beta}^*)$ in (6.16) is positive definite.

In the linear case defined by (6.13), provided that the matrix in (6.17) is nonsingular, the set of resistors is said to be *strictly passive* if the hybrid matrix H in (6.19) is positive definite.

The following result shows that, in the strictly passive framework, any other hybrid representation exists (locally in nonlinear cases), and that the strict passivity notions in Definition 6.1 are actually independent of the specific choice of this representation; that is, the positive definiteness assumption on the hybrid matrices above does not depend on the specific subsets α, β defining the partition of $\{1, \ldots, b_r\}$. This is consistent with the fact that the strict local passivity notion expresses that, for all $(v_r, i_r) \neq (v_r^*, i_r^*)$ satisfying $g(v_r, i_r) = 0$ and sufficiently close to (v_r^*, i_r^*), the incremental product $(v_r - v_r^*)^\mathsf{T}(i_r - i_r^*)$ is positive; note that the sign of this product near (v_r^*, i_r^*) equals that of

$$(i_{r_\alpha} - i_{r_\alpha}^*)^\mathsf{T}(v_{r_\alpha} - v_{r_\alpha}^*) + (v_{r_\beta} - v_{r_\beta}^*)^\mathsf{T}(i_{r_\beta} - i_{r_\beta}^*)$$

$$= ((i_{r_\alpha} - i_{r_\alpha}^*)^\mathsf{T} \ (v_{r_\beta} - v_{r_\beta}^*)^\mathsf{T}) H(i_{r_\alpha}^*, v_{r_\beta}^*) \begin{pmatrix} i_{r_\alpha} - i_{r_\alpha}^* \\ v_{r_\beta} - v_{r_\beta}^* \end{pmatrix}$$

and is therefore positive provided that the incremental hybrid matrix is positive definite. In the linear case the corresponding property holds globally. See also Remark 6.2 below.

Proposition 6.1. *If α and β define a partition of $\{1, \ldots, b_r\}$ which makes (6.14) nonsingular and the incremental hybrid matrix $H(i_{r_\alpha}^*, v_{r_\beta}^*)$ in (6.16) positive definite, then for any $\tilde{\alpha}$, $\tilde{\beta}$ defining another partition of $\{1, \ldots, b_r\}$ there exists a locally defined C^1 map $\tilde{\eta}$ such that (6.15) is equivalent to*

$$v_{r_{\tilde{\alpha}}} = \tilde{\eta}_1(i_{r_{\tilde{\alpha}}}, v_{r_{\tilde{\beta}}}) \tag{6.20a}$$

$$i_{r_{\tilde{\beta}}} = \tilde{\eta}_2(i_{r_{\tilde{\alpha}}}, v_{r_{\tilde{\beta}}}), \tag{6.20b}$$

with $\tilde{\eta}'(i_{r_{\tilde{\alpha}}}^*, v_{r_{\tilde{\beta}}}^*)$ positive definite.

In the linear setting, if α, β make (6.17) nonsingular and the hybrid matrix H in (6.19) positive definite, then for any other partition defined by $\tilde{\alpha}$, $\tilde{\beta}$ there exists a positive definite $\tilde{H}$ such that (6.18) is equivalent to

$$\begin{pmatrix} v_{r_{\tilde{\alpha}}} \\ i_{r_{\tilde{\beta}}} \end{pmatrix} = \tilde{H} \begin{pmatrix} i_{r_{\tilde{\alpha}}} \\ v_{r_{\tilde{\beta}}} \end{pmatrix}. \tag{6.21}$$

This means that, in a strictly passive circuit, one can choose *arbitrarily* the voltage or the current in every circuit branch as a variable for a hybrid description of the form (6.15) or (6.18); that is, all hybrid descriptions (as well as the extremal cases given by the voltage-controlled and current-controlled representations) are well-defined. Furthermore, we can speak of a strictly passive resistive network without the need to resort to a specific hybrid or voltage/current-controlled description.

6.2. Geometric index analysis and reduction of branch models

The key aspect in the proof of Proposition 6.1 is the following linear algebra result. We make use of the fact that a positive definite matrix is nonsingular.

Lemma 6.1. *Consider a linear mapping $y = Hx$ with $H \in \mathbb{R}^{b_r \times b_r}$ positive definite. Let γ and λ define any partition of $\{1, \ldots, b_r\}$, and assume that x_γ, y_γ (resp. x_λ, y_λ) denote the components of x, y indexed by γ (resp. λ). Then there exists a matrix $\tilde{H} \in \mathbb{R}^{b_r \times b_r}$ such that $y = Hx$ is equivalent to*

$$\begin{pmatrix} x_\gamma \\ y_\lambda \end{pmatrix} = \tilde{H} \begin{pmatrix} y_\gamma \\ x_\lambda \end{pmatrix}. \quad (6.22)$$

Moreover, $\tilde{H}$ is positive definite.

Proof. Assume, without loss of generality, that γ (resp. λ) are the first p (resp. last $b_r - p$) elements of $\{1, \ldots, b_r\}$. Write then $y = Hx$ as

$$\begin{pmatrix} y_\gamma \\ y_\lambda \end{pmatrix} = \begin{pmatrix} H_{11} & H_{12} \\ H_{21} & H_{22} \end{pmatrix} \begin{pmatrix} x_\gamma \\ x_\lambda \end{pmatrix}. \quad (6.23)$$

The positive definiteness of H makes the block H_{11} positive definite as well (see e.g. 7.1.2 in [137]), since

$$\begin{pmatrix} w^\mathsf{T} & 0 \end{pmatrix} \begin{pmatrix} H_{11} & H_{12} \\ H_{21} & H_{22} \end{pmatrix} \begin{pmatrix} w \\ 0 \end{pmatrix} = w^\mathsf{T} H_{11} w > 0$$

for any non-vanishing $w \in \mathbb{R}^p$. Hence, the submatrix H_{11} is nonsingular; this leads to $x_\gamma = H_{11}^{-1}(y_\gamma - H_{12}x_\lambda)$ and thereby $y_\lambda = H_{21}H_{11}^{-1}(y_\gamma - H_{12}x_\lambda) + H_{22}x_\lambda$, yielding an explicit expression for $\tilde{H}$ in (6.22), namely

$$\tilde{H} = \begin{pmatrix} \tilde{H}_{11} & \tilde{H}_{12} \\ \tilde{H}_{21} & \tilde{H}_{22} \end{pmatrix} = \begin{pmatrix} H_{11}^{-1} & -H_{11}^{-1}H_{12} \\ H_{21}H_{11}^{-1} & H_{22} - H_{21}H_{11}^{-1}H_{12} \end{pmatrix}. \quad (6.24)$$

Finally, the identities

$$\begin{pmatrix} x_\gamma^\mathsf{T} & x_\lambda^\mathsf{T} \end{pmatrix} \begin{pmatrix} H_{11} & H_{12} \\ H_{21} & H_{22} \end{pmatrix} \begin{pmatrix} x_\gamma \\ x_\lambda \end{pmatrix} = x_\gamma^\mathsf{T} y_\gamma + x_\lambda^\mathsf{T} y_\lambda = y_\gamma^\mathsf{T} x_\gamma + x_\lambda^\mathsf{T} y_\lambda$$

$$= \begin{pmatrix} y_\gamma^\mathsf{T} & x_\lambda^\mathsf{T} \end{pmatrix} \begin{pmatrix} \tilde{H}_{11} & \tilde{H}_{12} \\ \tilde{H}_{21} & \tilde{H}_{22} \end{pmatrix} \begin{pmatrix} y_\gamma \\ x_\lambda \end{pmatrix}$$

show that the positive definiteness of H is transferred to $\tilde{H}$. $\square$

Proof of Proposition 6.1. In the light of Lemma 6.1, let x_γ stand for the set of voltage and current variables present in $(i_{r_\alpha}, v_{r_\beta})$ but not in $(i_{r_{\tilde\alpha}}, v_{r_{\tilde\beta}})$; x_λ for the ones which are both in $(i_{r_\alpha}, v_{r_\beta})$ and $(i_{r_{\tilde\alpha}}, v_{r_{\tilde\beta}})$; y_γ for the variables in $(v_{r_\alpha}, i_{r_\beta})$ not belonging to $(v_{r_{\tilde\alpha}}, i_{r_{\tilde\beta}})$; and y_λ for the ones present both in $(v_{r_\alpha}, i_{r_\beta})$ and $(v_{r_{\tilde\alpha}}, i_{r_{\tilde\beta}})$. Note that, by construction, x_γ equals the set of variables in $(v_{r_{\tilde\alpha}}, i_{r_{\tilde\beta}})$ which do not belong to $(v_{r_\alpha}, i_{r_\beta})$ and, similarly, y_γ equals the variables $(i_{r_{\tilde\alpha}}, v_{r_{\tilde\beta}})$ which are not present in $(i_{r_\alpha}, v_{r_\beta})$.

In the linear case the result follows immediately from Lemma 6.1. Focus in the sequel on the nonlinear case. With the above-introduced notation, equation (6.15) can be written as

$$y_\gamma = \eta_\gamma(x_\gamma, x_\lambda) \qquad (6.25a)$$
$$y_\lambda = \eta_\lambda(x_\gamma, x_\lambda). \qquad (6.25b)$$

We need to show that (6.20) is locally well-defined; we will write this relation as $(x_\gamma, y_\lambda) = \tilde\eta(y_\gamma, x_\lambda)$.

Let H_{11} stand for the derivative $\frac{\partial \eta_\gamma}{\partial x_\gamma}$. As detailed in the proof of Lemma 6.1, from the positive definiteness of the incremental matrix H, it follows that H_{11} is positive definite and, in particular, nonsingular. This means that from (6.25a) we can write $x_\gamma = \tilde\eta_\gamma(y_\gamma, x_\lambda)$ for a locally defined map $\tilde\eta_\gamma$. Additionally, from (6.25b) we derive $y_\lambda = \eta_\lambda(\tilde\eta_\gamma(y_\gamma, x_\lambda), x_\lambda) = \tilde\eta_\lambda(y_\gamma, x_\lambda)$. From the implicit function theorem one gets

$$\frac{\partial \tilde\eta_\gamma}{\partial y_\gamma} = \left(\frac{\partial \eta_\gamma}{\partial x_\gamma}\right)^{-1}$$

$$\frac{\partial \tilde\eta_\gamma}{\partial x_\lambda} = -\left(\frac{\partial \eta_\gamma}{\partial x_\gamma}\right)^{-1} \frac{\partial \eta_\gamma}{\partial x_\lambda}$$

$$\frac{\partial \tilde\eta_\lambda}{\partial y_\gamma} = \frac{\partial \eta_\lambda}{\partial x_\gamma} \frac{\partial \tilde\eta_\gamma}{\partial y_\gamma} = \frac{\partial \eta_\lambda}{\partial x_\gamma}\left(\frac{\partial \eta_\gamma}{\partial x_\gamma}\right)^{-1}$$

$$\frac{\partial \tilde\eta_\lambda}{\partial x_\lambda} = \frac{\partial \eta_\lambda}{\partial x_\gamma} \frac{\partial \tilde\eta_\gamma}{\partial x_\lambda} + \frac{\partial \eta_\lambda}{\partial x_\lambda} = -\frac{\partial \eta_\lambda}{\partial x_\gamma}\left(\frac{\partial \eta_\gamma}{\partial x_\gamma}\right)^{-1} \frac{\partial \eta_\gamma}{\partial x_\lambda} + \frac{\partial \eta_\lambda}{\partial x_\lambda}.$$

The positive definiteness of the incremental hybrid matrix $\tilde\eta'(y_\gamma, x_\lambda)$ then follows from these expressions and (6.24) by writing $H_{11} = \frac{\partial \eta_\gamma}{\partial x_\gamma}$, as indicated above, and $H_{12} = \frac{\partial \eta_\gamma}{\partial x_\lambda}$, $H_{21} = \frac{\partial \eta_\lambda}{\partial x_\gamma}$, $H_{22} = \frac{\partial \eta_\lambda}{\partial x_\lambda}$. $\square$

Remark 6.2. In the setting of Proposition 6.1, for $\tilde\alpha$, $\tilde\beta$ defining any partition of $\{1, \ldots, b_r\}$, it is actually true that

$$\left(\frac{\partial g_r}{\partial v_{r_{\tilde\alpha}}} \frac{\partial g_r}{\partial i_{r_{\tilde\beta}}}\right)\bigg|_{(v_r^*, i_r^*)} \qquad (6.26)$$

in nonlinear cases, as well as

$$(D^{\omega,\tilde\alpha} \ E^{\omega,\tilde\beta}) \qquad (6.27)$$

6.2. Geometric index analysis and reduction of branch models

in linear problems, are nonsingular. This is an easy consequence of the following fact: if D and E are $b_r \times b_r$ matrices, with D nonsingular and $H = -D^{-1}E$ positive definite, then the exchange of the j-th columns of D and E result in two matrices $\tilde{D}$, $\tilde{E}$ with $\tilde{D}$ nonsingular and $\tilde{H} = -\tilde{D}^{-1}\tilde{E}$ positive definite. Indeed, the nonsingularity of $\tilde{D}$ follows from the identity $DH = -E$ and Cramer's rule, since the j-th diagonal entry of a positive definite matrix H is positive; the positive definiteness of $\tilde{H}$ follows from Lemma 6.1. The transition from (6.14) and (6.17) to (6.26) and (6.27) can be seen as a sequence of such column exchanges in $g'(v_r^*, i_r^*)$ or in the matrix $(D\ E)$ coming from (6.13), respectively; the nonsingularity of (6.14) is then equivalent to that of (6.26), the same holding for (6.17) and (6.27).

6.2.3 Multiport reduction

In the state formulation problem, as well as in the geometric index analysis performed in subsection 6.2.4, a key step is the elimination of the resistive voltages and currents from the models (6.1) and (6.3). If the source currents i_j, i_u and voltages v_j, v_u are also eliminated, this would yield a reduced model involving only reactive variables and excitation terms which, for the reasons detailed below, will be called a *multiport model*. We show here that, under a strict passivity assumption on resistors, this multiport reduction is always feasible around an operating point. Furthermore, the multiport model will preserve the geometric index of the original branch-oriented one. Focusing on nonlinear problems we state the results in a local way, although the linear analogs can be easily derived.

6.2.3.1 Multiport model

Since all circuits are assumed to be connected and well-posed (cf. page 206), we may choose a tree containing all voltage sources and no current source. Using the subscripts '1' and '2' for twigs and links, rewrite the model (6.3) as

$$C(v_c)v_c' = i_c \qquad (6.28a)$$
$$L(i_l)i_l' = v_l \qquad (6.28b)$$

$$0 = \begin{pmatrix} i_{c_1} \\ i_{l_1} \\ i_u \\ i_{r_1} \end{pmatrix} - F^\mathsf{T} \begin{pmatrix} i_{l_2} \\ i_{c_2} \\ i_j \\ i_{r_2} \end{pmatrix} \qquad (6.28c)$$

$$0 = \begin{pmatrix} v_{l_2} \\ v_{c_2} \\ v_j \\ v_{r_2} \end{pmatrix} + F \begin{pmatrix} v_{c_1} \\ v_{l_1} \\ v_u \\ v_{r_1} \end{pmatrix} \qquad (6.28d)$$

$$0 = g_r(v_r, i_r) \qquad (6.28e)$$
$$0 = i_j - i_s(t) \qquad (6.28f)$$
$$0 = v_u - v_s(t). \qquad (6.28g)$$

The matrix F will be split in blocks according to the different variables entering (6.28d) as follows:

$$F = \begin{pmatrix} F_{11} & F_{12} & F_{13} & F_{14} \\ F_{21} & F_{22} & F_{23} & F_{24} \\ F_{31} & F_{32} & F_{33} & F_{34} \\ F_{41} & F_{42} & F_{43} & F_{44} \end{pmatrix}. \qquad (6.29)$$

In particular, the block $\tilde{F} = F_{44}$ will be important in the elimination of resistive currents and voltages.

Now, the chance to write locally the resistive variables i_{r_1}, v_{r_2}, v_{r_1} and i_{r_2} in terms of reactive variables and the excitation terms via the implicit function theorem relies on the key assumption that the derivative matrix

$$M = \begin{pmatrix} I & 0 & 0 & -\tilde{F}^\mathsf{T} \\ 0 & I & \tilde{F} & 0 \\ \frac{\partial g_r}{\partial i_{r_1}} & \frac{\partial g_r}{\partial v_{r_2}} & \frac{\partial g_r}{\partial v_{r_1}} & \frac{\partial g_r}{\partial i_{r_2}} \end{pmatrix} \qquad (6.30)$$

is nonsingular at a resistive operating point (v_r^*, i_r^*). The first two sets of rows in (6.30) describe the derivatives with respect to i_{r_1}, v_{r_2}, v_{r_1} and i_{r_2} of the entries corresponding to i_{r_1} and v_{r_2} within (6.28c) and (6.28d), respectively, whereas the last rows comprise the corresponding derivatives of g_r in (6.28e). It is worth remarking that, if the nonsingularity of M in (6.30) holds at a resistive operating point (v_r^*, i_r^*), then the elimination procedure

6.2. Geometric index analysis and reduction of branch models

detailed below is locally feasible around *any* operating point (v^*, i^*) at t^* for which $v_r = v_r^*$ and $i_r = i_r^*$, since the entries of M depend only on the circuit topology and the resistive variables.

The invertibility of the matrix M in (6.30) plays a fundamental role in index analyses and in the state formulation problem, and it is shown in Theorem 6.1 to hold for problems with strictly locally passive resistors; cf. also 6.2.3.2. In problems with linear resistors, the last rows of M in (6.30) are defined in terms of the matrices D and E in (6.13) and the forthcoming procedure holds globally.

Provided that the nonsingularity of (6.30) is met, we can locally write

$$i_{r_1} = \alpha_1(v_{c_1}, v_{l_1}, i_{l_2}, i_{c_2}, v_s(t), i_s(t)) \tag{6.31a}$$

$$v_{r_2} = \alpha_2(v_{c_1}, v_{l_1}, i_{l_2}, i_{c_2}, v_s(t), i_s(t)) \tag{6.31b}$$

$$v_{r_1} = \alpha_3(v_{c_1}, v_{l_1}, i_{l_2}, i_{c_2}, v_s(t), i_s(t)) \tag{6.31c}$$

$$i_{r_2} = \alpha_4(v_{c_1}, v_{l_1}, i_{l_2}, i_{c_2}, v_s(t), i_s(t)), \tag{6.31d}$$

for certain locally defined mappings α_i; we have used $i_j = i_s(t)$, $v_u = v_s(t)$ from (6.28f), (6.28g). From these relations and (6.28c), (6.28d) we can also express locally i_u and v_j in terms of the same variables, that is

$$i_u = \alpha_5(v_{c_1}, v_{l_1}, i_{l_2}, i_{c_2}, v_s(t), i_s(t)) \tag{6.32a}$$

$$v_j = \alpha_6(v_{c_1}, v_{l_1}, i_{l_2}, i_{c_2}, v_s(t), i_s(t)). \tag{6.32b}$$

Using the expressions for v_{r_1} and i_{r_2} given in (6.31c) and (6.31d), the remaining equations of (6.28) yield a system of the form

$$C(v_c)v_c' = i_c \tag{6.33a}$$

$$L(i_l)i_l' = v_l \tag{6.33b}$$

$$0 = \begin{pmatrix} i_{c_1} \\ i_{l_1} \end{pmatrix} - \Psi_1(v_{c_1}, v_{l_1}, i_{l_2}, i_{c_2}, v_s(t), i_s(t)) \tag{6.33c}$$

$$0 = \begin{pmatrix} v_{l_2} \\ v_{c_2} \end{pmatrix} - \Psi_2(v_{c_1}, v_{l_1}, i_{l_2}, i_{c_2}, v_s(t), i_s(t)), \tag{6.33d}$$

for certain locally defined maps Ψ_1, Ψ_2. For later use, the expressions defining i_{c_1} and i_{l_1} in (6.33c) read

$$i_{c_1} = F_{11}^\mathsf{T} i_{l_2} + F_{21}^\mathsf{T} i_{c_2} + F_{31}^\mathsf{T} i_s(t) + F_{41}^\mathsf{T} \alpha_4 \tag{6.34a}$$

$$i_{l_1} = F_{12}^\mathsf{T} i_{l_2} + F_{22}^\mathsf{T} i_{c_2} + F_{32}^\mathsf{T} i_s(t) + F_{42}^\mathsf{T} \alpha_4, \tag{6.34b}$$

whereas v_{l_2} and v_{c_2} in (6.33d) are given by

$$v_{l_2} = -F_{11}v_{c_1} - F_{12}v_{l_1} - F_{13}v_s(t) - F_{14}\alpha_3 \tag{6.35a}$$

$$v_{c_2} = -F_{21}v_{c_1} - F_{22}v_{l_1} - F_{23}v_s(t) - F_{24}\alpha_3. \tag{6.35b}$$

Note that, for notational simplicity, we do not write explicitly the dependence of α_3 and α_4 on $(v_{c_1}, v_{l_1}, i_{l_2}, i_{c_2}, v_s(t), i_s(t))$.

It is worth emphasizing that, so far, the unique requirement on the tree used in this scheme is that it contains all voltage sources and no current source; in particular, we do not yet need the tree to be a proper or a normal one, as will be required at later stages. Note also that no specific hybrid description has been assumed for the set of resistors.

Remark 6.3. System (6.33) will be called a *multiport model*. Indeed, the maps Ψ_1 and Ψ_2 within equations (6.33c), (6.33d) can be understood to define b_c+b_l abstract relations involving the currents and voltages of certain b_c+b_l ports of a (possibly nonlinear) subnetwork which includes all resistors and sources, with a time dependence coming from the sources; cf. Figure 6.2.

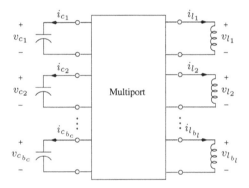

Fig. 6.2 Multiport model.

The connection of reactances at those ports leads to the dynamical system modeled by the DAE (6.33). This description is often assumed as a starting point in circuit analysis, cf. for instance [60, 244, 258]. See also [13, 269, 273]. Theorem 6.1 will show that this formulation is feasible for well-posed circuits with fully-coupled, strictly locally passive resistors at (v_r^*, i_r^*), provided that there exists at least one operating point (v^*, i^*) at t^* for which $v_r = v_r^*$ and $i_r = i_r^*$; this result will be a consequence of the invertibility in this context of the matrix M in (6.30). As indicated above, in nonlinear problems this formulation holds locally around *any* such operating point, whereas in linear settings the multiport model is globally defined.

6.2. Geometric index analysis and reduction of branch models

Theorem 6.1. *Assume that (6.1) and (6.3) model a well-posed, connected circuit with strictly locally passive resistors at a resistive operating point (v_r^*, i_r^*). For any tree containing all voltage sources and no current source, the matrix (6.30) is nonsingular and therefore the mappings (6.31) and (6.32) as well as the multiport model (6.33) are locally well-defined around any operating point for which $v_r = v_r^*$, $i_r = i_r^*$.*

Moreover, if the incremental matrices $C(v_c)$ and $L(i_l)$ are nonsingular, the geometric index of the branch-oriented models (6.1) and (6.3) equals the one of the multiport model (6.33).

The proof of Theorem 6.1 relies on the linear algebra property stated independently in Lemma 6.2. Recall that a real square matrix S is *skew-symmetric* if $S^\mathsf{T} = -S$.

Lemma 6.2. *Any matrix of the form*

$$\begin{pmatrix} I & S \\ D & I \end{pmatrix}, \qquad (6.36)$$

with S skew-symmetric and D definite, is nonsingular.

Proof. We make use of the fact that the matrix in (6.36) is nonsingular if and only if so it is the Schur complement (cf. Lemma 3.3 on p. 127) of the first identity block. This means that the nonsingularity of (6.36) amounts to that of $I - DS$.

In order to assess the nonsingularity of $I - DS$, assume that

$$(I - DS)w = 0. \qquad (6.37)$$

Premultiplying by $w^\mathsf{T} S^\mathsf{T}$ we get $0 = w^\mathsf{T} S^\mathsf{T} w - w^\mathsf{T} S^\mathsf{T} DSw = w^\mathsf{T} S^\mathsf{T} DSw$. Here we have used the fact that $w^\mathsf{T} S^\mathsf{T} w = 0$ for any skew-symmetric matrix S, since $S = -S^\mathsf{T}$ yields

$$w^\mathsf{T} S^\mathsf{T} w = (w^\mathsf{T} S^\mathsf{T} w)^\mathsf{T} = w^\mathsf{T} S w = -w^\mathsf{T} S^\mathsf{T} w.$$

The identity $w^\mathsf{T} S^\mathsf{T} D S w = 0$ and the definiteness of D then imply that $Sw = 0$, and by (6.37) this leads to $w = 0$, showing that $I - DS$ is indeed nonsingular. □

Proof of Theorem 6.1. Due to the strictly locally passive assumption on resistors at (v_r^*, i_r^*), Proposition 6.1 and Remark 6.2 imply that the block

$$K = \begin{pmatrix} \dfrac{\partial g_r}{\partial v_{r_1}} & \dfrac{\partial g_r}{\partial i_{r_2}} \end{pmatrix} \qquad (6.38)$$

in (6.30) is nonsingular at (v_r^*, i_r^*). Via the implicit function theorem, this defines a local hybrid description of resistors of the form

$$v_{r_1} = \eta_1(i_{r_1}, v_{r_2}) \qquad (6.39a)$$
$$i_{r_2} = \eta_2(i_{r_1}, v_{r_2}), \qquad (6.39b)$$

where it is worth recalling that the subscripts '1' and '2' stand for tree and cotree devices, respectively. For later use, set

$$H_{11} = \frac{\partial \eta_1}{\partial i_{r_1}}, \ H_{12} = \frac{\partial \eta_1}{\partial v_{r_2}}, \ H_{21} = \frac{\partial \eta_2}{\partial i_{r_1}}, \ H_{22} = \frac{\partial \eta_2}{\partial v_{r_2}},$$

the derivatives being evaluated at $(i_{r_1}^*, v_{r_2}^*)$.

Premultiply then the matrix M in (6.30) by

$$\begin{pmatrix} I & 0 & 0 \\ 0 & I & 0 \\ 0 & 0 & K^{-1} \end{pmatrix} \qquad (6.40)$$

to assess the nonsingularity of M in terms of

$$J = \begin{pmatrix} I & 0 & 0 & -\tilde{F}^{\mathsf{T}} \\ 0 & I & \tilde{F} & 0 \\ -H_{11} & -H_{12} & I & 0 \\ -H_{21} & -H_{22} & 0 & I \end{pmatrix}, \qquad (6.41)$$

where we have used the identity

$$\begin{pmatrix} H_{11} & H_{12} \\ H_{21} & H_{22} \end{pmatrix} = -\left(\frac{\partial g_r}{\partial v_{r_1}} \ \frac{\partial g_r}{\partial i_{r_2}} \right)^{-1} \left(\frac{\partial g_r}{\partial i_{r_1}} \ \frac{\partial g_r}{\partial v_{r_2}} \right)$$

which follows from the implicit function theorem. Note that J in (6.41) has the form depicted in (6.36) with

$$D = -H = -\begin{pmatrix} H_{11} & H_{12} \\ H_{21} & H_{22} \end{pmatrix}, \ S = \begin{pmatrix} 0 & -\tilde{F}^{\mathsf{T}} \\ \tilde{F} & 0 \end{pmatrix}.$$

From Proposition 6.1, the matrix H is positive definite, so that $D = -H$ is negative definite; the nonsingularity of J and therefore of M in (6.30) then follows from Lemma 6.2.

The index equivalence between the models (6.1), (6.3) and (6.28) and the multiport equations (6.33) is an immediate consequence of Proposition 3.7 (p. 128), since the elimination of $z_2 = (i_{r_1}, v_{r_2}, v_{r_1}, i_{r_2}, i_u, v_j, i_j, v_u)$ from (6.28) is a Schur reduction under the assumption that M in (6.30) is nonsingular; specifically, the mapping g_2 in Proposition 3.7 is defined by the expressions involving i_u, i_{r_1} in (6.28c) and v_j, v_{r_2} in (6.28d), together with (6.28e), (6.28f) and (6.28g). □

6.2. Geometric index analysis and reduction of branch models

Remark 6.4. The hybrid description (6.39) means that the voltages of resistive twigs and the currents of resistive links can be written in terms of the currents of twig resistors and the resistive link voltages. As shown above, the chance to use a hybrid description compatible with the chosen tree follows from the strict passivity assumption on resistors. Without such a passivity assumption, the existence of the hybrid description (6.39) is often required *a priori* (see [13] and 6.2.3.2 below). Note that the existence of this representation is, in general, independent of the nonsingularity of M in (6.30), which in this setting also follows from the passivity requirement.

6.2.3.2 Additional remarks on the invertibility of the matrix M

The nonsingularity of the matrix M in (6.30) is closely related with the so-called *associated resistive network* obtained by replacing twig reactances by independent voltage sources and link reactances by independent current sources (see [271, 272, 274] and references therein); mind that this term is often use in other senses (cf. for instance 2.4.12 in [128]). Note that every choice of a tree in the original circuit defines a different associated resistive network. For a given tree including all voltage sources and no current source, these replacements lead to a resistive circuit modeled by the algebraic equation

$$0 = \begin{pmatrix} i_{\tilde{u}} \\ i_u \\ i_{r_1} \end{pmatrix} - F^\mathsf{T} \begin{pmatrix} i_{\tilde{s}}(t) \\ i_s(t) \\ i_{r_2} \end{pmatrix} \tag{6.42a}$$

$$0 = \begin{pmatrix} v_{\tilde{j}} \\ v_j \\ v_{r_2} \end{pmatrix} + F \begin{pmatrix} v_{\tilde{s}}(t) \\ v_s(t) \\ v_{r_1} \end{pmatrix} \tag{6.42b}$$

$$0 = g_r(v_r, i_r), \tag{6.42c}$$

together with $i_j = i_s(t)$, $i_{\tilde{j}} = i_{\tilde{s}}(t)$, $v_u = v_s(t)$ and $v_{\tilde{u}} = v_{\tilde{s}}(t)$; we are using the tilde ˜ to refer to the newly introduced sources. The chance to express, locally around a given point satisfying (6.42), all resistive currents and voltages in terms of the excitations via the implicit function theorem relies on the nonsingularity of M. In a linear setting, the nonsingularity of this matrix is equivalent to the unique solvability of the associated resistive network.

Theorem 6.1 shows that any associated resistive network is uniquely solvable if the original circuit is well-posed and resistors are strictly passive; via Proposition 6.1, this property holds without the need to assume *a priori* the existence of a hybrid description of resistors compatible with any tree. Without such a strict passivity requirement on the resistors, however, the matrix M need not be invertible. In spite of the fact that our attention is mainly focused on problems with strictly passive resistors, a brief digression in this regard follows.

In such a non-passive context, the nonsingularity of M can be further discussed if we assume that, for some tree, there exists indeed a hybrid description of resistors of the form depicted in (6.39), as it is done in [13]. We are then led to assess the nonsingularity of the matrix J in (6.41). In particular, sufficient conditions for the nonsingularity of J are given, with a different notation, in [13] (see eqs. (127) and (128) in pp. 289-290), where it is required that either

$$I - H_{21}\tilde{F}^\mathsf{T} \tag{6.43}$$

and

$$I + H_{12}\tilde{F} + H_{11}\tilde{F}^\mathsf{T}(I - H_{21}\tilde{F}^\mathsf{T})^{-1}H_{22}\tilde{F} \tag{6.44}$$

are nonsingular, or so they are

$$I + H_{12}\tilde{F} \tag{6.45}$$

and

$$I - H_{21}\tilde{F}^\mathsf{T} + H_{22}\tilde{F}(I + H_{12}\tilde{F})^{-1}H_{11}\tilde{F}^\mathsf{T}. \tag{6.46}$$

These requirements can be easily understood by noticing that the Schur complement (cf. Lemma 3.3 on p. 127) of the upper-left identity blocks in J within (6.41) reads

$$\begin{aligned}\tilde{J} &= \begin{pmatrix} I & 0 \\ 0 & I \end{pmatrix} + \begin{pmatrix} H_{11} & H_{12} \\ H_{21} & H_{22} \end{pmatrix} \begin{pmatrix} 0 & -\tilde{F}^\mathsf{T} \\ \tilde{F} & 0 \end{pmatrix} \\ &= \begin{pmatrix} I + H_{12}\tilde{F} & -H_{11}\tilde{F}^\mathsf{T} \\ H_{22}\tilde{F} & I - H_{21}\tilde{F}^\mathsf{T} \end{pmatrix},\end{aligned} \tag{6.47}$$

which makes it possible to examine the nonsingularity of J, supporting index analyses or unique solvability properties of associated resistive networks, in terms of the somehow simpler matrix $\tilde{J}$ in (6.47). In particular, the fact that the nonsingularity of the pair of matrices (6.43)-(6.44) or that of (6.45)-(6.46) implies the nonsingularity of J becomes clear, since (6.44) and (6.46) are the Schur complements of (6.43) and (6.45), respectively, in $\tilde{J}$. Incidentally, none of these pairs of requirements is actually necessary for the nonsingularity of $\tilde{J}$; indeed, a two-branch example with $\tilde{F} = 1$, $H_{12} = -1$, $H_{21} = 1$, $H_{11} \neq 0 \neq H_{22}$ yields nonsingular matrices J and $\tilde{J}$ in (6.41) and (6.47) even though the ones in (6.43) and (6.45) are singular.

6.2.4 Index characterization

Using the results presented in subsections 3.4.7 and 6.2.3, we characterize below the geometric index of the branch-oriented model (6.1) and the tree-based reformulations (6.3) and (6.28) in problems with strictly locally passive resistors. The tractability index of these models is discussed in [234]. Mind that, according to the results of subsection 6.2.2, the strict passivity assumption avoids the need to assume *a priori* the existence of a hybrid description for resistors or the invertibility of the matrix M in (6.30).

Theorem 6.2. *Consider a well-posed, connected circuit with nonsingular incremental capacitance and inductance matrices C, L, and strictly locally passive resistors at all operating points.*

(1) The branch-oriented models (6.1) and (6.3), as well as the multiport model (6.33), have geometric index one if and only if the circuit has neither VC-loops nor IL-cutsets.

(2) Assuming additionally that the matrices C and L are positive definite, the existence of VC-loops and/or IL-cutsets makes systems (6.1), (6.3) and (6.33) index two.

Proof. According to Theorem 6.1 (p. 275), it is enough to prove the assertions for the multiport model (6.33), which is locally well-defined around any operating point and retains the index of the branch models (6.1), (6.3) and (6.28) for any tree containing all voltage sources and no current source, under the assumption that resistors are strictly locally passive; see also Remark 6.1 on p. 260. For item (1) we will assume that the tree leading to (6.33) is *proper*, whereas for item (2) it will be a *normal* tree (cf. page 206).

Assume first that there are neither VC-loops nor IL-cutsets, as stated in item (1). Letting then the tree be a proper one, in which all capacitors are twigs and all inductors are links, the multiport model (6.33) amounts to

$$C(v_c)v_c' = i_c \quad (6.48a)$$
$$L(i_l)i_l' = v_l \quad (6.48b)$$
$$0 = i_c - \Psi_1(v_c, i_l, v_s(t), i_s(t)) \quad (6.48c)$$
$$0 = v_l - \Psi_2(v_c, i_l, v_s(t), i_s(t)), \quad (6.48d)$$

which is an index one DAE since the derivative of (6.48c)-(6.48d) with respect to the algebraic variables i_c, v_l is obviously nonsingular.

The converse result in item (1) can be directly addressed in terms of the original branch model (6.1). Assume that this system is index one, so that the derivative of the mappings in (6.1c), (6.1d), (6.1e), (6.1f) and (6.1g) with respect to the algebraic variables i_c, i_r, i_j, i_u, v_l, v_r, v_j, v_u must be nonsingular at any operating point. Since i_c, i_u only enter (6.1c), it follows that $(A_c\ A_u)$ has full column rank, so that the circuit has no VC-loops according to Lemma 5.3 (p. 199). Analogously, since v_l, v_j are only present in (6.1d), the submatrix $(B_l\ B_j)$ must have full column rank, which according to Lemma 5.7 (p. 200) indicates that there are no IL-cutsets. Item (1) is then proved.

Focus in the sequel on item (2), and assume that the tree used in the derivation of (6.33) is a normal one. In a normal tree, the fundamental cutset defined by each twig inductor only has link inductors and current sources and, analogously, the fundamental loop defined by each link capacitor only involves twig capacitors and voltage sources (see page 207). This implies in particular that the blocks F_{22}, F_{42} and F_{24} within the matrix F in (6.29) vanish, and then the expressions (6.34b) and (6.35b) for i_{l_1} and v_{c_2} read

$$i_{l_1} = F_{12}^\mathsf{T} i_{l_2} + F_{32}^\mathsf{T} i_s(t) \qquad (6.49a)$$

$$v_{c_2} = -F_{21} v_{c_1} - F_{23} v_s(t). \qquad (6.49b)$$

Additionally, the relations $F_{24} = 0$ and $F_{42} = 0$ show that the variable i_{c_2} is not involved in the expression for i_{r_1} in (6.28c), nor is the variable v_{l_1} in the equation defining v_{r_2} within (6.28d). Therefore, the α_i-mappings in (6.31) are actually independent of v_{l_1} and i_{c_2}, so that (6.34a) and (6.35a) can be rewritten as

$$i_{c_1} = F_{11}^\mathsf{T} i_{l_2} + F_{21}^\mathsf{T} i_{c_2} + F_{31}^\mathsf{T} i_s(t) + F_{41}^\mathsf{T} \alpha_4(v_{c_1}, i_{l_2}, v_s(t), i_s(t))$$
$$= \beta_1(v_{c_1}, i_{l_2}, i_{c_2}, v_s(t), i_s(t)) \qquad (6.50)$$

$$v_{l_2} = -F_{11} v_{c_1} - F_{12} v_{l_1} - F_{13} v_s(t) - F_{14} \alpha_3(v_{c_1}, i_{l_2}, v_s(t), i_s(t))$$
$$= \beta_2(v_{c_1}, i_{l_2}, v_{l_1}, v_s(t), i_s(t)), \qquad (6.51)$$

where, with notational abuse, we have removed the dependence of α_3 and α_4 on v_{l_1} and i_{c_2}. Mind that β_1 does not depend on v_{l_1} and β_2 is independent of i_{c_2}; moreover, β_1 and β_2 are linear in i_{c_2} and v_{l_1}, respectively, with

$$\frac{\partial \beta_1}{\partial i_{c_2}} = F_{21}^\mathsf{T}, \quad \frac{\partial \beta_2}{\partial v_{l_1}} = -F_{12}. \qquad (6.52)$$

Using the expressions depicted in (6.50) and (6.51), we can then eliminate i_{c_1} and v_{l_2} from the multiport model (6.33). This is a Schur reduction

6.2. Geometric index analysis and reduction of branch models

which, according to Proposition 3.7, does not change the index. The resulting model is defined in terms of the variables $(v_{c_1}, v_{c_2}, i_{l_1}, i_{l_2}, v_{l_1}, i_{c_2})$ by

$$C(v_c)\begin{pmatrix} v'_{c_1} \\ v'_{c_2} \end{pmatrix} = \begin{pmatrix} \beta_1(v_{c_1}, i_{l_2}, i_{c_2}, v_s(t), i_s(t)) \\ i_{c_2} \end{pmatrix} \quad (6.53a)$$

$$L(i_l)\begin{pmatrix} i'_{l_1} \\ i'_{l_2} \end{pmatrix} = \begin{pmatrix} v_{l_1} \\ \beta_2(v_{c_1}, i_{l_2}, v_{l_1}, v_s(t), i_s(t)) \end{pmatrix} \quad (6.53b)$$

$$0 = i_{l_1} - F_{12}^T i_{l_2} - F_{32}^T i_s(t) \quad (6.53c)$$

$$0 = v_{c_2} + F_{21} v_{c_1} + F_{23} v_s(t), \quad (6.53d)$$

where it is worth emphasizing that at least one of the two equations (6.53c) or (6.53d) must necessarily be present in the model because of the existence of at least one IL-cutset or VC-loop.

System (6.53) is a Hessenberg DAE of the form depicted in (3.78), with dynamic variables $y = (v_{c_1}, v_{c_2}, i_{l_1}, i_{l_2})$ and algebraic variables $z = (v_{l_1}, i_{c_2})$. Note that, indeed, the twig inductor voltages v_{l_1} and the link capacitor currents i_{c_2} within z do not appear in the algebraic relations (6.53c)-(6.53d). The mappings h and g in (3.78) are here defined by

$$h(y, z, t) = \begin{pmatrix} C(v_c)^{-1} \begin{pmatrix} \beta_1(v_{c_1}, i_{l_2}, i_{c_2}, v_s(t), i_s(t)) \\ i_{c_2} \end{pmatrix} \\ L(i_l)^{-1} \begin{pmatrix} v_{l_1} \\ \beta_2(v_{c_1}, i_{l_2}, v_{l_1}, v_s(t), i_s(t)) \end{pmatrix} \end{pmatrix} \quad (6.54a)$$

$$g(y, t) = \begin{pmatrix} i_{l_1} - F_{12}^T i_{l_2} - F_{32}^T i_s(t) \\ v_{c_2} + F_{21} v_{c_1} + F_{23} v_s(t) \end{pmatrix}. \quad (6.54b)$$

According to the results in 3.4.7.3, for the Hessenberg system (6.53) to have geometric index two we need to check that the product $g_y h_z$ is nonsingular. In the light of (6.54), this product reads

$$\begin{pmatrix} 0 & 0 & I & -F_{12}^T \\ F_{21} & I & 0 & 0 \end{pmatrix} \begin{pmatrix} C^{-1}(v_c) \begin{pmatrix} 0 & F_{21}^T \\ 0 & I \end{pmatrix} \\ L^{-1}(i_l) \begin{pmatrix} I & 0 \\ -F_{12} & 0 \end{pmatrix} \end{pmatrix}, \quad (6.55)$$

where we have used (6.52) and the fact that β_1 and β_2 do not depend on v_{l_1} and i_{c_2}, respectively. The expression depicted in (6.55) can be rewritten as

$$\begin{pmatrix} 0 & 0 \\ F_{21} & I \end{pmatrix} C^{-1}(v_c) \begin{pmatrix} 0 & F_{21}^T \\ 0 & I \end{pmatrix} + \begin{pmatrix} I & -F_{12}^T \\ 0 & 0 \end{pmatrix} L^{-1}(i_l) \begin{pmatrix} I & 0 \\ -F_{12} & 0 \end{pmatrix}, \quad (6.56)$$

that is,

$$\begin{pmatrix} (I\ -F_{12}^{\mathsf{T}})\,L^{-1}(i_l)\begin{pmatrix} I \\ -F_{12} \end{pmatrix} & 0 \\ 0 & (F_{21}\ I)\,C^{-1}(v_c)\begin{pmatrix} F_{21}^{\mathsf{T}} \\ I \end{pmatrix} \end{pmatrix}, \qquad (6.57)$$

which is easily proved nonsingular since the positive definiteness assumptions on C and L make C^{-1} and L^{-1} positive definite as well. This shows that the Hessenberg model (6.53) is index two, and thereby the multiport equations (6.33) as well as the branch-oriented models (6.1) and (6.3) have geometric index two, thus completing the proof of item (2). □

Theorem 6.2 implicitly presumes the existence of at least one operating point (cf. subsection 6.2.1). In this regard, the index two setting raises the question of whether there exist operating points actually satisfying the hidden constraint for the Hessenberg DAE (6.53). This can be answered in the affirmative. Indeed, the hidden constraint $g_y h + g_t = 0$ reads

$$\begin{pmatrix} 0 & 0 \\ F_{21} & I \end{pmatrix} C^{-1}(v_c) \begin{pmatrix} \beta_1(v_{c_1},i_{l_2},i_{c_2},v_s(t),i_s(t)) \\ i_{c_2} \end{pmatrix} \\ + \begin{pmatrix} I & -F_{12}^{\mathsf{T}} \\ 0 & 0 \end{pmatrix} L^{-1}(i_l) \begin{pmatrix} v_{l_1} \\ \beta_2(v_{c_1},i_{l_2},v_{l_1},v_s(t),i_s(t)) \end{pmatrix} + \begin{pmatrix} -F_{32}^{\mathsf{T}} i_s'(t) \\ F_{23} v_s'(t) \end{pmatrix} = 0. \qquad (6.58)$$

Since β_1 does not depend on v_{l_1} and is linear in i_{c_2} and, similarly, β_2 is independent of i_{c_2} and linear in v_{l_1}, it follows that this system is linear in v_{l_1} and i_{c_2}. The nonsingularity of the product $g_y h_z$ proved above makes it possible to express $z = (v_{l_1}, i_{c_2})$ in terms of the remaining variables. Specifically, the first rows of (6.58) allow us to write

$$v_{l_1} = \zeta_1(v_{c_1}, i_{l_2}, v_s(t), i_s(t), i_s'(t)), \qquad (6.59)$$

whereas the last ones yield

$$i_{c_2} = \zeta_2(v_{c_1}, i_{l_2}, v_s(t), i_s(t), v_s'(t)). \qquad (6.60)$$

In turn, the conditions $F_{24} = 0$ and $F_{42} = 0$ in a normal tree indicate that i_{c_2} and v_{l_1} are not involved in the requirements on the configuration space imposed by the equations defining i_{r_1} and v_{r_2} within (6.28c) and (6.28d), respectively. Broadly speaking, this means that v_{l_1} and i_{c_2} can be freely assigned; since the existence of an operating point guarantees that the linear system defined by (6.28c) and (6.28d) with $i_r = i_r^*$, $i_j = i_s(t^*)$, $v_r = v_r^*$

and $v_u = v_s(t^*)$ has a solution, it follows that there actually exist operating points at t^* with v_{l_1}, i_{c_2} given by the relations depicted in (6.59)-(6.60).

Note also that, by Proposition 6.1, there would be no loss of generality in recasting Theorem 6.2 in terms of local voltage-controlled or current-controlled descriptions for resistors, with positive definite incremental conductance or resistance matrices, respectively.

6.2.5 State space reduction

The reduction approach supporting the geometric index characterization in Theorem 6.2 naturally yields local state equations modeling the circuit dynamics. Based on the semiexplicit form of the branch models (6.1) and (6.3) under the assumption of nonsingularity on $C(v_c)$, $L(i_l)$, and the Hessenberg form of (6.53) in the index two case, we may follow the ideas discussed in 3.4.7.2 and 3.4.7.3 to derive such state space models. This is done in Theorem 6.3 below, which can be understood as a restatement in DAE terms of the proper and normal tree methods [18, 35, 36].

For the sake of notational simplicity, we remove in (6.61) and (6.62) below the explicit dependence on t of the source terms $v_s(t)$, $i_s(t)$ and their derivatives $v'_s(t)$, $i'_s(t)$. The expressions defining the maps in (6.62) are detailed within the proof.

Theorem 6.3. *Under the assumptions of Theorem 6.2, the circuit dynamics are locally modeled by state space equations of the form*

$$v'_c = C(v_c)^{-1}\Psi_1(v_c, i_l, v_s, i_s) \tag{6.61a}$$

$$i'_l = L(i_l)^{-1}\Psi_2(v_c, i_l, v_s, i_s), \tag{6.61b}$$

in problems with geometric index one, and

$$v'_{c_1} = \tilde{C}(v_{c_1}, v_s) \begin{pmatrix} \beta_1(v_{c_1}, i_{l_2}, \zeta_2(v_{c_1}, i_{l_2}, v_s, i_s, v'_s), v_s, i_s) \\ \zeta_2(v_{c_1}, i_{l_2}, v_s, i_s, v'_s) \end{pmatrix} \tag{6.62a}$$

$$i'_{l_2} = \hat{L}(i_{l_2}, i_s) \begin{pmatrix} \zeta_1(v_{c_1}, i_{l_2}, v_s, i_s, i'_s) \\ \beta_2(v_{c_1}, i_{l_2}, \zeta_1(v_{c_1}, i_{l_2}, v_s, i_s, i'_s), v_s, i_s) \end{pmatrix} \tag{6.62b}$$

for circuits with geometric index two.

Proof. In the index one context, the reduction (6.61) is immediately obtained from (6.48). Note that this state space model has the form displayed in (3.77).

For the index two case, we use the Hessenberg form of (6.53) to derive a state space model of the form depicted in (3.82). The state variables

u will be defined by the twig capacitor voltages v_{c_1} and the link inductor currents i_{l_2}, whereas w will include the twig inductor currents i_{l_1} and the link capacitor voltages v_{c_2}. The mapping $w = \psi(u,t)$ used in the reduction of Hessenberg DAEs within 3.4.7.3 is now explicitly given by (6.53c) and (6.53d), namely

$$i_{l_1} = F_{12}^\mathsf{T} i_{l_2} + F_{32}^\mathsf{T} i_s(t) = \psi_1(i_{l_2}, i_s(t)) \tag{6.63a}$$

$$v_{c_2} = -F_{21} v_{c_1} - F_{23} v_s(t) = \psi_2(v_{c_1}, v_s(t)). \tag{6.63b}$$

Let us then insert these expressions within the capacitance and inductance matrices $C(v_c)$, $L(i_l)$ and split their inverses as

$$C((v_{c_1}, \psi_2(v_{c_1}, v_s(t))))^{-1} = \begin{pmatrix} \tilde{C}(v_{c_1}, v_s(t)) \\ \hat{C}(v_{c_1}, v_s(t)) \end{pmatrix}, \tag{6.64}$$

and

$$L((\psi_1(i_{l_2}, i_s(t)), i_{l_2}))^{-1} = \begin{pmatrix} \tilde{L}(i_{l_2}, i_s(t)) \\ \hat{L}(i_{l_2}, i_s(t)) \end{pmatrix}, \tag{6.65}$$

respectively. In both cases, the rows defining the ~ submatrices are those corresponding to tree reactances, whereas the ones entering the ^ submatrices correspond to cotree elements.

We also need to express the variables $z = (v_{l_1}, i_{c_2})$ in terms of v_{c_1} and i_{l_2}. Following again 3.4.7.3, this will be given by the mapping $z = \zeta(u,t)$, which comes from the hidden constraint $g_y h + g_t = 0$. The map ζ is in this case defined by the relations displayed in (6.59)-(6.60). The insertion of these expressions into the corresponding entries of (6.53a) and (6.53b) finally yields the local state space model (6.62). $\square$

Note that the explicit appearance of the derivatives $v'_s(t)$, $i'_s(t)$ of the excitation terms in (6.62) is consistent with the index two nature of the circuit configurations arising in item (2) of Theorem 6.2.

An index one example: Bashkow's circuit

Let us drive our attention again to Bashkow's circuit (Figure 6.1). The multiport model (6.33) for this circuit is obtained after eliminating from the equations discussed in subsection 6.1.2 the resistive branch variables using (6.5c), (6.5d), (6.7c) and (6.8), as well as the voltage and the current in the source by means of (6.7b) and the identity $i_j = i_0(t)$. This leads to

6.2. Geometric index analysis and reduction of branch models 285

$$C_1 v'_{c_1} = i_{c_1} \tag{6.66a}$$
$$C_2 v'_{c_2} = i_{c_2} \tag{6.66b}$$
$$C_3 v'_{c_3} = i_{c_3} \tag{6.66c}$$
$$L_1 i'_{l_1} = v_{l_1} \tag{6.66d}$$
$$L_2 i'_{l_2} = v_{l_2} \tag{6.66e}$$
$$0 = i_{c_1} - i_{l_1} + i_0(t) \tag{6.66f}$$
$$0 = i_{c_2} - i_{l_1} + G_3(v_{c_2} + v_{c_3}) \tag{6.66g}$$
$$0 = i_{c_3} + i_{l_2} + G_3(v_{c_2} + v_{c_3}) \tag{6.66h}$$
$$0 = v_{l_1} + v_{c_1} + v_{c_2} + R_4 i_{l_1} \tag{6.66i}$$
$$0 = v_{l_2} - v_{c_3} + R_5 i_{l_2}. \tag{6.66j}$$

This is an index one DAE, consistently with Theorem 6.2 and the absence of VC-loops and IL-cutsets. The state space equation given by (6.61) is

$$C_1 v'_{c_1} = i_{l_1} - i_0(t) \tag{6.67a}$$
$$C_2 v'_{c_2} = -G_3(v_{c_2} + v_{c_3}) + i_{l_1} \tag{6.67b}$$
$$C_3 v'_{c_3} = -G_3(v_{c_2} + v_{c_3}) - i_{l_2} \tag{6.67c}$$
$$L_1 i'_{l_1} = -v_{c_1} - v_{c_2} - R_4 i_{l_1} \tag{6.67d}$$
$$L_2 i'_{l_2} = v_{c_3} - R_5 i_{l_2}, \tag{6.67e}$$

which, as expected, coincides with the one derived in [18].

State equations of index two circuits

VC-loops. Consider the circuit discussed in Example 2 (pp. 229-230), displayed in Figure 6.3 together with a normal tree. According to Theorem 6.2, the presence of a VC-loop should make the branch and multiport models for this circuit index two, under positive definite assumptions on the circuit matrices which in this uncoupled problem hold if the individual capacitances, inductances and resistances are positive. This is indeed the case, as shown below; for the sake of brevity we work directly with the multiport model (6.33), which can be easily derived from the corresponding tree-based model of the form (6.3).

Using the normal tree shown in Figure 6.3, the multiport model for this

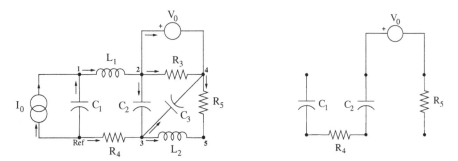

Fig. 6.3 Circuit with a VC-loop and normal tree.

circuit reads

$$C_1 v'_{c_1} = i_{c_1} \tag{6.68a}$$
$$C_2 v'_{c_2} = i_{c_2} \tag{6.68b}$$
$$C_3 v'_{c_3} = i_{c_3} \tag{6.68c}$$
$$L_1 i'_{l_1} = v_{l_1} \tag{6.68d}$$
$$L_2 i'_{l_2} = v_{l_2} \tag{6.68e}$$
$$0 = i_{c_1} - i_{l_1} + i_0(t) \tag{6.68f}$$
$$0 = i_{c_2} - i_{c_3} - i_{l_1} - i_{l_2} \tag{6.68g}$$
$$0 = v_{l_1} + v_{c_1} + v_{c_2} + R_4 i_{l_1} \tag{6.68h}$$
$$0 = v_{l_2} + v_{c_2} + R_5 i_{l_2} - v_0(t) \tag{6.68i}$$
$$0 = v_{c_3} + v_{c_2} - v_0(t). \tag{6.68j}$$

Note that the capacitor C_3 is located within a link, owing to the existence of the VC-loop defined by C_2 and C_3 together with the voltage source. The fundamental loop equation for the VC-loop is (6.68j). With respect to the circuit and the tree considered in Figure 6.1, note that now the twig capacitor C_2 defines a cutset together with the link reactances L_1, C_3 and L_2, so that (6.68g) replaces (6.5a). The fundamental loop defined by the link inductor L_2 also changes with respect to the one in Figure 6.1.

Now, twig capacitor currents and link inductor voltages can be eliminated as in (6.50) and (6.51). This yields the Hessenberg model (6.53),

6.2. Geometric index analysis and reduction of branch models

which for this example reads

$$C_1 v'_{c_1} = i_{l_1} - i_0(t) \tag{6.69a}$$
$$C_2 v'_{c_2} = i_{l_1} + i_{l_2} + i_{c_3} \tag{6.69b}$$
$$C_3 v'_{c_3} = i_{c_3} \tag{6.69c}$$
$$L_1 i'_{l_1} = -v_{c_1} - v_{c_2} - R_4 i_{l_1} \tag{6.69d}$$
$$L_2 i'_{l_2} = -v_{c_2} - R_5 i_{l_2} + v_0(t) \tag{6.69e}$$
$$0 = v_{c_2} + v_{c_3} - v_0(t), \tag{6.69f}$$

where it is worth remarking the presence of the algebraic variable i_{c_3}, corresponding to the link capacitor C_3. The hidden constraint for this index two DAE is easily derived from (6.69b), (6.69c) and (6.69f), and can be written as (cf. (6.60))

$$i_{c_3} = -\frac{C_3}{C_2 + C_3}(i_{l_1} + i_{l_2}) + \frac{C_2 C_3}{C_2 + C_3} v'_0(t).$$

This expression finally leads to the state space model

$$C_1 v'_{c_1} = i_{l_1} - i_0(t) \tag{6.70a}$$
$$(C_2 + C_3) v'_{c_2} = i_{l_1} + i_{l_2} + C_3 v'_0(t) \tag{6.70b}$$
$$L_1 i'_{l_1} = -v_{c_1} - v_{c_2} - R_4 i_{l_1} \tag{6.70c}$$
$$L_2 i'_{l_2} = -v_{c_2} - R_5 i_{l_2} + v_0(t). \tag{6.70d}$$

The state dimension of this equation (four) equals the order of complexity of the circuit, which in the absence of IL-cutsets is defined by the number of reactances minus the number of independent VC-loops.

Note that the leading coefficients of (6.70) do not vanish since all reactances are positive and then $C_1 \neq 0 \neq C_2 + C_3$ and $L_1 \neq 0 \neq L_2$; in this regard, it is worth mentioning that in problems without IL-cutsets the proofs of the index two cases in Theorems 6.2 and 6.3 hold if the capacitance matrix C is positive definite and the inductance matrix L is just nonsingular.

IL-cutsets. If all circuit matrices are positive definite, the existence of IL-cutsets also leads to index two models within the branch-oriented approach, as illustrated below. The procedure discussed above allows one to derive a state space equation also in this setting. Let us then consider the circuit in Example 3 (pp. 231-232), displayed in Figure 6.4 together with a normal tree, assuming again that all capacitances, inductances and conductances are positive.

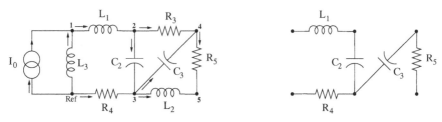

Fig. 6.4 Circuit with an IL-cutset and normal tree.

The inductor L_1 is now located in the tree, due to the existence of the IL-cutset defined by L_1, L_3 and the current source. The reader can check that the normal tree depicted in Figure 6.4 leads to the multiport model

$$C_2 v'_{c_2} = i_{c_2}$$
$$C_3 v'_{c_3} = i_{c_3}$$
$$L_1 i'_{l_1} = v_{l_1}$$
$$L_2 i'_{l_2} = v_{l_2}$$
$$L_3 i'_{l_3} = v_{l_3}$$
$$0 = i_{c_2} - i_{l_3} + G_3(v_{c_2} + v_{c_3}) - i_0(t)$$
$$0 = i_{c_3} + i_{l_2} + G_3(v_{c_2} + v_{c_3})$$
$$0 = i_{l_1} - i_{l_3} - i_0(t)$$
$$0 = v_{l_2} - v_{c_3} + R_5 i_{l_2}$$
$$0 = v_{l_3} + v_{l_1} + v_{c_2} + R_4(i_{l_3} + i_0(t)).$$

The Hessenberg DAE (6.53), which includes the algebraic variable v_{l_1} coming from the twig inductor, reads

$$C_2 v'_{c_2} = -G_3(v_{c_2} + v_{c_3}) + i_{l_3} + i_0(t)$$
$$C_3 v'_{c_3} = -G_3(v_{c_2} + v_{c_3}) - i_{l_2}$$
$$L_1 i'_{l_1} = v_{l_1}$$
$$L_2 i'_{l_2} = v_{c_3} - R_5 i_{l_2}$$
$$L_3 i'_{l_3} = -v_{c_2} - R_4(i_{l_3} + i_0(t)) - v_{l_1}$$
$$0 = i_{l_1} - i_{l_3} - i_0(t),$$

and the hidden constraint for this index two equation can be checked to yield, as in (6.59),

$$v_{l_1} = -\frac{L_1}{L_1 + L_3}(v_{c_2} + R_4(i_{l_3} + i_0(t))) + \frac{L_1 L_3}{L_1 + L_3} i'_0(t),$$

6.2. Geometric index analysis and reduction of branch models

which finally leads to the state space model

$$C_2 v'_{c_2} = -G_3 v_{c_2} - G_3 v_{c_3} + i_{l_3} + i_0(t) \quad (6.71a)$$
$$C_3 v'_{c_3} = -G_3 v_{c_2} - G_3 v_{c_3} - i_{l_2} \quad (6.71b)$$
$$L_2 i'_{l_2} = v_{c_3} - R_5 i_{l_2} \quad (6.71c)$$
$$(L_1 + L_3) i'_{l_3} = -v_{c_2} - R_4 i_{l_3} - R_4 i_0(t) - L_1 i'_0(t). \quad (6.71d)$$

The order of complexity (four) of the circuit now equals the number of reactances minus the number of independent IL-cutsets. Again, the leading coefficients do not vanish since the capacitances and inductances are positive and thus $C_2 \neq 0 \neq C_3$, $L_2 \neq 0 \neq L_1 + L_3$. In problems without VC-loops the index two characterization in Theorem 6.2 and the state space derivation in Theorem 6.3 are actually feasible with nonsingular capacitance and positive definite inductance, in consonance with the requirements in this example.

6.2.6 Controlled sources

The results discussed above can be extended to a broad family of circuits including controlled sources. Specifically, the geometric index characterization and the state space reduction introduced in subsections 6.2.4 and 6.2.5, respectively, can be adapted in order to apply to circuits with controlled sources under certain restrictions on the controlling variables of these sources and on the network topology. These ideas are detailed below.

Current and voltage sources will now be assumed to be defined by certain relations of the form

$$i_j = g_j(v, i, t) \quad (6.72a)$$
$$v_u = g_u(v, i, t) \quad (6.72b)$$

where i_j and v_u stand for the currents and voltages in current and voltage sources, respectively. These relations accommodate in particular *independent* sources, which correspond to certain components of i_j and v_u which are governed by explicit functions of time, to be denoted by $i_s(t)$ and $v_s(t)$. The remaining sources, associated with the remaining components of i_j and v_u, are controlled by certain branch voltages and/or currents.

Letting i_u, v_j describe the currents and voltages in all voltage and current sources, respectively, the general branch-oriented model for circuits with controlled sources is defined by the DAE

$$C(v_c)v'_c = i_c \tag{6.73a}$$
$$L(i_l)i'_l = v_l \tag{6.73b}$$
$$0 = A_c i_c + A_l i_l + A_r i_r + A_j i_j + A_u i_u \tag{6.73c}$$
$$0 = B_c v_c + B_l v_l + B_r v_r + B_j v_j + B_u v_u \tag{6.73d}$$
$$0 = g_r(v_r, i_r) \tag{6.73e}$$
$$i_j = g_j(v, i, t) \tag{6.73f}$$
$$v_u = g_u(v, i, t). \tag{6.73g}$$

Assume now that the circuit has neither I-cutsets nor V-loops; in the present framework, this means that there are neither cutsets formed by independent and/or controlled current sources, nor loops formed by independent and/or controlled voltage sources. Choosing a tree in which all voltage sources are twigs and all current sources are links, the corresponding tree-based model reads

$$C(v_c)v'_c = i_c \tag{6.74a}$$
$$L(i_l)i'_l = v_l \tag{6.74b}$$
$$0 = \begin{pmatrix} i_{c_1} \\ i_{l_1} \\ i_u \\ i_{r_1} \end{pmatrix} - F^\mathsf{T} \begin{pmatrix} i_{l_2} \\ i_{c_2} \\ i_j \\ i_{r_2} \end{pmatrix} \tag{6.74c}$$

$$0 = \begin{pmatrix} v_{l_2} \\ v_{c_2} \\ v_j \\ v_{r_2} \end{pmatrix} + F \begin{pmatrix} v_{c_1} \\ v_{l_1} \\ v_u \\ v_{r_1} \end{pmatrix} \tag{6.74d}$$

$$0 = g_r(v_r, i_r) \tag{6.74e}$$
$$i_j = g_j(v, i, t) \tag{6.74f}$$
$$v_u = g_u(v, i, t), \tag{6.74g}$$

where, as usual, the subscripts '**1**' and '**2**' refer to tree and cotree elements, respectively,

The key assumption making it feasible to extend the results of subsections 6.2.4 and 6.2.5 to circuits with controlled sources is defined in terms of the controlling variables allowed for them. Specifically, we will assume that equations (6.74c)-(6.74g) are such that i_j and v_u can be eventually written in terms of capacitor voltages, inductor currents, and voltages and currents in independent (voltage and current, respectively) sources. This means that

6.2. Geometric index analysis and reduction of branch models

the defining relations (6.74f) and (6.74g) for the sources, together with the circuit topology, make it possible to write

$$i_j = \gamma_j(v_c, i_l, v_s(t), i_s(t)) \qquad (6.75a)$$
$$v_u = \gamma_u(v_c, i_l, v_s(t), i_s(t)). \qquad (6.75b)$$

We will elaborate further on this point in 6.2.6.3 below.

6.2.6.1 Index one

Let us suppose that the circuit has neither VC-loops nor IL-cutsets, where the excluded sources are both independent and controlled ones. The tree leading to (6.74) can be then assumed to be a proper one.

Proceeding as in subsection 6.2.3, under a strict passivity assumption on resistors (or in the broader setting discussed in 6.2.3.2), resistive voltages and currents can be explicitly written in terms of (v_c, i_l, v_u, i_j) and in turn, using (6.75), in terms of $(v_c, i_l, v_s(t), i_s(t))$. Therefore, the derivation of the model (6.48) is still feasible in this context and the circuit is immediately seen to be index one; a state space reduction of the form (6.61) also follows in a trivial manner.

6.2.6.2 Index two

Assume now that the circuit has VC-loops and/or IL-cutsets. The tree used in the formulation of (6.74) can be then taken to be a normal one.

Suppose additionally that no controlled voltage source enters a VC-loop and, analogously, that controlled current sources are absent from IL-cutsets. In this situation, the normal tree can be chosen in a way such that the fundamental cutset defined by each twig inductor only has link inductors and independent current sources, and the fundamental loop defined by each link capacitor only involves twig capacitors and independent voltage sources. This means that the analogs of (6.49a)-(6.49b) can be written as

$$i_{l_1} = F_{12}^\mathsf{T} i_{l_2} + \left(F_{32}^\mathsf{T}\right)^* i_s(t) \qquad (6.76a)$$
$$v_{c_2} = -F_{21} v_{c_1} - (F_{23})^* v_s(t), \qquad (6.76b)$$

where $\left(F_{32}^\mathsf{T}\right)^*$ and $(F_{23})^*$ are the submatrices of F_{32}^T and F_{23} defined by the columns corresponding to independent sources.

Again, under a strict passivity assumption on resistors, or in the framework of 6.2.3.2, resistive voltages and currents can be written in terms of $(v_{c_1}, i_{l_2}, v_s(t), i_s(t))$, as in the proof of the index two case in Theorem 6.2

and using additionally (6.75) and (6.76). This implies that i_{c_1} and v_{l_2} can still be expressed in the form

$$i_{c_1} = \beta_1(v_{c_1}, i_{l_2}, i_{c_2}, v_s(t), i_s(t)) \tag{6.77}$$

$$v_{l_2} = \beta_2(v_{c_1}, i_{l_2}, v_{l_1}, v_s(t), i_s(t)), \tag{6.78}$$

as in (6.50)-(6.51), the identities (6.52) also holding in the present context. Hence, the circuit equations can be written as the Hessenberg system of size two

$$C(v_c)\begin{pmatrix} v'_{c_1} \\ v'_{c_2} \end{pmatrix} = \begin{pmatrix} \beta_1(v_{c_1}, i_{l_2}, i_{c_2}, v_s(t), i_s(t)) \\ i_{c_2} \end{pmatrix} \tag{6.79a}$$

$$L(i_l)\begin{pmatrix} i'_{l_1} \\ i'_{l_2} \end{pmatrix} = \begin{pmatrix} v_{l_1} \\ \beta_2(v_{c_1}, i_{l_2}, v_{l_1}, v_s(t), i_s(t)) \end{pmatrix} \tag{6.79b}$$

$$0 = i_{l_1} - F_{12}^\mathsf{T} i_{l_2} - \left(F_{32}^\mathsf{T}\right)^* i_s(t) \tag{6.79c}$$

$$0 = v_{c_2} + F_{21} v_{c_1} + (F_{23})^* v_s(t). \tag{6.79d}$$

The index two condition on this DAE again relies on the nonsingularity of the matrix (6.55), which holds exactly as detailed in the proof of Theorem 6.2 (p. 282). Note also that a state space reduction analogous to the one presented in Theorem 6.3 can be derived from the Hessenberg model (6.79).

6.2.6.3 *Assumptions on controlling variables for the sources*

The assumption that the voltages and currents in controlled sources can be expressed in the form displayed in (6.75) is met in practice by many circuits. The reader should not erroneously understand that only the voltages of capacitors and independent voltage sources and the currents of inductors and independent current sources can be controlling variables; the key idea is that not only these variables, but also those which can be expressed in terms of them using the network equations, may act as controlling variables for the sources.

In particular, the controlling variables accommodated in conditions 1 and 2(a) of the MNA index characterization stated in Theorem 4.1 of [87] (cf. Tables I-IV there) fall in this framework. For instance, VCCS's (note that [87] presumes that all controlled sources fall in the four categories defined on p. 211 above) are assumed in Table III of [87] to be controlled by the voltages of capacitors, all kind of voltage sources, and branches that form a loop with capacitors and voltage sources. By Kirchhoff's voltage law, the latter implies that these controlling voltages can be written in terms of those coming from the capacitors and the voltage sources in the loop. This means that the currents of voltage-controlled current sources can be eventually expressed as

$$i_j^{\text{vccs}} = g_j^{\text{vccs}}(v_c, v_u). \qquad (6.80)$$

Analogously, the remaining sources are assumed (in Tables IV, I, and II, respectively) to be governed by relations that can be written in the form

$$i_j^{\text{cccs}} = g_j^{\text{cccs}}(i_l, i_s(t), \tilde{\gamma}_r(v_c, v_u), \check{g}_j(v_c, v_u)) \qquad (6.81)$$

$$v_u^{\text{vcvs}} = g_u^{\text{vcvs}}(v_c, v_s(t), \hat{g}_u(i_l, i_s(t), \hat{\gamma}_r(v_c, v_s(t)), \hat{g}_j(v_c, v_s(t)))) \qquad (6.82)$$

$$v_u^{\text{ccvs}} = g_u^{\text{ccvs}}(i_l, i_s(t), \tilde{\gamma}_r(v_c, v_s(t), \tilde{g}_u(v_c, v_s(t))), \tilde{g}_j(v_c, v_s(t), \tilde{g}_u(v_c, v_s(t)))). (6.83)$$

From the expressions depicted in (6.82) and (6.83) it is clear that the voltages in all voltage sources can be written in terms of $(v_c, i_l, v_s(t), i_s(t))$ and, subsequently, the same holds for the currents in current sources (cf. (6.80) and (6.81)).

Therefore, the assumption that i_j and v_u can be eventually written in the form displayed in (6.75) is met by the above-mentioned sources. As detailed in [87, 292], this accommodates in particular the controlled sources arising in MOSFET transistors, used in many electronic applications. The extension of these results to circuits including broader families of controlled sources is however an open problem.

6.3 Qualitative properties

Qualitative features of nonlinear electrical circuits are often addressed via state models; see [60, 64, 65, 88, 103–105, 128, 201, 204, 286, 287] and references therein. Many results involving equilibria were established via state space formulations in the 1970s; a compilation of these results can be found in the survey paper [60]. As it happens in index analyses, qualitative properties of equilibria can be assessed in terms of the topology of the circuit and the electrical features of the devices. Topological aspects were already emphasized in [60, 64, 65].

Nevertheless, in spite of the wide scope of the state space framework discussed in these references and particularly in [60], a drawback of this approach is that the topological conditions arising in qualitative analyses are somehow merged with those supporting a state formulation. Indeed, the use of state models in [60] requires *local solvability* hypotheses (see specifically Theorems 4 and 12 there) which can be seen as local index one conditions on a multiport model (cf. [244]), and therefore restrict the results to these index one configurations. As a simple example, a C-loop, displayed for instance in many MOS transistor models [287, 292], would put the circuit out of the scope of this framework. Other related results, such as Theorem 10 in [64], are also explicitly restricted to circuit configurations not displaying C-loops or L-cutsets.

The DAE setting accommodates circuit models in great generality, allowing for a precise distinction between "qualitative" topological conditions and those supporting state space reductions. In the study of equilibria carried out in this Section, we will hence clarify the nature of the different topological conditions arising in the analysis, stating the results in a way such that they apply to index one and index two configurations. Under some passivity and reciprocity requirements on the circuit matrices, the qualitative analysis will rely on certain graph configurations involving VL-loops, IC-cutsets, VCL-loops and ICL-cutsets. The results must then be stated in a manner which applies in particular to well-posed circuits with VC-loops (including C-loops) and/or IL-cutsets (including L-cutsets).

We will split the topological conditions supporting different qualitative features, specifically the nonsingularity, hyperbolicity, and exponential stability of equilibrium points. These linear stability features are expressed in terms of certain properties of the matrix pencil spectrum, namely on the conditions $\lambda \neq 0$, $\mathrm{Re}\,\lambda \neq 0$ and $\mathrm{Re}\,\lambda < 0$ for the pencil eigenvalues λ. Note that the qualitative analysis is possible in terms of matrix pencil theory because of the geometric index characterization carried out in subsection 6.2.4 above, which makes it possible to apply directly the results of Section 3.5 and, specifically, Theorem 3.5 (p. 131). This way we improve on the results of [252], where the tractability index analysis requires an *ad hoc* support of the use of the matrix pencil spectrum (cf. Theorem 5 there).

Other qualitative aspects of nonlinear circuits will not be touched upon here. Somehow related issues concerning DC operating points are addressed in [103–105], where operating points *potential stability* and *instability* are defined and analyzed in terms of the DC equations only. Broadly speaking, these papers classify an operating point of a given DC circuit as unstable if no insertion of parallel capacitors and series inductors results in a stable equilibrium of the corresponding dynamic circuit. Find details in the above-mentioned references. Several results based upon Lyapunov function methods can be found in [64, 88, 128, 287] and references therein, whereas qualitative features of certain circuit families are tackled in [98, 155, 257, 286, 287]. Different qualitative properties of nonlinear circuit dynamics can be studied via the geometric ap-

6.3. Qualitative properties

proach stemming from the work of Brayton and Moser [28, 29]; later results in this direction can be found in [77, 118, 119, 268, 302, 303]. Jump phenomena, impasse points and related singularities are addressed in [61, 62, 233, 236, 246, 258, 288]. Chaotic effects in state space models of nonlinear circuits have been exhaustively analyzed in the literature; see [69, 265, 266] as a sample.

6.3.1 Equilibria of DC circuits

Assume that a given circuit has no controlled sources and that the independent ones are DC sources; this means that the circuit is driven by constant excitations $i_s(t) = I_s$ and $v_s(t) = V_s$ with $I_s \in \mathbb{R}^{b_j}$, $V_s \in \mathbb{R}^{b_u}$. For simplicity, let us substitute $i_j = I_s$, $v_u = V_s$ within equations (6.1c)-(6.1d) of the branch-oriented model (6.1) to derive

$$C(v_c)v'_c = i_c \tag{6.84a}$$
$$L(i_l)i'_l = v_l \tag{6.84b}$$
$$0 = A_c i_c + A_l i_l + A_r i_r + A_j I_s + A_u i_u \tag{6.84c}$$
$$0 = B_c v_c + B_l v_l + B_r v_r + B_j v_j + B_u V_s \tag{6.84d}$$
$$0 = g_r(v_r, i_r). \tag{6.84e}$$

This is a quasilinear DAE of the form $A(x)x' = f(x)$, where x is the semistate vector defined by all branch currents and voltages (except for the currents and voltages in current and voltage sources, respectively). For later use, we will order the variables in x as $(v_c,\ i_l,\ i_c,\ v_l,\ i_r,\ v_r,\ i_u,\ v_j)$. The matrix-valued mapping $A(x)$ can be written as

$$A(x) = \begin{pmatrix} C(v_c) & 0 & 0 & 0 & 0 & 0 & 0 \\ 0 & L(i_l) & 0 & 0 & 0 & 0 & 0 \\ 0 & 0 & 0 & 0 & 0 & 0 & 0 \\ 0 & 0 & 0 & 0 & 0 & 0 & 0 \\ 0 & 0 & 0 & 0 & 0 & 0 & 0 \end{pmatrix}, \tag{6.85}$$

and the vector field $f(x)$ reads

$$f(x) = \begin{pmatrix} i_c \\ v_l \\ -A_c i_c - A_l i_l - A_r i_r - A_j I_s - A_u i_u \\ -B_c v_c - B_l v_l - B_r v_r - B_j v_j - B_u V_s \\ -g_r(v_r, i_r) \end{pmatrix}, \tag{6.86}$$

where the − sign in the last entries is aimed at later convenience. In the sequel we assume that a given x^* is an equilibrium point of (6.84), namely, that $f(x^*) = 0$.

Remark 6.5. In the light of (6.84a) and (6.84b), the equilibrium condition requires that $i_c = 0$ and $v_l = 0$. This is equivalent to open-circuiting capacitors, so that $i_c = 0$, and short-circuiting inductors, which makes $v_l = 0$. From the point of view of circuit theory, this is the usual procedure to compute the so-called *DC operating points* of the network. The relations which result from performing the substitutions $i_c = 0$ and $v_l = 0$ in (6.84c) and (6.84d), respectively, together with (6.84e), are known as the *DC equations* of the circuit.

Given an equilibrium x^* of (6.84), our goal in this Section is to characterize topologically certain properties of the spectrum of the matrix pencil $\lambda A(x^*) - f'(x^*)$, that is,

$$\sigma(\{A(x^*), -f'(x^*)\}) = \{\lambda \in \mathbb{C} \ / \ \det(\lambda A(x^*) - f'(x^*)) = 0\}. \quad (6.87)$$

Following Theorem 3.5, in circuits with a well-defined geometric index this spectrum will describe linear stability properties of the equilibrium point x^*. Certainly, these circuits include in particular the ones arising in Theorem 6.2. Recalling that the semistate variables x are ordered as $(v_c, i_l, i_c, v_l, i_r, v_r, i_u, v_j)$, in the light of (6.85) and (6.86) we have

$$\lambda A(x^*) - f'(x^*) = \begin{pmatrix} \lambda C & 0 & -I & 0 & 0 & 0 & 0 & 0 \\ 0 & \lambda L & 0 & -I & 0 & 0 & 0 & 0 \\ 0 & A_l & A_c & 0 & A_r & 0 & A_u & 0 \\ B_c & 0 & 0 & B_l & 0 & B_r & 0 & B_j \\ 0 & 0 & 0 & 0 & E & D & 0 & 0 \end{pmatrix}, \quad (6.88)$$

where C and L stand for $C(v_c^*)$ and $L(i_l^*)$, respectively, whereas

$$D = \frac{\partial g_r}{\partial v_r}(v_r^*, i_r^*), \ E = \frac{\partial g_r}{\partial i_r}(v_r^*, i_r^*).$$

We will assume that the set of resistors is strictly locally passive at (v_r^*, i_r^*) (cf. Definition 6.1); it follows from Proposition 6.1 (p. 268) that all hybrid definitions of the resistors are locally well-defined and yield positive definite incremental hybrid matrices H. In particular, the incremental conductance and resistance matrices $G = -E^{-1}D$, $R = -D^{-1}E$ are well-defined and positive definite. Recall that full coupling is allowed within the sets of capacitors, inductors and resistors.

6.3. Qualitative properties   297

Remark 6.6. The expression depicted in (6.88) shows that ill-posed circuits yield a singular matrix pencil. Indeed, the existence of a V-loop implies that A_u does not have maximal column rank (cf. Lemma 5.3), making (6.88) singular for any λ. Analogously, an I-cutset prevents B_j from having maximal column rank, according to Lemma 5.7, and thereby also makes (6.88) singular. In problems with strictly locally passive resistors and positive definite matrices C, L the converse is also true, that is, a singular matrix pencil (6.88) necessarily corresponds to an ill-posed circuit; see Remark 6.11 after Theorem 6.5.

Remark 6.7. At several points we will compute the matrix pencil spectrum at equilibrium using models slightly different from (6.84). Specifically, we will use tree-based formulations, as well as hybrid descriptions of resistors of the form (6.39) instead of (6.84e). Proceeding as in Remark 6.1 (p. 260) and in the proof of Theorem 6.1 (p. 275), respectively, these reformulations can be easily seen to result in a premultiplication of $\lambda A(x^*) - f'(x^*)$ in (6.88) by a nonsingular matrix, which does not change the pencil spectrum nor affects the invertibility of $f'(x^*)$.

6.3.2 *Nonsingularity*

In different contexts, it is of interest to assess if a given equilibrium x^* of the circuit equations (6.84) is *nonsingular* in the sense that the Jacobian matrix $f'(x^*)$ is invertible. In nonlinear problems, the nonsingularity of $f'(x^*)$ guarantees the isolation of this equilibrium and makes it well-conditioned for Newton-based computations (see for instance [63, 105, 160, 161, 245] and references therein). In linear cases, the nonsingularity of the matrix coming from the right-hand side of (6.84) implies the existence of a unique equilibrium for the circuit dynamics.

Restricting the attention to problems with strictly locally passive resistors (or strictly passive resistors in linear settings), this nonsingularity requirement can be examined topologically, as shown in Theorem 6.4 below. Specifically, the nonsingularity of equilibria will rely on the absence of VL-loops and IC-cutsets. Circuits with inductor-only loops and capacitor-only cutsets were examined in [201] and, later, in [119, 120]; Theorem 6.4 can be seen as a restatement, in the DAE setting, of a property already discussed in these references.

It is worth emphasizing that these topological conditions are entirely independent of those characterizing the index in Theorem 6.2, so that the statement in Theorem 6.4 applies to index one and index two configurations. Note also that this result is consistent with the fact that a VL-loop or an IC-cutset would yield an ill-posed resistive circuit after short-circuiting inductors and open-circuiting capacitors (cf. Remark 6.5), due to the presence of a V-loop or an I-cutset, respectively, in this resistive circuit.

Theorem 6.4. *Assume that (6.84) models a well-posed, connected circuit with strictly locally passive resistors. An equilibrium point x^* of (6.84) is nonsingular if and only if the circuit has neither VL-loops nor IC-cutsets.*

Proof. Assume first that the matrix $f'(x^*)$ is nonsingular. This is equivalent to the nonsingularity of the matrix

$$\begin{pmatrix} 0 & 0 & I & 0 & 0 & 0 & 0 & 0 \\ 0 & 0 & 0 & I & 0 & 0 & 0 & 0 \\ 0 & A_l & A_c & 0 & A_r & 0 & A_u & 0 \\ B_c & 0 & 0 & B_l & 0 & B_r & 0 & B_j \\ 0 & 0 & 0 & 0 & E & D & 0 & 0 \end{pmatrix}.$$

It follows immediately that $(A_l \; A_u)$ must have full column rank, and according to Lemma 5.3 this rules out VL-loops in the circuit. Similarly, $(B_c \; B_j)$ must have full column rank, which precludes IC-cutsets in the light of Lemma 5.7.

Conversely, assume that the circuit contains neither VL-loops nor IC-cutsets. From Lemma 5.1 it follows that there exists a tree which contains all voltage sources and inductors and neither current sources nor capacitors. We may then rewrite (6.84) in the form

$$C(v_c)v_c' = i_c \tag{6.89a}$$

$$L(i_l)i_l' = v_l \tag{6.89b}$$

$$0 = \begin{pmatrix} i_l \\ i_u \\ i_{r_1} \end{pmatrix} - F^\mathsf{T} \begin{pmatrix} i_c \\ I_s \\ i_{r_2} \end{pmatrix} \tag{6.89c}$$

$$0 = \begin{pmatrix} v_c \\ v_j \\ v_{r_2} \end{pmatrix} + F \begin{pmatrix} v_l \\ V_s \\ v_{r_1} \end{pmatrix} \tag{6.89d}$$

$$0 = g_r(v_r, i_r), \tag{6.89e}$$

where, as usual, the subscripts '1' and '2' are used to specify twig and link elements, respectively. Note that, as indicated in Remark 6.7, this transformation just involves the premultiplication of (6.1) by a nonsingular matrix and hence does not affect the invertibility of the matrix $f'(x^*)$.

6.3. Qualitative properties

Splitting the matrix F in blocks as

$$F = \begin{pmatrix} F_{11} & F_{12} & F_{13} \\ F_{21} & F_{22} & F_{23} \\ F_{31} & F_{32} & F_{33} \end{pmatrix}, \tag{6.90}$$

and writing for simplicity the columns below according to the order of variables i_c, v_l, i_l, i_u, v_c, v_j, i_{r_1}, v_{r_2}, v_{r_1}, i_{r_2}, the derivative of the right-hand side of (6.89) is nonsingular if and only if so it is

$$\begin{pmatrix} I & 0 & 0 & 0 & 0 & 0 & 0 & 0 & 0 & 0 \\ 0 & I & 0 & 0 & 0 & 0 & 0 & 0 & 0 & 0 \\ -F_{11}^T & 0 & I & 0 & 0 & 0 & 0 & 0 & 0 & -F_{31}^T \\ -F_{12}^T & 0 & 0 & I & 0 & 0 & 0 & 0 & 0 & -F_{32}^T \\ 0 & F_{11} & 0 & 0 & I & 0 & 0 & 0 & F_{13} & 0 \\ 0 & F_{21} & 0 & 0 & 0 & I & 0 & 0 & F_{23} & 0 \\ -F_{13}^T & 0 & 0 & 0 & 0 & 0 & I & 0 & 0 & -F_{33}^T \\ 0 & F_{31} & 0 & 0 & 0 & 0 & 0 & I & F_{33} & 0 \\ 0 & 0 & 0 & 0 & 0 & 0 & -H_{11} & -H_{12} & I & 0 \\ 0 & 0 & 0 & 0 & 0 & 0 & -H_{21} & -H_{22} & 0 & I \end{pmatrix}, \tag{6.91}$$

where the matrix H (which is split in blocks) comes from the resistive hybrid description (6.39) associated with the tree chosen above (cf. Remark 6.7). Note that we have written the row corresponding to i_{r_1} coming from (6.89c) right before the one associated with v_{r_2} from (6.89d) for the sake of clarity in the sequel.

Looking at the first two rows and at the third, forth, fifth and sixth columns, it is clear that the nonsingularity of (6.91) amounts to that of

$$\begin{pmatrix} I & 0 & 0 & -F_{33}^T \\ 0 & I & F_{33} & 0 \\ -H_{11} & -H_{12} & I & 0 \\ -H_{21} & -H_{22} & 0 & I \end{pmatrix}, \tag{6.92}$$

which is a nonsingular matrix, according to Lemma 6.2 (p. 275). Therefore, the absence of VL-loops and IC-cutsets indeed makes $f'(x^*)$ a nonsingular matrix. □

Remark 6.8. The absence of VL-loops and IC-cutsets in Theorem 6.4 is somehow reminiscent of the index characterization in terms of VC-loops and IL-cutsets stated in Theorem 6.2. A notion of *reactive duality* underlies this similarity. Indeed, the substitution of capacitors by inductors and vice-versa in a given electrical circuit results in a spectral transformation of the form $\lambda \to \lambda^{-1}$; in particular, VL-loops or IC-cutsets yielding null eigenvalues (mind that the nonsingularity of $f'(x^*)$ is equivalent to the property $0 \notin \sigma(\{A(x^*), -f'(x^*)\})$) lead to VC-loops or IL-cutsets which increase the multiplicity of the infinite eigenvalue in the circuit resulting from these substitutions. Find details in [248].

Remark 6.9. If, in the setting of Theorem 6.4, there are neither VC-loops nor IL-cutsets, and the capacitance and inductance matrices C, L are positive definite, then zero eigenvalues owing to VL-loops or IC-cutsets are shown in Theorem 4 of [248] to be simple or semisimple for the linearization of a local state space circuit model at equilibrium. Recall that, given a matrix M, the index of an eigenvalue λ is the smallest positive integer κ for which $\text{rk}\,(\lambda I - M)^\kappa = \text{rk}\,(\lambda I - M)^{\kappa+1}$; eigenvalues of algebraic multiplicity one are called *simple*, and those with algebraic multiplicity greater than one but with index one are called *semisimple*. This is of interest e.g. in the analysis of local bifurcation phenomena in electrical circuits, since this property precludes for instance Takens-Bogdanov bifurcations [110] which are based on the existence of null eigenvalues with index two.

6.3.3 Hyperbolicity and exponential stability

Assume that the circuit model (6.84) has a well-defined geometric index at a given equilibrium point x^*. According to the results in Section 3.5, the equilibrium x^* may then be said to be *hyperbolic* if the matrix pencil spectrum (6.87) has no purely imaginary eigenvalues, that is, if $\text{Re}\,\lambda \neq 0$ for all eigenvalues λ. As shown in Theorem 6.4 above, the absence of null eigenvalues in problems with strictly locally passive resistors is equivalent to the absence of VL-loops and IC-cutsets, so that the hyperbolicity problem in this setting requires studying the existence of non-vanishing, purely imaginary eigenvalues. The importance of these stem from the fact that, in linear problems, they correspond to natural frequencies yielding proper oscillations, whereas in the nonlinear context a pair of purely imaginary eigenvalues may be responsible for Hopf bifurcations [204] leading as well to nonlinear oscillations.

We provide in item (1) of Theorem 6.5 below sufficient conditions to guarantee that $\text{Re}\,\lambda \neq 0$ for all pencil eigenvalues λ. In a certain sense, the topological conditions arising in this problem will be mild extensions of the topological requirements for nonsingularity depicted in Theorem 6.4. Via Proposition 6.2, VC-loops and IL-cutsets (yielding index two config-

6.3. Qualitative properties

urations) can be accommodated in the analysis, so that the topological conditions in Theorem 6.5 will be again independent of the ones arising in the index characterization presented in Theorem 6.2.

Supported also on the results of Section 3.5, in hyperbolic problems strict local passivity assumptions on the circuit matrices will naturally yield the *exponential stability* of the equilibrium. This is given by the condition $\text{Re}\,\lambda < 0$ for all eigenvalues in the pencil spectrum (6.87); cf. item (2) of Theorem 6.5.

Proposition 6.2. *Let K and J be two sets of edges of a connected digraph $\mathcal{G}$, with $K \subseteq J$.*

(1) If all cutsets in J are contained in K, then $\ker A_{\mathcal{G}-J}^\mathsf{T} = \ker A_{\mathcal{G}-K}^\mathsf{T}$ or, equivalently,

$$w^\mathsf{T} A_{\mathcal{G}-J} = 0 \Rightarrow w^\mathsf{T} A_{J-K} = 0. \qquad (6.93)$$

(2) If all loops in J are contained in K, then $\ker B_{\mathcal{G}-J}^\mathsf{T} = \ker B_{\mathcal{G}-K}^\mathsf{T}$ or, equivalently,

$$w^\mathsf{T} B_{\mathcal{G}-J} = 0 \Rightarrow w^\mathsf{T} B_{J-K} = 0. \qquad (6.94)$$

Proof. Item (1) is proved in [252] (Proposition 4.4). In turn, item (2) is a restatement of Proposition 4.5 in [252], which shows that, if all loops in J are contained in K, then

$$w_1^\mathsf{T} A_K^\mathsf{T} + w_2^\mathsf{T} A_{J-K}^\mathsf{T} = 0 \Rightarrow w_2 = 0. \qquad (6.95)$$

From this result we can check that (6.94) holds by means of the identity $BA^\mathsf{T} = 0$ (cf. Lemma 5.9) written in the form

$$B_J A_J^\mathsf{T} + B_{\mathcal{G}-J} A_{\mathcal{G}-J}^\mathsf{T} = 0. \qquad (6.96)$$

Indeed, assume that $w^\mathsf{T} B_{\mathcal{G}-J} = 0$. From (6.96) we derive $w^\mathsf{T} B_J A_J^\mathsf{T} = 0$, that is

$$w^\mathsf{T} B_K A_K^\mathsf{T} + w^\mathsf{T} B_{J-K} A_{J-K}^\mathsf{T} = 0.$$

Writing

$$w_1^\mathsf{T} = w^\mathsf{T} B_K, \ w_2^\mathsf{T} = w^\mathsf{T} B_{J-K},$$

from the relation depicted in (6.95) we get $w_2^\mathsf{T} = w^\mathsf{T} B_{J-K} = 0$, as we aimed to show. $\square$

Below we will use the fact that, if M is a positive definite real matrix, and denoting as $u^\star$ the conjugate transpose of u, then $u^\star(M+M^\mathsf{T})u$ is real and positive for any non-vanishing complex vector u. This can be easily checked by writing $u = u_1 + iu_2$ for real vectors u_1, u_2; indeed,

$$u^\star(M + M^\mathsf{T})u = u_1^\mathsf{T}(M + M^\mathsf{T})u_1 + u_2^\mathsf{T}(M + M^\mathsf{T})u_2 > 0$$

since the purely imaginary terms cancel each other due to the symmetry of $M + M^\mathsf{T}$. It then follows that

$$u^\star(M + M^\mathsf{T})u = 0 \Rightarrow u = 0 \qquad (6.97)$$

for any positive definite matrix M. Recall that in the definition of a positive definite matrix (p. 211) we do not require it to be symmetric.

Theorem 6.5. *Assume that (6.84) models a well-posed, connected circuit with strictly locally passive resistors and symmetric and nonsingular incremental capacitance and inductance matrices C, L. Let x^* be an equilibrium point of (6.84).*

(1) All eigenvalues λ in the matrix pencil spectrum (6.87) verify $\operatorname{Re}\lambda \neq 0$ if at least one of the following two pairs of topological conditions holds.

  (a) There are neither VL-loops nor ICL-cutsets (except for IL-cutsets).

  (b) There are neither VCL-loops (except for VC-loops) nor IC-cutsets.

(2) If, additionally, C and L are positive definite, and at least one of the two pairs of conditions (a) and (b) above holds, then all eigenvalues verify $\operatorname{Re}\lambda < 0$.

Proof. The equations defining a non-vanishing $w = (w_c, w_l, w_a, w_b, w_r)$ as a left-eigenvector for (6.88) are

$$\lambda w_c^\mathsf{T} C + w_b^\mathsf{T} B_c = 0 \qquad (6.98\text{a})$$
$$\lambda w_l^\mathsf{T} L + w_a^\mathsf{T} A_l = 0 \qquad (6.98\text{b})$$
$$-w_c^\mathsf{T} + w_a^\mathsf{T} A_c = 0 \qquad (6.98\text{c})$$
$$-w_l^\mathsf{T} + w_b^\mathsf{T} B_l = 0 \qquad (6.98\text{d})$$
$$w_a^\mathsf{T} A_r + w_r^\mathsf{T} E = 0 \qquad (6.98\text{e})$$
$$w_b^\mathsf{T} B_r + w_r^\mathsf{T} D = 0 \qquad (6.98\text{f})$$
$$w_a^\mathsf{T} A_u = 0 \qquad (6.98\text{g})$$
$$w_b^\mathsf{T} B_j = 0, \qquad (6.98\text{h})$$

6.3. Qualitative properties

or, eliminating w_c^T and w_l^T via (6.98c)-(6.98d),

$$\lambda w_a^\mathsf{T} A_c C + w_b^\mathsf{T} B_c = 0 \tag{6.99a}$$
$$\lambda w_b^\mathsf{T} B_l L + w_a^\mathsf{T} A_l = 0 \tag{6.99b}$$
$$w_a^\mathsf{T} A_r + w_r^\mathsf{T} E = 0 \tag{6.99c}$$
$$w_b^\mathsf{T} B_r + w_r^\mathsf{T} D = 0 \tag{6.99d}$$
$$w_a^\mathsf{T} A_u = 0 \tag{6.99e}$$
$$w_b^\mathsf{T} B_j = 0. \tag{6.99f}$$

Let us transpose (6.99a) and premultiply the resulting equation by $w_a^\star A_c$, where as indicated above $w_a^\star$ stands for the conjugate transpose $\overline{w_a}^\mathsf{T}$. Since C is symmetric, this yields

$$\lambda w_a^\star A_c C A_c^\mathsf{T} w_a + w_a^\star A_c B_c^\mathsf{T} w_b. \tag{6.100}$$

Analogously, multiplying the conjugate of (6.99b) by $B_l^\mathsf{T} w_b$ we obtain

$$\overline{\lambda} w_b^\star B_l L B_l^\mathsf{T} w_b + w_a^\star A_l B_l^\mathsf{T} w_b. \tag{6.101}$$

In turn, multiplying the conjugate of (6.99c) by $B_r^\mathsf{T} w_b$ leads to

$$w_a^\star A_r B_r^\mathsf{T} w_b + w_r^\star E B_r^\mathsf{T} w_b = 0,$$

which, using $w_r^\star = -w_b^\star B_r D^{-1}$ from (6.99d) and the identity $R = -D^{-1}E$ for the incremental resistance matrix, is transformed into

$$w_a^\star A_r B_r^\mathsf{T} w_b + w_b^\star B_r R B_r^\mathsf{T} w_b = 0. \tag{6.102}$$

Finally, multiplying the conjugate of (6.99e) by $B_u^\mathsf{T} w_b$ and premultiplying the transpose of (6.99f) by $w_a^\star A_j$ we get

$$w_a^\star A_u B_u^\mathsf{T} w_b = 0 \tag{6.103}$$
$$w_a^\star A_j B_j^\mathsf{T} w_b = 0. \tag{6.104}$$

The sum of (6.100), (6.101), (6.102), (6.103) and (6.104) yields

$$\lambda w_a^\star A_c C A_c^\mathsf{T} w_a + \overline{\lambda} w_b^\star B_l L B_l^\mathsf{T} w_b + w_b^\star B_r R B_r^\mathsf{T} w_b = 0, \tag{6.105}$$

where we have used $A_c B_c^\mathsf{T} + A_l B_l^\mathsf{T} + A_r B_r^\mathsf{T} + A_u B_u^\mathsf{T} + A_j B_j^\mathsf{T} = AB^\mathsf{T} = 0$ (cf. Lemma 5.9 on p. 201) to cancel the terms leading to $w_a^\star AB^\mathsf{T} w_b$.

Since C and L are symmetric, the conjugate transpose of (6.105) reads

$$\overline{\lambda} w_a^\star A_c C A_c^\mathsf{T} w_a + \lambda w_b^\star B_l L B_l^\mathsf{T} w_b + w_b^\star B_r R^\mathsf{T} B_r^\mathsf{T} w_b = 0. \tag{6.106}$$

The sum of (6.105) and (6.106) results in

$$2\mathrm{Re}\,\lambda(w_a^\star A_c C A_c^\mathsf{T} w_a + w_b^\star B_l L B_l^\mathsf{T} w_b) + w_b^\star B_r(R + R^\mathsf{T})B_r^\mathsf{T} w_b = 0. \tag{6.107}$$

Now, to prove the statement in item (1), assume that $\operatorname{Re}\lambda = 0$, so that (6.107) amounts to

$$w_b^* B_r (R + R^{\mathsf{T}}) B_r^{\mathsf{T}} w_b = 0.$$

According to Proposition 6.1 (p. 268), R is positive definite and hence from (6.97) we derive $w_b^{\mathsf{T}} B_r = 0$. It then follows that $w_r = 0$, according to (6.99d), and from (6.99c) we derive also $w_a^{\mathsf{T}} A_r = 0$. Note additionally that $w_a^{\mathsf{T}} A_u = 0$ and $w_b^{\mathsf{T}} B_j = 0$, as depicted in (6.99e) and (6.99f), respectively.

Assume first that the topological requirements displayed in (a) are met. The fact that the only ICL-cutsets are IL-cutsets means, according to item (1) of Proposition 6.2 above, that $w_a^{\mathsf{T}} A_r = 0$, $w_a^{\mathsf{T}} A_u = 0$ imply $w_a^{\mathsf{T}} A_c = 0$. By (6.99a) the latter yields $w_b^{\mathsf{T}} B_c = 0$. Since $w_b^{\mathsf{T}} B_r = 0$ and $w_b^{\mathsf{T}} B_j = 0$, the absence of VL-loops yields $w_b = 0$, following Lemma 5.8 on p. 200. In the light of (6.99b) we then get $w_a^{\mathsf{T}} A_l = 0$ and, since no I-cutset may be present, Lemma 5.4 (p. 199) means that $w_a = 0$. We have then arrived at the contradiction that the left eigenvector w vanishes since $w_r = 0$, as indicated above, and $w_c = 0$, $w_l = 0$ because of (6.98c) and (6.98d).

Let us now suppose that the topological conditions in (b) hold. Now, the identities $w_b^{\mathsf{T}} B_r = 0$ and $w_b^{\mathsf{T}} B_j = 0$ together with the fact that all VCL-loops are VC-loops yield $w_b^{\mathsf{T}} B_l = 0$, using item (2) of Proposition 6.2. It then follows from (6.99b) that $w_a^{\mathsf{T}} A_l = 0$. Together with $w_a^{\mathsf{T}} A_r = 0$, $w_a^{\mathsf{T}} A_u = 0$ and the absence of IC-cutsets, this yields $w_a = 0$ according to Lemma 5.4. From (6.99a) we then derive $w_b^{\mathsf{T}} B_c = 0$. The absence of V-loops implies that $w_b = 0$ (cf. Lemma 5.8). Jointly with $w_r = 0$ and $w_c = 0$, $w_l = 0$ from (6.98c) and (6.98d), this would lead to the contradiction $w = 0$.

Finally, to prove item (2) we use again the relation (6.107). Assuming $\operatorname{Re}\lambda > 0$, the symmetry and positive definiteness of C, L, together with the positive definiteness of R, imply that $w_a^{\mathsf{T}} A_c = 0$, $w_b^{\mathsf{T}} B_l = 0$, $w_b^{\mathsf{T}} B_r = 0$. The latter implies that $w_r = 0$ and then $w_a^{\mathsf{T}} A_r = 0$ (cf. (6.99d) and (6.99c), respectively). Note also that $w_a^{\mathsf{T}} A_u = 0$, $w_b^{\mathsf{T}} B_j = 0$ as stated in (6.99e), (6.99f). Additionally, from (6.99a) and (6.99b) we obtain $w_b^{\mathsf{T}} B_c = 0$ and $w_a^{\mathsf{T}} A_l = 0$, respectively. The well-posedness of the circuit and Lemmas 5.4 and 5.8 then would yield $w_a = 0$ and $w_b = 0$, so that $w_c = 0$ and $w_l = 0$; together with $w_r = 0$, these properties contradict the assumption that w does not vanish.

This means that, without making use of any topological conditions, the positive definiteness assumptions yield $\operatorname{Re}\lambda \leq 0$. Combining this result with the topological criteria stated in (a) or (b), which guarantee that

$\operatorname{Re}\lambda \neq 0$, we finally derive that the condition $\operatorname{Re}\lambda < 0$ holds for all eigenvalues. $\square$

Remark 6.10. The proof of item (1) above also shows that the absence of ICL-cutsets alone is enough to preclude purely imaginary, non-vanishing eigenvalues in the circuit; cf. Theorem 4.3 in [252], where an analogous proof is given for MNA models. Indeed, the above-derived conditions $w_a^T A_r = 0$, $w_a^T A_u = 0$ would yield in this setting $w_a = 0$. This leads to $w_b^T B_c = 0$ in the light of (6.99a), and the non-vanishing of λ in (6.99b) together with the nonsingularity of L yields $w_b^T B_l = 0$. Using $w_b^T B_r = 0$, $w_b^T B_j = 0$ and the absence of V-loops we obtain $w_b = 0$. The identities $w_r = 0$, $w_c = 0$ and $w_l = 0$ are derived as above. The same happens if the circuit has no VCL-loops.

Remark 6.11. In turn, the proof of item (2) shows that, in cases with strictly locally passive resistors and positive definite matrices C, L, a singular matrix pencil (6.88) necessarily corresponds to an ill-posed circuit (see Remark 6.6 on page 297). Indeed, if the matrix pencil is singular, any $\lambda \in \mathbb{C}$ must admit a nontrivial left-eigenvector w (which may certainly depend on λ) for (6.88). The reasoning in the proof of item (2) shows that those verifying $\operatorname{Re}\lambda > 0$ yield the conditions $w_a^T(A_c \; A_l \; A_r \; A_u) = 0$ and $w_b^T(B_c \; B_l \; B_r \; B_j) = 0$ for the components w_a, w_b of the eigenvector $w = (w_c, w_l, w_a, w_b, w_r)$, where w_a and w_b cannot vanish simultaneously; the case $w_a \neq 0$ indicates that an I-cutset is present, whereas the condition $w_b \neq 0$ expresses the existence of a V-loop.

From the results of Section 3.5 it follows that the requirements stated in Theorem 6.5 support the hyperbolicity or exponential stability of equilibria in circuits with a well-defined geometric index. Recall, in this regard, the geometric index characterization of branch-oriented circuit models presented in Theorem 6.2.

Example

A simple illustration of the results stated in Theorems 6.4 and 6.5 is provided by the circuit displayed in Figure 6.5, which includes a DC voltage source with voltage V_s, a DC current source injecting a current I_s, as well as two (uncoupled) inductors with inductances L_1, L_2, a capacitor with capacitance C, a linear resistor with conductance G and a diode, denoted by D, with a current-voltage characteristic given by $i_d = \gamma(v_d) = i_0(e^{v_d/v_0} - 1)$. The constants i_0 and v_0 are positive, and so they are L_1, L_2, C and G.

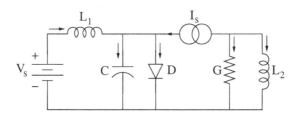

Fig. 6.5   Nonlinear DC circuit.

The circuit has a unique DC operating point, defined by the conditions $v_c = V_s$, $i_{l_1} = \gamma(V_s) - I_s$, $i_{l_2} = -I_s$, $v_d = V_s$, $i_d = \gamma(V_s)$, $v_g = i_g = 0$, $i_u = I_s - \gamma(V_s)$, $v_j = -V_s$ and $i_c = v_{l_1} = v_{l_2} = 0$. We are denoting by v_g and i_g the voltage and the current in the linear resistor, whereas i_u and v_j stand for the current and the voltage in the voltage and current sources, respectively. Mind that the convention on reference directions implies that i_u flows out of the "−" pole in the source. This operating point defines an equilibrium for the circuit dynamics (cf. (6.108) below). For later use, denote by G_d the (positive) incremental conductance $\gamma'(V_s)$ at equilibrium.

Topologically, the absence of VL-loops and IC-cutsets makes the equilibrium nonsingular, according to Theorem 6.4. This rules out null eigenvalues from the matrix pencil spectrum associated with this circuit. Additionally, even though the circuit displays a VCL-loop defined by the voltage source, the capacitor and the inductor L_1, there are no ICL-cutsets and hence, following item (a) of Theorem 6.5, the pencil is hyperbolic, namely, all pencil eigenvalues verify $\operatorname{Re}\lambda \neq 0$. Moreover, all of them satisfy $\operatorname{Re}\lambda < 0$ because of the positiveness of L_1, L_2, C, G and G_d, thereby guaranteeing the asymptotic stability of the equilibrium point.

The reader can easily verify these assertions using the DAE

$$Cv_c' = -\gamma(v_c) + i_{l_1} + I_s \qquad (6.108\text{a})$$
$$L_1 i_{l_1}' = -v_c + V_s \qquad (6.108\text{b})$$
$$L_2 i_{l_2}' = v_{l_2} \qquad (6.108\text{c})$$
$$0 = i_{l_2} + Gv_{l_2} + I_s, \qquad (6.108\text{d})$$

which models the circuit dynamics. Indeed, the eigenvalues of the pencil coming from the linearization of (6.108) at equilibrium are given by

$$\lambda_{1,2} = \frac{-G_d}{2C} \pm \sqrt{\left(\frac{G_d}{2C}\right)^2 - \frac{1}{L_1 C}}, \quad \lambda_3 = \frac{-1}{L_2 G}, \qquad (6.109)$$

and it is straightforward to check that all of them satisfy $\operatorname{Re}\lambda < 0$.

6.3. Qualitative properties

A non-hyperbolic configuration. It is interesting to note that, open-circuiting the diode, we get a circuit displaying a VCL-loop and an ICL-cutset (cf. Figure 6.6). In this situation, neither the pair of conditions in (a) nor those in (b) within Theorem 6.5 are met and, therefore, the hyperbolicity of the equilibrium can no longer be guaranteed. Mind the fact that item (1) in Theorem 6.5 only provides *sufficient* conditions for the hyperbolicity of the pencil. In this case, the eigenvalues are given by the expressions depicted in (6.109) with $G_d = 0$, namely,

$$\lambda_{1,2} = \pm i\sqrt{\frac{1}{L_1 C}}, \quad \lambda_3 = \frac{-1}{L_2 G}, \tag{6.110}$$

so that the pencil is actually non-hyperbolic in this problem. The pair of purely imaginary eigenvalues $\lambda_{1,2}$ characterizes the frequency of the proper oscillations in the linear circuit obtained after removing the diode.

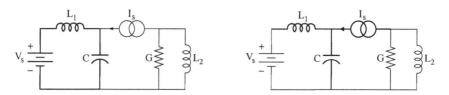

Fig. 6.6 A VCL-loop and an ICL-cutset resulting from the removal of the diode.

Index two. Finally, the fact that these qualitative results are independent of the topological conditions arising in index analyses (cf. Theorem 6.2 on page 279) can be easily illustrated by making $G = 0$ in (6.108). This corresponds to open-circuiting the linear resistor in Figure 6.5, as displayed in Figure 6.7.

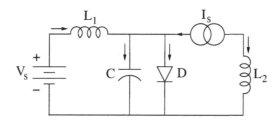

Fig. 6.7 Open-circuiting G yields an index two configuration.

The IL-cutset defined by the current source and the inductor L_2 makes the DAE (6.108) index two. The equilibrium conditions in this circuit do not change with respect to the ones discussed above, but now the dynamical behavior is two-dimensional, the eigenvalues of the linearization being given by the pair $\lambda_{1,2}$ in (6.109). The hyperbolicity of the equilibrium point is now explained by the fact that the unique ICL-cutset is an IL-cutset, so that item (a) in Theorem 6.5 still applies in this index two setting. Additionally, the removal of the diode results in a non-hyperbolic index two configuration, with a pair of eigenvalues again defined by $\lambda_{1,2}$ in (6.110).

Bibliography

[1] R. Abraham, J. E. Marsden and T. Ratiu, *Manifolds, Tensor Analysis, and Applications*, Springer-Verlag, 1988.

[2] G. Alí, A. Bartel, M. Günther and C. Tischendorf, Elliptic partial differential-algebraic multiphysics models in electrical network design, *Math. Models Methods Appl. Sci.* **13** (2003) 1261-1278.

[3] H. Amann, *Ordinary Differential Equations*, Walter de Gruyter, 1990.

[4] B. Andrásfai, *Introductory Graph Theory*, Akadémiai Kiadó, Budapest, 1977.

[5] B. Andrásfai, *Graph Theory: Flows, Matrices*, Adam Hilger, 1991.

[6] V. I. Arnol'd, *Ordinary Differential Equations*, MIT Press, 1973.

[7] V. I. Arnol'd, *Geometrical Methods in the Theory of Ordinary Differential Equations*, Springer-Verlag, 1988.

[8] U. M. Ascher, H. Chin and S. Reich, Stabilization of DAEs and invariant manifolds, *Numer. Math.* **67** (1994) 131-149.

[9] U. M. Ascher and P. Lin, Sequential regularization methods for higher index DAEs with constraint singularities: The linear index-2 case, *SIAM J. Numer. Anal.* **33** (1996) 1921-1940.

[10] U. M. Ascher and L. R. Petzold, *Computer Methods for Ordinary Differential Equations and Differential-Algebraic Equations*, SIAM, 1998.

[11] S. Ayasun, C. O. Nwankpa and H. G. Kwatny, Computation of singular and singularity induced bifurcation points of differential-algebraic power system model, *IEEE Trans. Circ. Sys. I* **51** (2004) 1525-1538.

[12] A. Backes, *Extremalbedingungen für Optimierungsprobleme mit Algebro-Differentialgleichungen*, PhD Thesis, Inst. Math., Humboldt University, Berlin, 2005.

[13] N. Balabanian and T. A. Bickart, *Electrical Network Theory*, John Wiley & Sons, New York, 1969.

[14] K. Balla, G. A. Kurina and R. März, Index criteria for differential algebraic equations arising from linear-quadratic optimal control problems, *J. Dyn. Control Syst.* **12** (2006) 289-311.

[15] K. Balla and V. H. Linh, Adjoint pairs of differential-algebraic equations and Hamiltonian systems, *Appl. Numer. Math.* **53** (2005) 131-148.

[16] K. Balla and R. März, A unified approach to linear differential-algebraic equations and their adjoint equations, *Z. Anal. Anwendungen* **21** (2002) 783-802.

[17] F. Barone, R. Grassini and G. Mendella, A unified approach to constrained mechanical systems as implicit differential equations, *Ann. Inst. Henri Poincar, Phys. Théor.* **70** (1999) 515-546.

[18] T. R. Bashkow, The A matrix, new network description, *IRE Trans. Circuit Theory* **4** (1957) 117-119.

[19] R. E. Beardmore, Stability and bifurcation properties of index-1 DAEs, *Numer. Algorithms* **19** (1998) 43-53.

[20] R. E. Beardmore, Double singularity-induced bifurcation points and singular Hopf bifurcations, *Dynamics and Stability of Systems* **15** (2000) 319-342.

[21] R. E. Beardmore, The singularity-induced bifurcation and its Kronecker normal form, *SIAM J. Matrix Anal. Appl.* **23** (2001) 126-137.

[22] R. E. Beardmore and R. Laister, The flow of a DAE near a singular equilibrium, *SIAM J. Matrix Anal. Appl.* **24** (2002) 106-120.

[23] R. E. Beardmore, R. Laister and A. Peplow, Trajectories of a DAE near a pseudo-equilibrium, *Nonlinearity* **17** (2004) 253-279.

[24] A. Ben-Israel and T. N. E. Greville, *Generalized Inverses: Theory and Applications*, Springer-Verlag, 2003.

[25] M. Bodestedt and C. Tischendorf, PDAE models of integrated circuits and index analysis, *Math. Comput. Model. Dyn. Syst.* **13** (2007) 1-17.

[26] B. Bollobás, *Modern Graph Theory*, Springer-Verlag, 1998.

[27] W. M. Boothby, *An Introduction to Differentiable Manifolds and Riemannian Geometry*, Academic Press, 1986.

[28] R. K. Brayton and J. K. Moser, A theory of nonlinear networks, I, *Quarterly of Applied Mathematics* **22** (1964) 1-33.

[29] R. K. Brayton and J. K. Moser, A theory of nonlinear networks, II, *Quarterly of Applied Mathematics* **22** (1964) 81-104.

[30] K. E. Brenan, S. L. Campbell and L. R. Petzold, *Numerical Solution of Initial-Value Problems in Differential-Algebraic Equations*, SIAM, 1996.

[31] F. Brickell and R. S. Clark, *Differentiable Manifolds. An Introduction*, Van Nostrand Reinhold, 1970.

[32] D. P. Brown, Derivative-explicit differential equations for RLC graphs, *Journal of the Franklin Institute* **275** (1963) 503-514.

[33] P. N. Brown, A. C. Hindmarsh and L. R. Petzold, Consistent initial condition calculation for differential-algebraic systems, *SIAM J. Sci. Comput.* **19** (1998) 1495-1512.

[34] J. W. Bruce, G. J. Fletcher and F. Tari, Bifurcations of implicit differential equations, *Proc. Royal Soc. Edimburgh A* **130** (2000) 485-506.

[35] P. R. Bryant, The order of complexity of electrical networks, *Proceedings of the IEE, Part C* **106** (1959) 174-188.

[36] P. R. Bryant, The explicit form of Bashkow's A matrix, *IRE Trans. Circuit Theory* **9** (1962) 303-306.

[37] P. R. Bryant and J. Tow, The A-matrix of linear passive reciprocal networks, *Journal of the Franklin Institute* **293** (1972) 401-419.

[38] K. Burrage and L. R. Petzold, On order reduction for Runge-Kutta methods applied to differential/algebraic systems and to stiff systems of ODEs, *SIAM J. Numer. Anal.* **27** (1990) 447-456.
[39] S. L. Campbell, *Singular Systems of Differential Equations*, Pitman, 1980.
[40] S. L. Campbell, *Singular Systems of Differential Equations II*, Pitman, 1982.
[41] S. L. Campbell, One canonical form for higher index linear time varying singular systems, *Cir. Sys. Signal Process.* **2** (1983) 311-326.
[42] S. L. Campbell, A general form for solvable linear time varying singular systems of differential equations, *SIAM J. Math. Anal.* **18** (1987) 1101-1115.
[43] S. L. Campbell, Least squares completions for nonlinear differential algebraic equations, *Numer. Math.* **65** (1993) 77-94.
[44] S. L. Campbell, Numerical methods for unstructured higher index DAEs *Ann. Numer. Math.* **1** (1994) 265-277.
[45] S. L. Campbell and C. W. Gear, The index of general nonlinear DAEs, *Numer. Math.* **72** (1995) 173-196.
[46] S. L. Campbell and E. Griepentrog, Solvability of general differential algebraic equations, *SIAM J. Sci. Comput* **16** (1995) 257-270.
[47] S. L. Campbell, R. Hollenbeck, K. Yeomans and Y. Zhong, Mixed symbolic-numeric computations with general DAEs I: System properties, *Numer. Algorithms* **19** (1998) 73-83.
[48] S. L. Campbell and W. Marszalek, DAEs arising from traveling wave solutions of PDEs, *J. Comput. Appl. Math.* **82** (1997) 41-58.
[49] S. L. Campbell and W. Marszalek, Mixed symbolic-numeric computations with general DAEs II: An applications case study, *Numer. Algorithms* **19** (1998) 85-94.
[50] S. L. Campbell and R. März, Direct transcription solution of high index optimal control problems and regular Euler-Lagrange equations, *J. Comput. Appl. Math.* **202** (2007) 186-202.
[51] S. L. Campbell and E. Moore, Constraint preserving integrators for general nonlinear higher index DAEs, *Numer. Math.* **69** (1995) 383-399.
[52] S. L. Campbell, E. Moore and Y. Zhong, Constraint preserving integrators for unstructured higher index DAEs, *Z. Angew. Math. Mech.* **76**, Suppl. 1 (1996) 83-86.
[53] S. L. Campbell, N. K. Nichols and W. J. Terrell, Duality, observability, and controllability for linear time-varying descriptor systems, *Cir. Sys. Signal Process.* **10** (1991) 455-470.
[54] S. L. Campbell and L. R. Petzold, Canonical forms and solvable singular systems of differential equations, *SIAM J. Alg. Disc. Met.* **4** 517-521 (1983).
[55] S. L. Campbell and J. W. Silverstein, A nonlinear system with singular vector field near equilibria, *Applicable Analysis* **12** (1981) 57-71.
[56] H. Cendra and M. Etchechoury, Desingularization of implicit analytic differential equations, *J. Phys. A: Math. Gen.* **39** (2006) 10975-11001.
[57] W. Chen, M. Saif and B. Shafai, Fault diagnosis in a class of differential-algebraic systems, *Proc. 2004 American Control Conf.*, 4398-4402, 2004.

[58] J. Choma Jr, *Electrical Networks. Theory and Analysis*, John Wiley & Sons, New York, 1985.

[59] S. N. Chow and J. K. Hale, *Methods of Bifurcation Theory*, Springer-Verlag, 1982.

[60] L. O. Chua, Dynamic nonlinear networks: state-of-the-art, *IEEE Trans. Circ. Sys.* **27** (1980) 1059-1087.

[61] L. O. Chua and A. D. Deng, Impasse points, I: Numerical aspects, *Internat. J. Circuit Theory Appl.* **17** (1989) 213-235.

[62] L. O. Chua and A. D. Deng, Impasse points, II: Analytical aspects, *Internat. J. Circuit Theory Appl.* **17** (1989) 271-282.

[63] L. O. Chua, C. A. Desoer and E. S. Kuh, *Linear and Nonlinear Circuits*, McGraw-Hill, 1987.

[64] L. O. Chua and D. N. Green, A qualitative analysis of the behavior of dynamic nonlinear networks: Stability of autonomous networks, *IEEE Trans. Circ. Sys.* **23** (1976) 355-379.

[65] L. O. Chua and D. N. Green, Graph-theoretic properties of dynamic nonlinear networks, *IEEE Trans. Circ. Sys.* **23** (1976) 292-312.

[66] L. O. Chua and P. M. Lin, *Computer-Aided Analysis of Electronic Circuits: Algorithms and Computational Techniques*, Prentice-Hall, 1975.

[67] L. O. Chua and H. Oka, Normal forms for constrained nonlinear differential equations, Part I: Theory, *IEEE Trans. Circ. Sys.* **35** (1988) 881-901.

[68] L. O. Chua and H. Oka, Normal forms for constrained nonlinear differential equations, Part II: Bifurcation, *IEEE Trans. Circ. Sys.* **36** (1989) 71-88.

[69] L. O. Chua, Ch. W. Wu, A. Huang, and G.-Q. Zhong, A universal circuit for studying and generating chaos. I: Routes to chaos, *IEEE Trans. Circ. Sys. I* **40** (1993) 732-744.

[70] K. D. Clark, A structural form for higher-index semistate equations. I. Theory and applications to circuit and control theory, *Lin. Alg. Appl.* **98** (1988) 169-197.

[71] E. A. Coddington and R. Carlson, *Linear Ordinary Differential Equations*, SIAM, 1997.

[72] A. A. Davydov, Normal form of a differential equation, not solvable for the derivative, in a neighborhood of a singular point, *Funct. Anal. Appl.* **19** (1985) 81-89; *Funktsional'nyi Analiz i Ego Prilozheniya* **19** (1985) 1-10.

[73] M. Delgado-Téllez and A. Ibort, On the geometry and topology of singular optimal control problems and their solutions, *Discrete Contin. Dyn. Syst.* 2003, Suppl. Vol., 223-233 (2003).

[74] A. Demir, Floquet theory and non-linear perturbation analysis for oscillators with differential-algebraic equations, *Internat. J. Circuit Theory Appl.* **28** (2000) 163-185.

[75] A. Dervisoglu, Bashkow's A-matrix for active RLC networks, *IEEE Trans. Circuit Theory* **11** (1964) 404-406.

[76] A. Dervisoglu, The realization of the A-matrix of a certain class of RLC networks, *IEEE Trans. Circuit Theory* **13** (1966) 164-170.

[77] Ch. A. Desoer and F. F. Wu, Trajectories of nonlinear RLC networks: A geometric approach, *IEEE Trans. Circuit Theory* **19** (1972) 562-571.

[78] J. Dieudonné, *Foundations of Modern Analysis*, Academic Press, 1969.
[79] P. A. M. Dirac, Generalized Hamiltonian dynamics, *Can. J. Math.* **2** (1950) 129-148.
[80] P. A. M. Dirac, Generalized Hamiltonian dynamics, *Proc. Royal Soc. London A* **246** (1958) 326-332.
[81] P. A. M. Dirac, *Lectures on Quantum Mechanics*, Yeshiva University, 1964; Dover, 2001.
[82] B. Dziurla and R. Newcomb, The Drazin inverse and semi-state equations, *Proc. Intl. Symp. Math. Theory of Networks and Systems*, 283-289, 1979.
[83] E. Eich-Soellner and C. Führer, *Numerical Methods in Multibody Dynamics*, B. G. Teubner, 1998.
[84] A. Encinas and R. Riaza, Tree-based characterization of low index circuit configurations without passivity restrictions, *Internat. J. Circuit Theory Appl.* **36** (2008) 135-160.
[85] D. Estévez-Schwarz, A step-by-step approach to compute a consistent initialization for the MNA, *Internat. J. Circuit Theory Appl.* **30** (2002) 1-16.
[86] D. Estévez Schwarz and R. Lamour, The computation of consistent initial values for nonlinear index-2 differential-algebraic equations, *Numer. Algorithms* **26** (2001) 49-75.
[87] D. Estévez-Schwarz and C. Tischendorf, Structural analysis of electric circuits and consequences for MNA, *Internat. J. Circuit Theory Appl.* **28** (2000) 131-162.
[88] M. Fosséprez, *Non-Linear Circuits: Qualitative Analysis of Non-Linear, Non-Reciprocal Circuits,* John Wiley & Sons, 1992.
[89] L. R. Foulds, *Graph Theory Applications*, Springer, 1992.
[90] F. R. Gantmacher, *The Theory of Matrices*, Chelsea, 1959.
[91] F. R. Gantmacher, *The Theory of Matrices*, Vol. 2, Chelsea, 1959.
[92] B. García-Celayeta, *Stability for Differential-Algebraic Equations*, PhD Thesis, Universidad Pública de Navarra, 1998.
[93] C. W. Gear, The simultaneous numerical solution of differential-algebraic equations, *IEEE Trans. Circuit Theory* **18** (1971) 89-95.
[94] C. W. Gear, Differential-algebraic equation index transformations, *SIAM J. Sci. Stat. Comput.* **9** (1988) 39-47.
[95] C. W. Gear and L. R. Petzold, Differential/algebraic systems and matrix pencils, in B. Kagstrom and A. Ruhe (eds.), *Matrix Pencils*, Lect. Notes Maths., Vol. 973, 75-89, Springer-Verlag, 1983.
[96] C. W. Gear and L. R. Petzold, ODE methods for the solution of differential/algebraic systems, *SIAM J. Numer. Anal.* **21** (1984) 716-728.
[97] M. Gerdts and C. Büskens, Computation of consistent initial values for optimal control problems with DAE systems of higher index, *Z. Angew. Math. Mech.* **81** Suppl. 2 (2001) 249-250.
[98] L. B. Goldgeisser and M. M. Green, On the topology and number of operating points of MOSFET circuits, *IEEE Trans. Circ. Sys. I* **48** (2001) 218-221.
[99] M. Golubitsky and V. Guillemin, *Stable Mappings and Their Singularities*, Springer-Verlag, 1973.

[100] M. Golubitsky and D. G. Schaeffer, *Singularities and Groups in Bifurcation Theory I*, Springer-Verlag, 1985.
[101] X. Gràcia and J. M. Pons, Constrained systems: A unified geometric approach, *Intl. J. Theor. Phys.* **30** (1991) 511-516.
[102] X. Gràcia and J. M. Pons, A generalized geometric framework for constrained systems, *Diff. Geom. Appl.* **2** (1992) 223-247.
[103] M. M. Green and A. N. Willson Jr, How to identify unstable dc operating points, *IEEE Trans. Circ. Sys. I* **39** (1992) 820-832.
[104] M. M. Green and A. N. Willson Jr, (Almost) half on any circuit's operating points are unstable, *IEEE Trans. Circ. Sys. I* **41** (1994) 286-293.
[105] M. M. Green and A. N. Willson Jr, An algorithm for identifying unstable operating points using SPICE, *IEEE Trans. Comput. Aid. Des. Cir. Sys.* **14** (1995) 360-370.
[106] E. Griepentrog, Index reduction methods for differential-algebraic equations, Seminar Report 92-1, Inst. Math., Humboldt University, 14-29, 1992.
[107] E. Griepentrog and R. März, *Differential-Algebraic Equations and Their Numerical Treatment*, Teubner, Leipzig, 1986.
[108] E. Griepentrog and R. März, Basic properties of some differential-algebraic equations, *Z. Anal. Anwendungen* **8** (1989) 25-40.
[109] D. M. Gritsis, C. C. Pantelides and R. W. H. Sargent, Optimal control of systems described by index two differential-algebraic equations, *SIAM J. Sci. Comput.* **16** (1995) 1349-1366.
[110] J. Guckenheimer and P. Holmes, *Nonlinear Oscillations, Dynamical Systems and Bifurcations of Vector Fields,* Springer-Verlag, 1983.
[111] M. Günther, A PDAE model for interconnected linear RLC networks, *Math. Comp. Model. Dyn. Sys.* **7** (2001) 189-203.
[112] M. Günther and U. Feldmann, The DAE-index in electric circuit simulation, *Math. Comp. Simul.* **39** (1995) 573-582.
[113] M. Günther and U. Feldmann, CAD-based electric-circuit modeling in industry. I: Mathematical structure and index of network equations, *Surv. Math. Ind.* **8** (1999) 97-129.
[114] M. Günther and U. Feldmann, CAD-based electric-circuit modeling in industry. II: Impact of circuit configurations and parameters, *Surv. Math. Ind.* **8** (1999) 131-157.
[115] M. Günther and P. Rentrop, The differential-algebraic index concept in electric circuit simulation, *Z. Angew. Math. Mech.* **76** S. 1 (1996) 91-94.
[116] M. Günther and P. Rentrop, Numerical simulation of electrical circuits, *Mitt. Ges. Angew. Math. Mech* **23** (2000) 51-77.
[117] G. D. Hachtel, R. K. Brayton and F. G. Gustafson, The sparse tableau approach to network analysis and design, *IEEE Trans. Circ. Sys.* **18** (1971) 101-113.
[118] B. C. Haggman and P. R. Bryant, Solutions of singular constrained differential equations: A generalization of circuits containing capacitor-only loops and inductor-only cutsets, *IEEE Trans. Circ. Sys.* **31** (1984) 1015-1029.
[119] B. C. Haggman and P. R. Bryant, Geometric properties of nonlinear networks containing capacitor-only cutsets and/or inductor-only loops. Part

I: Conservation laws, *Cir. Sys. Signal Process.* **5** (1986) 279-319.
[120] B. C. Haggman and P. R. Bryant, Geometric properties of nonlinear networks containing capacitor-only cutsets and/or inductor-only loops. Part II: Symmetries, *Cir. Sys. Signal Process.* **5** (1986) 435-448.
[121] E. Hairer, C. Lubich and M. Roche, *The Numerical Solution of Differential-Algebraic Systems by Runge-Kutta Methods*, Lect. Notes Maths., Vol. 1409, Springer-Verlag, 1989.
[122] E. Hairer and G. Wanner, *Solving Ordinary Differential Equations II: Stiff and Differential-Algebraic Problems*, Springer-Verlag, 1996.
[123] J. K. Hale, *Ordinary Differential Equations*, 2nd ed., Robert E. Krieger Publishing Co., 1980.
[124] M. Hanke and R. Lamour, Consistent initialization for nonlinear index-2 differential–algebraic equation: Large sparse systems in MATLAB, *Numer. Algorithms* **32** (2003) 67-85.
[125] B. Hansen, Linear time-varying differential-algebraic equations being tractable with the index k, Preprint 246, Inst. Math., Humboldt University, Berlin, 1990.
[126] B. Hansen, Computing consistent initial values for nonlinear index-2 differential- algebraic equations, Seminar Report 92-1, Humboldt University, Berlin, 142-157, 1992.
[127] P. Hartman, *Ordinary Differential Equations*, SIAM, 2002.
[128] M. Hasler and J. Neirynck, *Nonlinear Circuits*, Artech House, 1986.
[129] I. Higueras and B. García-Celayeta, Runge-Kutta methods for DAEs. A new approach, *J. Comput. Appl. Math.* **111** (1999) 49-61.
[130] I. Higueras and R. März, Differential algebraic equations with properly stated leading terms, *Comp. Math. Appl.* **48** (2004) 215-235.
[131] I. Higueras, R. März and C. Tischendorf, Stability preserving integration of index-1 DAEs, *Appl. Numer. Math.* **45** (2003) 175-200.
[132] I. Higueras, R. März and C. Tischendorf, Stability preserving integration of index-2 DAEs, *Appl. Numer. Math.* **45** (2003) 201-229.
[133] I. Higueras and T. Roldán, Starting algorithms for a class of RK methods for index-2 DAEs, *Comp. Math. Appl.* **49** (2005) 1081-1099.
[134] D. J. Hill and I. M. Y. Mareels, Stability theory for differential/algebraic systems with application to power systems, *IEEE Trans. Circ. Sys.* **37** (1990) 1416-1423.
[135] C. W. Ho, A. E. Ruehli and P. A. Brennan, The modified nodal approach to network analysis, *IEEE Trans. Circ. Sys.* **22** (1975) 504-509.
[136] A. M. Hodge and R. W. Newcomb, Semistate theory and analog VLSI design, *IEEE Circuits and Systems Magazine* **2** (2002) 30-51.
[137] R. A. Horn and Ch. R. Johnson, *Matrix Analysis*, Cambridge University Press, 1985.
[138] M. Hou, Fault detection and isolation for descriptor systems, in R. J. Patton, P. M. Frank and R. N. Clark (eds.), *Issues of Fault Diagnosis for Dynamic Systems*, pp. 115-144, Springer, 2000.
[139] A. Ilchmann and V. Mehrmann, A behavioral approach to time-varying linear systems. II: Descriptor systems, *SIAM J. Control Optimization* **44**

(2005) 1748-1765.
[140] L. O. Jay, Symplectic partitioned Runge-Kutta methods for constrained Hamiltonian systems, *SIAM J. Numer. Anal.* **33** (1996) 368-387.
[141] L. O. Jay, Structure preservation for constrained dynamics with super partitioned additive Runge-Kutta methods, *SIAM J. Sci. Comput.* **20** (1999) 416-446.
[142] L. V. Kalachev and R. E. O'Malley Jr, The regularization of linear differential-algebraic equations, *SIAM J. Math. Anal.* **27** (1996) 258-273.
[143] M. Knorrenschild, Differential/algebraic equations as stiff ordinary differential equations, *SIAM J. Numer. Anal.* **29** (1992) 1694-1715.
[144] L. Kronecker, Algebraische Reduction der Schaaren bilinearer Formen, *Sitzungsberichte Akad. Wiss. Berlin* 1890, 1225-1237; *Leopold Kronecker's Werke*, Chelsea, 1968, 139-155.
[145] E. S. Kuh and R. A. Rohrer, The state-variable approach to network analysis, *Proceedings of the IEEE* **53** (1965) 672-686.
[146] A. Kumar and P. Daoutidis, *Control of Nonlinear Differential Algebraic Equation Systems with Applications to Chemical Processes,* Chapman and Hall, 1999.
[147] P. Kunkel and V. Mehrmann, Canonical forms for linear differential-algebraic equations with variable coefficients, *J. Comput. Appl. Math.* **56** (1994) 225-251.
[148] P. Kunkel and V. Mehrmann, Local and global invariants of linear differential-algebraic equations and their relation, *E. Trans. Numerical Analysis* **4** (1996) 138-157.
[149] P. Kunkel and V. Mehrmann, Regular solutions of nonlinear differential-algebraic equations and their numerical determination, *Numer. Math.* **79** (1998) 581-600.
[150] P. Kunkel and V. Mehrmann, Analysis of over- and underdetermined nonlinear differential-algebraic systems with application to nonlinear control problems, *Math. Control Signals Syst.* **14** (2001) 233-256.
[151] P. Kunkel and V. Mehrmann, *Differential-Algebraic Equations. Analysis and Numerical Solution*, EMS, 2006.
[152] P. Kunkel, V. Mehrmann and W. Rath, Analysis and numerical solution of control problems in descriptor form, *Math. Control Signals Syst.* **14** (2001) 29-61.
[153] G. A. Kurina, Linear-quadratic discrete optimal control problems for descriptor systems in Hilbert space, *J. Dyn. Control Syst.* **10** (2004) 365-375.
[154] G. A. Kurina and R. März, On linear-quadratic optimal control problems for time-varying descriptor systems, *SIAM J. Control Optimization* **42** (2004) 2062-2077.
[155] J. C. Lagarias and L. Trajkovic, Bounds for the number of DC operating points of transistor circuits, *IEEE Trans. Circ. Sys. I* **46** (1999) 1216-1221.
[156] R. Lamour, Index determination and calculation of consistent initial values for DAEs, *Comp. Math. Appl.* **50** (2005) 1125-1140.
[157] R. Lamour, R. März and C. Tischendorf, *Projector Based DAE Analysis*, in preparation, 2008.

[158] R. Lamour, R. März and R. Winkler, How Floquet theory applies to index 1 differential algebraic equations, *J. Math. Anal. Appl.* **217** (1998) 372-394.
[159] R. Lamour, R. März and R. Winkler, Stability of periodic solutions of index-2 differential algebraic equations, *J. Math. Anal. Appl.* **279** (2003) 475-494.
[160] J. Lee and H. D. Chiang, Constructive homotopy methods for finding all or multiple DC operating points of nonlinear circuits and systems, *IEEE Trans. Circ. Sys. I* **48** (2001) 35-50.
[161] J. Lee and H. D. Chiang, Convergent regions of the Newton homotopy method for nonlinear systems: Theory and computational applications, *IEEE Trans. Circ. Sys. I* **48** (2001) 51-66.
[162] B. Leimkuhler, L. R. Petzold and C. W. Gear, Approximation methods for the consistent initialization of differential-algebraic equations, *SIAM J. Numer. Anal.* **28** (1991) 205-226.
[163] B. Leimkuhler and S. Reich, *Simulating Hamiltonian Dynamics*, Cambridge University Press, 2004.
[164] F. Li and P. Y. Woo, A new method for establishing state equations: the branch replacement and augmented node-voltage equation approach, *Cir. Sys. Signal Process.* **21** (2002) 149-161.
[165] J. Llibre and J. Sotomayor, Structural stability of constrained polynomial systems, *Bull. Lond. Math. Soc.* **30** (1998) 589-595.
[166] J. Llibre, J. Sotomayor and M. Zhitomirskii, Impasse bifurcations of constrained systems, in A. Galves et al. (eds.), *Differential Equations and Dynamical Systems*, Fields Inst. Commun. 31 (2002) 235-255.
[167] P. Lötstedt and L. Petzold, Numerical solution of nonlinear differential equations with algebraic constraints. I: Convergence results for backward differentiation formulas, *Math. Comput.* **46** (1986) 491-516.
[168] D. G. Luenberger, Dynamic equations in descriptor form, *IEEE Trans. Aut. Control* **22** (1977) 312-321.
[169] D. G. Luenberger, Time-invariant descriptor systems, *Automatica* **14** (1978) 473-485.
[170] G. Marmo, G. Mendella and W. M. Tulczyjew, Symmetries and constants of the motion for dynamics in implicit form, *Ann. Inst. Henri Poincar, Phys. Théor.* **57** (1992) 147-166.
[171] G. Marmo, G. Mendella and W. M. Tulczyjew, Constrained Hamiltonian systems as implicit differential equations, *J. Phys. A: Math. Gen.* **30** (1997) 277-293.
[172] W. Marszalek, *Analysis of Partial Differential Algebraic Equations*, PhD Thesis, North Carolina State University, 1997.
[173] W. Marszalek, T. Amdeberhan and R. Riaza, Singularity crossing phenomena in DAEs: a two-phase fluid flow application case study, *Comp. Math. Appl.* **49** (2005) 303-319.
[174] W. Marszalek and S. L. Campbell, DAEs arising from traveling wave solutions of PDEs II, *Comp. Math. Appl.* **37** (1999) 15-34.
[175] W. Marszalek and Z. W. Trzaska, Singularity-induced bifurcations in electrical power systems, *IEEE Trans. Power Systems* **20** (2005) 312-320.

[176] W. S. Martinson and P. I. Barton, Index and characteristic analysis of linear PDAE systems, *SIAM J. Sci. Comput.* **24** (2002) 905-923.

[177] R. März, Multistep methods for initial value problems in implicit differential-algebraic equations, *Beiträge Numer. Math.* **12** (1984) 107-123.

[178] R. März, On difference and shooting methods for boundary value problems in differential-algebraic equations, *Z. Ang. Math. Mech.* **64** (1984) 463-473.

[179] R. März, A matrix chain for analyzing differential algebraic equations, Preprint 162, Inst. Math., Humboldt University, Berlin, 1987.

[180] R. März, Index-2 differential-algebraic equations, *Results in Mathematics* **15** (1989) 149-171.

[181] R. März, Some new results concerning index-3 differential-algebraic equations, *J. Math. Anal. Appl* **140** (1989) 177-199.

[182] R. März, Numerical methods for differential algebraic equations, *Acta Numerica 1992*, 141-198 (1992).

[183] R. März, On quasilinear index 2 differential-algebraic equations, Seminar Report 92-1, Inst. Math., Humboldt University, Berlin, 39-60, 1992.

[184] R. März, Practical Lyapunov stability criteria for differential algebraic equations, *Banach Center Publications* **29** (1994) 245-266.

[185] R. März, Canonical projectors for linear differential algebraic equations, *Comp. Math. Appl.* **31** (1996) 121-135.

[186] R. März, Criteria for the trivial solution of differential algebraic equations with small nonlinearities to be asymptotically stable, *J. Math. Anal. Appl.* **225** (1998) 587-607.

[187] R. März, Adjoint equations of differential-algebraic systems and optimal control problems, *Proc. Inst. Math., NAS Belarus*, **7** (2001) 88-97.

[188] R. März, Differential algebraic equations anew, *Appl. Numer. Math.* **42** (2002) 315-335.

[189] R. März, The index of linear differential algebraic equations with properly stated leading terms, *Results in Mathematics* **42** (2002) 308-338.

[190] R. März, Differential algebraic systems with properly stated leading term and MNA equations, in *Modeling, Simulation, and Optimization of Integrated Circuits* (Oberwolfach, 2001), *Int. Ser. Numer. Math.* **146** (2003) 135-151.

[191] R. März, Solvability of linear differential algebraic equations with properly stated leading terms, *Results in Mathematics* **45** (2004) 88-105.

[192] R. März, Fine decouplings of regular differential algebraic equations, *Results in Mathematics* **46** (2004) 57-72.

[193] R. März, Characterizing differential algebraic equations without the use of derivative arrays, *Comp. Math. Appl.* **50** (2005) 1141-1156.

[194] R. März, D. Estévez-Schwarz, U. Feldmann, S. Sturtzel and C. Tischendorf, Finding beneficial DAE structures in circuit simulation, in W. Jäger et al. (eds.), *Mathematics - Key Technology for the Future,* Springer, pp. 413-428, 2003.

[195] R. März and R. Riaza, Linear differential-algebraic equations with properly stated leading term: Regular points, *J. Math. Anal. Appl.* **323** (2006) 1279-1299.

[196] R. März and R. Riaza, Linear differential-algebraic equations with properly stated leading term: A-critical points, Math. Comp. Model. Dyn. Sys. 13 (2007) 291-314.
[197] R. März and R. Riaza, Linear differential-algebraic equations with properly stated leading term: B-critical points, Preprint 07-9, Inst. Math., Humboldt University, Berlin, 2007.
[198] R. März and C. Tischendorf, Recent results in solving index-2 differential-algebraic equations in circuit simulation, SIAM J. Sci. Comput. 18 (1997) 139-159.
[199] R. März and E. B. Weinmüller, Solvability of boundary value problems for systems of singular differential-algebraic equations, SIAM J. Math. Anal. 24 (1993) 200-215.
[200] J. N. Mather, Stability of C^∞ mappings. III: Finitely determined map germs, Publ. Math., Inst. Hautes Etud. Sci. 35 (1968) 127-156.
[201] T. Matsumoto, L. O. Chua and A. Makino, On the implications of capacitor-only cutsets and inductor-only loops in nonlinear networks, IEEE Trans. Circ. Sys. 26 (1979) 828-845.
[202] M. Medved, Normal forms of implicit and observed implicit differential equations, Riv. Mat. Pura ed Appl. 10 (1991) 95-107.
[203] M. Medved, Qualitative properties of generalized vector fields, Riv. Mat. Pura ed Appl. 15 (1994) 7-31.
[204] A. I. Mees and L. O. Chua, The Hopf bifurcation theorem and its applications to nonlinear oscillations in circuits and systems, IEEE Trans. Circ. Sys. 26 (1979) 235-254.
[205] G. Mendella, G. Marmo and W. M. Tulczyjew, Integrability of implicit differential equations, J. Phys. A: Math. Gen. 28 (1995) 149-163.
[206] M. M. Milić and P. M. Simic, General linear nonstationary passive systems: Derivation of the explicit state model, Cir. Sys. Signal Process. 6 (1987) 299-314.
[207] P. K. Moore, L. R. Petzold and Y. Ren, Regularization of index-1 differential-algebraic equations with rank-deficient constraints, Comp. Math. Appl. 35 (1998) 43-61.
[208] M. C. Muñoz-Lecanda and N. Román-Roy, Implicit quasilinear differential systems: A geometrical approach, Electron. J. Differ. Equ. 1999 (1999) No. 10, 1-33.
[209] S. Natarajan, A systematic method for obtaining state equations using MNA, IEE Proceedings-G 138 (1991) 341-346.
[210] R. W. Newcomb, The semistate description of nonlinear time-variable circuits, IEEE Trans. Circ. Sys. 28 (1981) 62-71.
[211] R. W. Newcomb and B. Dziurla, Some circuits and systems applications of semistate theory, Cir. Sys. Signal Process. 8 (1989) 235-260.
[212] R. Nikoukhah, A new methodology for observer design and implementation, IEEE Trans. Aut. Control 43 (1998) 229-234.
[213] T. Ohtsuki and H. Watanabe, State-variable analysis of RLC networks containing nonlinear coupling elements, IEEE Trans. Circuit Theory 16 (1969) 26-38.

[214] R. E. O'Malley Jr. and L. V. Kalachev, Regularization of nonlinear differential-algebraic equations, *SIAM J. Math. Anal.* **25** (1994) 615-629.

[215] C. C. Pantelides, The consistent initialization of differential algebraic systems, *SIAM J. Sci. Stat. Comput.* **9** (1988) 213-231.

[216] L. R. Petzold, Differential-algebraic equations are not ODEs, *SIAM J. Sci. Stat. Comput.* **3** (1982) 367-384.

[217] L. R. Petzold, A description of DASSL: A differential/algebraic system solver, in R.S. Stepleman *et al.* (eds), *Scientific Computing*, pp. 65-68, North-Holland, 1983.

[218] L. R. Petzold, Y. Ren and T. Maly, Regularization of higher-index differential-algebraic equations with rank-deficient constraints, *SIAM J. Sci. Comput.* **18** (1997) 753-774.

[219] P. J. Rabier, Implicit differential equations near a singular point, *J. Math. Anal. Appl.* **144** (1989) 425-449.

[220] P. J. Rabier and W. C. Rheinboldt, A general existence and uniqueness theory for implicit differential-algebraic equations, *Differential and Integral Equations* **4** (1991) 563-582.

[221] P. J. Rabier and W. C. Rheinboldt, A geometric treatment of implicit differential-algebraic equations, *J. Diff. Equations* **109** (1994) 110-146.

[222] P. J. Rabier and W. C. Rheinboldt, On impasse points of quasi-linear differential-algebraic equations, *J. Math. Anal. Appl.* **181** (1994) 429-454.

[223] P. J. Rabier and W. C. Rheinboldt, On the computation of impasse points of quasi-linear differential-algebraic equations, *Math. Comp.* **62** (1994) 133-154.

[224] P. J. Rabier and W. C. Rheinboldt, Finite difference methods for time-dependent, linear differential-algebraic equations, *Appl. Math. Letters* **7** (1994) 29-34.

[225] P. J. Rabier and W. C. Rheinboldt, Classical and generalized solutions of time-dependent linear differential-algebraic equations, *Lin. Alg. Appl.* **245** (1996) 259-293.

[226] P. J. Rabier and W. C. Rheinboldt, Time-dependent linear DAEs with discontinuous inputs, *Lin. Alg. Appl.* **247** (1996) 1-29.

[227] P. J. Rabier and W. C. Rheinboldt, *Nonholonomic Motion of Rigid Mechanical Systems from a DAE Viewpoint*, SIAM, 2000.

[228] P. J. Rabier and W. C. Rheinboldt, Theoretical and numerical analysis of differential-algebraic equations, in P. G. Ciarlet *et al.* (eds.), *Handbook of Numerical Analysis*, Vol. VIII, pp. 183-540, North Holland/Elsevier, 2002.

[229] S. Reich, On a geometrical interpretation of differential-algebraic equations, *Cir. Sys. Signal Process.* **9** (1990) 367-382.

[230] S. Reich, On an existence and uniqueness theory for nonlinear differential-algebraic equations, *Cir. Sys. Signal Process.* **10** (1991) 343-359.

[231] S. Reich, On the local qualitative behavior of differential-algebraic equations, *Cir. Sys. Signal Process.* **14** (1995) 427-443.

[232] T. Reis and C. Tischendorf, Frequency domain methods and decoupling of linear infinite dimensional differential algebraic systems, *J. Evol. Equ.* **5** (2005) 357-385.

[233] G. Reißig, Differential-algebraic equations and impasse points, *IEEE Trans. Circ. Sys. I* **43** (1996) 122-133.
[234] G. Reißig, The index of the standard circuit equations of passive RLCTG-networks does not exceed 2, *Proc. ISCAS'98*, Vol. 3, 419-422, 1998.
[235] G. Reißig, Extension of the normal tree method, *Internat. J. Circuit Theory Appl.* **27** (1999) 241-265.
[236] G. Reißig and H. Boche, On singularities of autonomous implicit ordinary differential equations, *IEEE Trans. Circ. Sys. I* **50** (2003) 922-931.
[237] G. Reißig, W. S. Martinson and P. I. Barton, Differential-algebraic equations of index 1 may have an arbitrarily high structural index, *SIAM J. Sci. Comput.* **21** (2000) 1987-1990.
[238] W. C. Rheinboldt, Differential-algebraic systems as differential equations on manifolds, *Math. Comput.* **43** (1984) 473-482.
[239] W. C. Rheinboldt, On the existence and uniqueness of solutions of nonlinear semi-implicit differential-algebraic equations, *Nonlinear Analysis TMA* **16** (1991) 647-661.
[240] R. Riaza, Stability issues in regular and non-critical singular DAEs, *Acta Appl. Math.* **73** (2002) 301-336.
[241] R. Riaza, On the Singularity-Induced Bifurcation Theorem, *IEEE Trans. Aut. Control* **47** (2002) 1520-1523.
[242] R. Riaza, Singular bifurcations in higher index differential-algebraic equations, *Dynamical Systems* **17** (2002) 243-261.
[243] R. Riaza, Double SIB points in differential-algebraic systems, *IEEE Trans. Aut. Control* **48** (2003) 1625-1629.
[244] R. Riaza, A matrix pencil approach to the local stability analysis of nonlinear circuits, *Internat. J. Circuit Theory Appl.* **32** (2004) 23-46.
[245] R. Riaza, Attraction domains of degenerate singular equilibria in quasilinear ODEs, *SIAM J. Math. Anal.* **36** (2004) 678-690.
[246] R. Riaza, Singularity-induced bifurcations in lumped circuits, *IEEE Trans. Circ. Sys. I* **52** (2005) 1442-1450.
[247] R. Riaza, On the local classification of smooth maps induced by Newton's method, *J. Diff. Equations* **217** (2005) 377-392.
[248] R. Riaza, Time-domain properties of reactive dual circuits, *Internat. J. Circuit Theory Appl.* **34** (2006) 317-340.
[249] R. Riaza, S. L. Campbell and W. Marszalek, On singular equilibria of index-1 DAEs, *Cir. Sys. Signal Process.* **19** (2000) 131-157.
[250] R. Riaza and R. März, Linear index-1 DAEs: regular and singular problems, *Acta Appl. Math.* **84** (2004) 29-53.
[251] R. Riaza and R. März, A simpler construction of the matrix chain defining the tractability index of linear DAEs, *Appl. Math. Letters* **21** (2008) 326-331.
[252] R. Riaza and C. Tischendorf, Qualitative features of matrix pencils and DAEs arising in circuit dynamics, *Dynamical Systems* **22** (2007) 107-131.
[253] R. Riaza and J. Torres-Ramírez, Nonlinear circuit modeling via nodal methods, *Internat. J. Circuit Theory Appl.* **33** (2005) 281-305.
[254] R. Riaza and P. J. Zufiria, Stability of singular equilibria in quasilinear

implicit differential equations, *J. Diff. Equations* **171** (2001) 24-53.
[255] R. Riaza and P. J. Zufiria, Discretization of implicit ODEs for singular root-finding problems, *J. Comput. Appl. Math.* **140** (2002) 695-712.
[256] R. Riaza and P. J. Zufiria, Differential-algebraic equations and singular perturbation methods in recurrent neural learning, *Dynamical Systems* **18** (2003) 89-105.
[257] A. Sarmiento-Reyes, A novel method to predict both, the upper bound on the number and the stability of DC operating points of transistor circuits, *Proc. ISCAS'95*, Vol. 1, pp. 101-104, 1995.
[258] S. S. Sastry and C. A. Desoer, Jump behavior of circuits and systems, *IEEE Trans. Circ. Sys.* **28** (1981) 1109-1124.
[259] O. Schein, *Stochastic Differential-Algebraic Equations in Circuit Simulation*, Shaker Verlag, 1999.
[260] J. Schropp, Ordinary differential equations, differential algebraic equations and their use in optimization, *Proc. EQUADIFF'99*, Vol. 2, pp. 928-933, World Scientific, 2000.
[261] W. M. Seiler, Numerical analysis of constrained Hamiltonian systems and the formal theory of differential equations, *Math. Comp. Simul.* **45** (1998) 561-576.
[262] W. M. Seiler, Numerical integration of constrained Hamiltonian systems using Dirac brackets, *Math. Comp.* **68** (1999) 661-681.
[263] W. M. Seiler, Involution and constrained dynamics I: the Dirac approach, *J. Phys. A: Math. Gen.* **28** (1995) 4431-4451.
[264] M. Selva Soto and C. Tischendorf, Numerical analysis of DAEs from coupled circuit and semiconductor simulation, *Appl. Numer. Math.* **53** (2005) 471-488.
[265] L. P. Shilnikov, A. L. Shilnikov, D. V. Turaev and L. O. Chua, *Methods of Qualitative Theory in Nonlinear Dynamics I*, World Scientific, 1998.
[266] L. P. Shilnikov, A. L. Shilnikov, D. V. Turaev and L. O. Chua, *Methods of Qualitative Theory in Nonlinear Dynamics II*, World Scientific, 2001.
[267] B. Simeon, MBSPACK – numerical integration software for constrained mechanical motion, *Surv. Math. Ind.* **5** (1995) 169-202.
[268] S. Smale, On the mathematical foundations of electrical circuit theory, *J. Diff. Geometry* **7** (1972) 193-210.
[269] H. C. So, On the hybrid description of a linear n-port resulting from the extraction of arbitrarily specified elements, *IEEE Trans. Circuit Theory* **12** (1965) 381-387.
[270] G. Söderlind and L. Wang, Evaluating numerical ODE/DAE methods, algorithms and software, *J. Comput. Appl. Math.* **185** (2006) 244-260.
[271] A. M. Sommariva, On a specific substitution theorem, *Internat. J. Circuit Theory Appl.* **26** (1998) 509-512.
[272] A. M. Sommariva, On a specific substitution theorem: Further results, *Internat. J. Circuit Theory Appl.* **27** (1999) 277-281.
[273] A. M. Sommariva, State-space equations of regular and strictly topologically degenerate linear lumped time-invariant networks: the multiport method, *Internat. J. Circuit Theory Appl.* **29** (2001) 435-453.

[274] A. M. Sommariva, State-space equations of regular and strictly topologically degenerate linear lumped time-invariant networks: the implicit tree-tableau method, *IEEE Proc. Circ. Sys.* **8** (2001) 1139-1141.
[275] H. von Sosen, *Folds and Bifurcations in the Solutions of Semi-Explicit Differential-Algebraic Equations,* PhD. Thesis, Part I. California Institute of Technology, 1994.
[276] J. Sotomayor, Structurally stable differential systems of the form $A(x)x' = F(x)$, *Diff. Eqs. Dyn. Syst.* **5** (1997) 415-422.
[277] J. Sotomayor and M. Zhitomirskii, Impasse singularities of differential systems of the form $A(x)x' = F(x)$, *J. Diff. Equations* **169** (2001) 567-587.
[278] R. J. Spiteri, U. M. Ascher and D. K. Pai, Numerical solution of differential systems with algebraic inequalities arising in robot programming, *Proc. IEEE Intl. Conf. Robotics and Automation,* Vol. 3, 2373-2380, 1995.
[279] R. J. Spiteri, D. K. Pai and U. M. Ascher, Programming and control of robots by means of differential algebraic inequalities, *IEEE Trans. Robotics and Automation* **16** (2000) 135-145.
[280] T. E. Stern, On the equations of nonlinear networks, *IEEE Trans. Circuit Theory* **13** (1966) 74-81.
[281] T. Stykel, On criteria for asymptotic stability of differential-algebraic equations, *Z. Angew. Math. Mech.* **82** (2002) 147-158.
[282] T. Stykel, Gramian-based model reduction for descriptor systems, *Math. Control Signals Syst.* **16** (2004) 297-319.
[283] J. L. Synge, The fundamental theorem of electrical networks, *Quarterly of Applied Mathematics* **9** (1951) 113-127.
[284] A. Szatkowski, Generalized dynamical systems: differentiable dynamic complexes and differential dynamic systems, *Int. J. Systems Sci.* **21** (1990) 1631-1657.
[285] A. Szatkowski, Geometric characterization of singular differential algebraic equations, *Int. J. Systems Sci.* **23** (1992) 167-186.
[286] M. Tadeusiewicz, A method for identification of asymptotically stable equilibrium points of a certain class of dynamic circuits, *IEEE Trans. Circ. Sys. I* **46** (1999) 1101-1109.
[287] M. Tadeusiewicz, Global and local stability of circuits containing MOS transistors, *IEEE Trans. Circ. Sys. I* **48** (2001) 957-966.
[288] F. Takens, Constrained equations; a study of implicit differential equations and their discontinuous solutions, Lect. Notes Maths., Vol. 525, 143-234, Springer-Verlag, 1976.
[289] F. Tari, Geometric properties of the integral curves of an implicit differential equation, *Discr. Cont. Dyn. Sys.* **17** (2007) 349-364.
[290] G. Thomas, The problem of defining the singular points of quasi-linear differential-algebraic systems, *Theor. Comput. Science* **187** (1997) 49-79.
[291] C. Tischendorf, On the stability of solutions of autonomous index-1 tractable and quasilinear index-2 tractable DAEs, *Cir. Sys. Sig. Proc.* **13** (1994) 139-154.
[292] C. Tischendorf, Topological index calculation of DAEs in circuit simulation, *Surv. Math. Ind.* **8** (1999) 187-199.

[293] C. Tischendorf, Coupled systems of differential algebraic and partial differential equations in circuit and device simulation. Modeling and numerical analysis, Habilitationsschrift, Inst. Math., Humboldt University, Berlin, 2003.

[294] O. Tosun and A. Dervisoglu, Formulation of state equations in active RLC networks, *IEEE Trans. Circ. Sys.* **21** (1974) 36-38.

[295] A. T. Vemuri, M. M. Polycarpou and A. R. Ciric, Fault diagnosis of differential-algebraic systems, *IEEE Trans. Systems, Man, and Cybernetics A* **31** (2001) 143-152.

[296] V. Venkatasubramanian, *A Taxonomy of the Dynamics of Large Differential Algebraic Systems such as the Power System*, PhD. Thesis, Washington University, 1992.

[297] V. Venkatasubramanian, Singularity induced bifurcation and the van der Pol oscillator, *IEEE Trans. Circ. Sys. I* **41** (1994) 765-769.

[298] V. Venkatasubramanian, H. Schättler and J. Zaborszky, Local bifurcations and feasibility regions in differential-algebraic systems, *IEEE Trans. Aut. Control* **40** (1995) 1992-2013.

[299] J. Vlach and K. Singhal, *Computer Methods for Circuits Analysis and Design*, Kluwer, 1994.

[300] W. Wasow, *Asymptotic Expansions for Ordinary Differential Equations*, John Wiley & Sons, 1965.

[301] K. Weierstrass, Zur Theorie der bilinearen und quadratischen Formen, *Monatsberichte Akad. Wiss. Berlin* 1868, 310-338; *Mathematische Werke*, II, Georg Olms Verlag and Johnson Reprint Co., 1967, 19-44.

[302] L. Weiss and W. Mathis, A Hamiltonian formulation for complete nonlinear RLC-networks, *IEEE Trans. Circ. Sys. I* **44** (1997) 843-846.

[303] L. Weiss, W. Mahtis and L. Trajkovic, A generalization of Brayton-Moser's mixed potential function, *IEEE Trans. Circ. Sys. I* **45** (1998) 423-427.

[304] R. Winkler, On simple impasse points and their numerical computation, Preprint 94-15, Inst. Math., Humboldt University, Berlin, 1994.

[305] R. Winkler, Stochastic differential algebraic equations of index 1 and applications in circuit simulation, *J. Comput. Appl. Math.* **163** (2004) 435-463.

[306] T. Yamada and D. G. Luenberger, Generic controllability theorems for descriptor systems, *IEEE Trans. Aut. Control* **30** (1985) 144-152.

[307] T. Yamada and D. G. Luenberger, Algorithms to verify generic causality and controllability of descriptor systems, *IEEE Trans. Aut. Control* **30** (1985) 874-880.

[308] L. Yang and Y. Tang, An improved version of the singularity-induced bifurcation theorem, *IEEE Trans. Aut. Control* **46** (2001) 1483-1486.

[309] M. Zhitomirskii, Local normal forms for constrained systems on 2-manifolds, *Bol. Soc. Bras. Mat.* **24** (1993) 211-232.

[310] P. J. Zufiria and R. S. Guttalu, On the role of singularities in Branin's method from dynamic and continuation perspectives, *Appl. Math. Comp.* **130** (2002) 593-618.

Index

{1, 2}-inverse, 39

Π-projectors, 44, 45, 49

active circuit, 196, 211
admissible Π-projector sequence, 46
admissible P-projector sequence, 41
   equivalence to Π-sequence, 46
algebraic variables, 17
algebraically nice DAE, 41
algebraically nice point, 74
associated resistive network, 277
asymptotic stability, 130, 133
Augmented Nodal Analysis (ANA), 214
   index, 220, 238

backward impasse point, *see* impasse point
bifurcation, 160, 166, 300
boundary singularity, 169, 170
branch, 203
branch current, 203
branch voltage, 203
branch-oriented model, 258
   index, 279
   tree-based formulation, 260

C^1-space, 36
C-loop, 206, 232, 247, 294
canonical projector, 33
capacitance matrix, 208

capacitive block, 245
   non-degenerate, 246
capacitive tree, 232, 247
capacitor, 207
   linear, 208
   voltage-controlled, 207
Cauchy-Binet formula, 237
characteristic values
   geometric sense, 107
   of a nice at level k DAE, 41
   of a regular linear DAE, 43
   of a regular point, 74
charge, 207
charge-oriented model, 212
chart, 90
chord, 198
codimension, 91
complete decoupling, 53
completion, 136
conductance matrix, 210
conductance products, 238, 247
configuration space, 265
conjugacy, 116
connected component, 197
connected graph, 197
consistent initial value, 4
constrained system, 2
contact equivalence, 112
continuation, 145, 147, 156
controlled source, 210, 289–293
conventional model, 212
core-rank, 28

cotree, 198
coupling, 207–209
current source, 210, 289
current-controlled current source (CCCS), 211
current-controlled voltage source (CCVS), 211
cut space, 201
cutset, 197
　characterization, 199, 200, 301
cutset matrix, 200
　reduced, 201
cycle space, 201
cyclomatic number, 200

DC circuit, 295
DC equations, 296
DC operating point, 296
DC source, 210
decoupling
　index one, 31, 37
　properly stated form, 51
　singular, 149, 156
　standard form, 78
　time-invariant DAEs, 80
derivative array, 135
descriptor system, 3
diffeomorphism, 91
differential, 92
differential-algebraic equation (DAE), 2
differential-algebraic system, 2
differentiation index, 6, 136
　one, 85, 136
　two, 88, 136
　zero, 136
digraph, 196
　underlying graph, 196
diode, 209
directed graph, see digraph
directed subgraph, 197
dynamic variables, 17
dynamical degree of freedom, 28

edge, 196
eigenvalue (of a matrix pencil), 28

embedding, 92
enlarged system, 12
equilibrium point, 130, 296
　nonsingular, 297
　regular, 130
excess capacitors, 207
excess inductors, 207
explicit ODE, 1
exponential stability, 130, 301

fine decoupling, 52
flux, 208
forest, 198
forward impasse point, see impasse point
fundamental cutset, 203, 259
fundamental loop, 203, 259
fundamental matrices, 203

generalized inverse, 39
generalized vector field, 19, 94
geometric index, 7, 106
　independence of reduction operators, 112
　invariance, 112
　nonautonomous DAEs, 123, 124
　of branch-oriented circuit models, 279
　of fully nonlinear DAEs, 134
　of linear DAEs, 82
　one, 102, 124
　Rabier and Rheinboldt sense, 96, 111
　two, 104
　zero, 98, 123
graph, 196
gyrator, 209

harmless singularity, 152, 179
Hessenberg DAE, 19
　size k, 19
　size three, 19
　size two, 18
　　autonomous, 88, 120
　　circuit model, 281
　　nonautonomous, 125

Index

hidden constraint, 88, 121, 126
hybrid analysis, 257
hybrid description of resistors, *see* resistor
hybrid matrix, 267
hyperbolicity, 130, 300

I singularity
  general quasilinear DAEs, 180
  quasilinear ODEs, 166
  semiexplicit index one DAEs, 184
I-cutset, 206, 210
IC-cutset, 294, 297, 302
ICL-cutset, 294, 302, 305
ideal transformer, 209
IK singularity
  general quasilinear DAEs, 180
  quasilinear ODEs, 168
  semiexplicit index one DAEs, 184
IL-cutset, 206, 220, 232, 238, 247, 279, 294, 302
ill-posed circuit, 206
immersion, 92
impasse point
  general quasilinear DAEs, 180
    backward, 180
    forward, 180
  quasilinear ODEs, 163
    backward, 166
    forward, 166
  semiexplicit index one DAEs, 183
    backward, 183
    forward, 183
implicit ODE, 1
implicitly defined resistors, *see* resistor
incidence matrix, 198
  reduced, 199
incremental capacitance matrix, 208
incremental conductance matrix, 209
incremental hybrid matrix, 267
incremental inductance matrix, 208
incremental resistance matrix, 210
independent source, 210
index, 5–8
  differentiation, 6, 136

geometric, 7, 82, 106
Kronecker, 5, 27
perturbation, 8
strangeness, 8
structural, 8
tractability, 6, 42, 75
inductance matrix, 209
inductor, 208
  current-controlled, 208
  linear, 209
infinite eigenvalue, 28
inherent ODE
  index one, 31, 37
  properly stated form, 52
  scalarly implicit, 149, 156
  standard form, 78
  time-invariant DAEs, 80
initial value, 4
inner singularity, 169, 170
input-output description, 13–15, 32, 38

Josephson junction, 208

K singularity
  general quasilinear DAEs, 180
  quasilinear ODEs, 168
  semiexplicit index one DAEs, 183
Kirchhoff laws, 204–206
Kronecker canonical form, 27
Kronecker index, 5, 27
  characterization via projectors, 30, 34, 80

L-cutset, 206, 294
linear DAE, 5, 6, 25
  time-invariant, 5, 25
    homogeneous, 26
  time-varying, 6
    analytic, 158
    properly stated form, 15, 25
    standard form, 16, 25
link, 198
local equivalence, 112
  of one-step reductions, 114
  of singular reductions, 176

of state space reductions, 116
local parametrization, 90
locally passive, 211
locally regular DAE, 110
loop, 197
   characterization, 199, 200, 301
loop matrix, 200
   reduced, 200
lumped circuit, 195
Lyapunov function, 294

M-projectors, 44, 45, 49
matrix chain
   based on Π-projectors, 45
   based on P-projectors, 40
   equivalence of P- and Π-chains, 46
matrix pencil, 27
   spectrum, 28
mixed analysis, 257
Modified Nodal Analysis (MNA), 215
   index, 232, 247
MOSFET, 293
multiport model, 274, 275
   index, 279
mutual inductance, 209

nice DAE, 41
nice point, 74
nilpotency index, see Kronecker index
nodal analysis, 212
nodal capacitance matrix, 215
nodal conductance matrix, 237
node, 203
node potential, 205
Node Tableau Analysis (NTA), 213
   index, 220
non-passive, 211, 236, 278
noncritical singularity
   general quasilinear DAEs, 180
   quasilinear ODEs, 163
   semiexplicit index one DAEs, 183
normal tree, 206

Ohm's law, 210
open circuit, 211
operating point, 265

order of complexity, 262
orthogonal projector, 30

P-projectors, 34, 39
partial differential-algebraic equation (PDAE), 4, 195
passive, 195, 211, 219
path, 197
perturbation index, 8
positive definite, 211
positive semidefinite, 211
preadmissible Π-projector sequence, 46
preadmissible P-projector sequence, 41
projector, 30
proper tree, 206
properly stated leading term, 38
pseudoequilibrium, 166, 184

Q-projectors, 34, 40
qualitative properties, 9, 130, 293
quasilinear DAE, 19
   singular points, 168–171
quasilinear ODE, 19, 162

R-projector, 39
rank of a digraph, 199
rank of a smooth mapping, 92
reactance, see reactive element
reactive duality, 299
reactive element, 203
reciprocal, 211
reduced cutset matrix, 201
reduced incidence matrix, 199
reduced loop matrix, 200
reduced ODE, 7, 87, 107
   Hessenberg DAE of size two, 89, 121, 126
   linear DAEs, 82
   semiexplicit index one DAE, 86, 119, 125
reduction
   of linear DAEs, 82
   of quasilinear DAEs, 107
   one-step, 101

singular, 175–178
reference direction, 203
reference node, 199, 205
reflexive generalized inverse, 39
regular DAE, 42, 110
regular manifold, 110
regular matrix pencil, 27
regular point
   of a linear DAE, 74
   of a quasilinear DAE, 106
   k-regular point, 105
   0-regular point, 98
   1-regular point, 104
regularization, 12
resistance matrix, 210
resistive operating point, 265
resistor, 209
   current-controlled, 210
   hybrid description, 263, 276
   implicitly defined, 258, 266
      linear, 266
   linear, 210
   voltage-controlled, 209

Schur complement, 127
Schur reduction, 127–129, 276
self-inductance, 209
self-loop, 197
semi-implicit DAE, 187
semi-singular point, 168, 184
semiexplicit DAE, 17
   Hessenberg form, 19
   index one, 85–87, 119, 125
   nonautonomous, 124
   singularities, 181–184
semilinear DAE, 17
semisimple eigenvalue, 300
semistate model, 1, 193
semistate system, 3
short circuit, 211
simple eigenvalue, 300
singular matrix pencil, 27
singular perturbation problem, 12
singular point, *see* singularity
singularity, 9
   of linear DAEs, 140

harmless, 152
independence of projectors, 142
invariance, 142
type k-A, 141, 155
type k-B, 141, 155
type 0, 141, 154
of quasilinear DAEs, 170
   k-singular point, 170
   0-singular point, 169
   harmless, 179
   independence of reduction operators, 171
   invariance, 171
   last step, 161
of quasilinear ODEs, 162–168
singularity crossing, 166, 185
singularity-induced bifurcation, 160
skew-symmetric matrix, 275
smooth manifold, 90
smooth mapping, 91
solution, 4
solution manifold, 110
   Hessenberg DAE of size two, 89
   semiexplicit index one DAE, 86
solution set, 4
spanning tree, *see* tree
standard form linear DAE, 16, 25
state dimension, 28
state formulation problem, 255, 256, 262–264
state space model, 1, 262, 263, 283
strangeness index, 8
strictly locally passive, 211
   implicitly defined resistors, 267
strictly passive, 211
   implicitly defined resistors, 267
structural index, 8
subgraph, 197
subimmersion, 93
submanifold, 91
submersion, 92

tangent bundle, 92
tangent space, 91
topological properties of circuits, 204

tractability index, 6, 42, 75
  characterization via Π-projectors, 48
  independence of projectors, 43
  of ANA circuit models, 220, 238
  of MNA circuit models, 232, 247
  of NTA circuit models, 220
  of properly stated linear DAEs, 42
  of quasilinear DAEs, 218
  of standard form linear DAEs, 75
  one, 37
transistor, 293
tree, 197
tree-based model, 259
  index, 279
twig, 198

underlying ODE, 6, 87
  Hessenberg DAE of size two, 88
  semiexplicit index one DAE, 85

uniform differentiation index, 136

V-loop, 206, 210
V-tree, 247
VC-loop, 206, 220, 232, 238, 247, 279, 294, 302
VC-tree, 238
VCL-loop, 294, 302, 305
vertex, 196
VL-loop, 294, 297, 302
voltage source, 210, 289
voltage-controlled current source (VCCS), 211
voltage-controlled voltage source (VCVS), 211

well-posed circuit, 206, 210